Y0-AAA-510

de Gruyter Series in Nonlinear Analysis and Applications 5

Pavel Drábek
Alois Kufner
Francesco Nicolosi

Quasilinear Elliptic Equations with Degenerations and Singularities

Walter de Gruyter · Berlin · New York 1997

Authors

Pavel Drábek
Dept. of Mathematics
University of West Bohemia
Univerzitní 22
306 14 Pilsen
Czech Republic

Alois Kufner
Mathematical Institute
Czech Academy of Sciences
Žitná 25
115 67 Prague 1
Czech Republic

Francesco Nicolosi
Dept. of Mathematics
University of Catania
Viale A. Doria 6
95125 Catania
Italy

1991 Mathematics Subject Classification: 35-02; 35J70; 35J65; 35J60; 35D05; 35B45, 35A25

Keywords: Quasilinear elliptic equations, boundary value problems, degeneration, singularity, weak solution

⊗ Printed on acid-free paper which falls within the guidelines of the ANSI to ensure permanence and durability.

Library of Congress Cataloging-in-Publication Data

Drabek, P. (Pavel), 1953–
Quasilinear elliptic equations with degenerations and singularities / by Pavel Drábek, Alois Kufner, Francesco Nicolosi.
p. cm. – (De Gruyter series in nonlinear analysis and applications, ISSN 0941-813X ; 5)
Includes bibliographical references and index.
ISBN 3-11-015490-0 (alk. paper)
1. Differential equations, Elliptic–Numerical solutions. 2. Boundary value problems–Numerical solutions. 3. Bifurcation theory. I. Kufner, Alois. II. Nicolosi, Francesco, 1938– . III. Title. IV. Series.
QA377.D76 1997
515'.353–dc21 97-17293
CIP

Die Deutsche Bibliothek – Cataloging-in-Publication Data

Drábek, Pavel:
Quasilinear elliptic equations with degenerations and singularities / by Pavel Drábek ; Alois Kufner ; Francesco Nicolosi. – Berlin ; New York : de Gruyter, 1997
(De Gruyter series in nonlinear analysis and applications ; 5)
ISBN 3-11-015490-0

ISSN 0941-813 X

Typeset using the authors' TEX files: I. Zimmermann, Freiburg. Printing: Gerike GmbH, Berlin. Binding: Lüderitz & Bauer GmbH, Berlin. Cover design: Thomas Bonnie, Hamburg.

To our wives
Dana, Zlata and Tina
for patience, understanding
and permanent support

Preface

Boundary value problems for elliptic equations, more precisely, the concept of *weak* (generalized) solutions, have their background in applications (namely, in the variational approach connected with the critical level of a certain energy functional as well as in numerical methods like FEM etc.). This type of approach is closely related to the concept of Sobolev spaces and is well elaborated for both linear and nonlinear equations.

In various applications, we can meet boundary value problems for elliptic equations whose ellipticity is "disturbed" in the sense that some degeneration or singularity appears. This "bad" behaviour can be caused by the *coefficients* of the corresponding differential operator as well as by the *solution itself.* The so-called p-Laplacian is a prototype of such an operator and its character can be interpreted as a degeneration or as a singularity of the classical (linear) Laplace operator (with $p = 2$). There are several very concrete problems from practice which lead to such differential equations, e.g. from glaceology, non-Newtonian fluid mechanics, flows through porous media, differential geometry, celestial mechanics, climatology, petroleum extraction, reaction-diffusion problems, etc.

In this book, we concentrate mainly on nonlinear (or, more precisely, *quasilinear*) problems since linear problems have been investigated more frequently, and we will study both types of appearance of "disturbance" — coming from the coefficients and coming from the solution. It turns out that the apparatus of weighted Sobolev spaces is the natural tool for investigation of such problems and that not only differential operators with degeneration, but also operators with singularities can be treated in the same way.

As a matter of fact, the combination of properties of weighted Sobolev spaces with abstract methods of nonlinear functional analysis makes it possible to obtain *existence results* for such boundary value problems and to study basic *spectral properties and bifurcations* of a certain class of typical nonlinear differential operators.

The book consists of two somewhat different parts — Chapter 2 forming the first part, Chapters 3 and 4 the second. But first of all, in *Introduction*, for the reader's orientation it is shown how the investigation of degenerated and singular boundary value problems via weighted Sobolev spaces appears as a natural extension of the nowadays classical approach to (strongly) elliptic equations via the theory of monotone operators in classical (nonweighted) Sobolev spaces.

Chapter 1 is of auxiliary character and we recommend the reader to proceed from the introduction directly to part one or two and to consult Chapter 1 mainly in case he needs some information about the tools used in the book, i.e. the properties of the corresponding (weighted) function spaces, namely imbedding theorems and special superposition (Nemytskij) operators on these spaces, and the substantial results about the existence of solutions of abstract operator equations (as the degree theory for nonlinear operators).

Chapter 2 is concerned with the first aim of the authors — to deal with disturbed elliptic problems when the disturbance is caused by the coefficients. We provide a survey on the existence of weak solutions, picking out the main ideas and the main methods, comparing various approaches and, last but not least, presenting a number of examples showing the advantages and disadvantages of the methods in question. We distinguish between second order equations, where truncation techniques can be used and where the Leray–Lions theorem is the main functional analytic tool, and higher order boundary value problems where a more general theory of the topological degree of monotone mappings is used. Also, we try to illustrate the mutual interaction of the parameters appearing in the boundary value problems, in particular the growth properties describing the nonlinearity, and the weight functions describing the degeneration or singularity. Here we have collected results obtained by the authors, partially in collaboration with other colleagues, and published in several papers.

Chapters 3 and 4 deal with the second aim of the book, namely to collect some results connected with the up to now less investigated field of spectral analysis of certain nonlinear boundary value problems where the disturbance is given mainly by the solution itself. Here we concentrate on the (perturbed as well as nonperturbed) p-Laplacian because it can serve as a very useful and transparent model case for some more general quasilinear differential operators. In Chapter 3 we deal with problems containing the p-Laplacian and its more general perturbed version. The case of a bounded domain is considered here and questions connected with the existence and properties of the first eigenvalue, the maximum principle, the existence and the bifurcation of positive solutions are studied. Chapter 4 is devoted to similar problems but on the whole space $\mathbb{R}^N$. Many principal troubles arise in this case and so it should be pointed out that these results are not only trivial extensions of the previous ones. Let us emphasize that, in particular, our problems have a non-variational structure and that very fine apriori estimates enable us to deal with very general equations by use of functional analytic approach. Unlike the first part, where a survey is given, the results of this second part — which are mainly due to the research done by the first author — can be used as a starting point for further investigation of such spectral problems. It should be mentioned that the topic dealt with in these two chapters was studied rarely in the past but many papers of numerous authors have appeared recently.

Thus, the book addresses readers who wish to become acquainted with modern methods of solving some special types of boundary value problems (graduate students, applied mathematicians and people from related fields of science and engineering etc.) as well as specialists in differential equations who are looking for new problems and approaches in this important topic.

The authors would like to express their gratitude to Prof. Bronius Kvedaras from Institute of Mathematics and Informatics, Vilnius, Dr. Milan Kučera from Czech Academy of Sciences, Prague and Dr. Jiří Bouchala from Technical University, Ostrava whose valuable comments helped to improve the book, to colleagues who helped to produce the manuscript, namely Ms. Iva Sulková and Ing. Miloš Brejcha from the University of West Bohemia in Plzeň, and also to Dr. Jiří Jarník from Charles University, Praha who improved their English. It also should be mentioned that the work on the book was made possible by the support of the Grant Agency of the Czech Republic, Grant No. 201/94/0008, as well as by the support of the Italian G. N. A. F. A., CNR, Grant No. 341/91. This help is gratefully acknowledged.

Plzeň, Praha, Catania *The authors*

Spring 1997

Contents

List of symbols, theorems, definitions, assumptions, examples

List of symbols

Let us point out, that we will deal here only with *real* functions of one ore more *real* variables; c or c_i will always denote positive constants.

$\mathbb{R}$	the set of real numbers
$\mathbb{N}$	the set of natural numbers
$\mathbb{R}^N$	the N-dimensional Euclidean space of points $x = (x_1, x_2, \dots, x_N)$
Ω	a domain in $\mathbb{R}^N$ (i.e., an open and connected set)
$\partial\Omega$	the boundary of Ω
$\Omega(u > 0) = \{x \in \Omega;\ u(x) > 0\}$	
$\alpha = (\alpha_1, \alpha_2, \dots, \alpha_N)$	an (N-dimensional) multiindex (i.e., with components α_i which are nonnegative integers)
$\lvert\alpha\rvert = \alpha_1 + \alpha_2 + \dots + \alpha_N$	the length of the multiindex α
$D^\alpha = \dfrac{\partial^{\lvert\alpha\rvert}}{\partial x_1^{\alpha_1} \partial x_2^{\alpha_2} \dots \partial x_N^{\alpha_N}}$	(partial) derivative of order $\lvert\alpha\rvert$
p	usually $1 < p < \infty$
$p' = \dfrac{p}{p-1}$,	i.e. $\dfrac{1}{p} + \dfrac{1}{p'} = 1$
$p^* = \dfrac{Np}{N-p}$	with $1 < p < N$
$p^* = +\infty$	with $p \geq N$
$m = \dfrac{(N+k)!}{N!k!}$	the number of all N-dimensional multiindices of length k
$\nabla^j u = \{D^\alpha u;\ \lvert\alpha\rvert = j\}$	the gradient of order $j \in \mathbb{N}$ of a function u
$\Delta_p, \tilde{\Delta}_p$	the p-Laplacian (p-Laplace operator, $p > 1$), pp. 10, 11
X^*	the adjoint (dual) space to the space X, p. 31
$\langle \cdot, \cdot \rangle$	the duality between X and X^*, p. 40
Deg	the (topological) degree (of a mapping), p. 33
Ind	the index (of a mapping), p. 34
$\hookrightarrow$	the symbol of the *continuous* imbedding, p. 21
$\hookrightarrow\hookrightarrow$	the symbol of the *compact* imbedding, p. 21
BVP	boundary value problem, p. 39
CAR	the class of Carathéodory functions, p. 19
$h(x, \zeta)$	p. 71
$h(x, u(x))$	p. 72

$s^+ = \max\{s, 0\}$	the positive part of $s \in \mathbb{R}$
$s^- = \max\{-s, 0\}$	the negative part of $s \in \mathbb{R}$
$\operatorname{supp} u$	support of a function u
a.e.	almost everywhere (with respect to the Lebesgue measure)
$\diamond$	the end of an example
$\square$	the end of a proof

Various particular function spaces

$C^{m,\lambda}(\overline{\Omega})$	a space of smooth functions, p. 20
$C_0^\infty(\Omega)$	the space of infinitely differentiable functions with compact support, p. 20
$C(\Omega, \omega)$	a weighted space of continuous functions with norm $\|\cdot\|_\omega^*$, p. 71
$H^{k-1,p}(\Omega, \omega)$	a special weighted space, p. 81
$L^p(\Omega)$ (or L^p)	Lebesgue space, p. 18
$L^p_{\mathrm{loc}}(\Omega)$	p. 18
$L^p(\Omega, w)$ (or $L^p(w)$)	weighted Lebesgue space, p. 18
$w(x)$ or $w_\alpha(x)$	weights or weight functions, i.e., functions measurable and positive a.e.
$W^{k,2}(\Omega)$	Sobolev (Hilbert) space with norm $\|\cdot\|_{k,2}$, p. 5
$W^{k,2}(\Omega, w)$	weighted Sobolev (Hilbert) space with norm $\|\cdot\|_{k,2,w}$ and weight $w = \{w_\alpha\,;\ \lvert\alpha\rvert \le k\}$, pp. 6, 7
$W^{k,p}(\Omega)$	Sobolev (Banach) space with norm $\|\cdot\|_{k,p}$, pp. 8, 20
$W^{k,p}_{\mathrm{loc}}(\Omega)$	p. 20
$W^{k,p}(\Omega, w)$	weighted Sobolev (Banach) space with norm $\|\cdot\|_{k,p,w}$, pp. 13, 23
$W_0^{k,p}(\Omega)$	a subset of $W^{k,p}(\Omega)$ with norm $\vert\vert\vert\cdot\vert\vert\vert_{k,p}$, p. 21
$W_M^{k,p}(\Omega, w)$	a subset of $W^{k,p}(\Omega, w)$, p. 24
$W_0^{k,p}(\Omega, w)$	a subset of $W^{k,p}(\Omega, w)$, p. 23
$W^{1,p}(\nu, \Omega)$	a special weighted Sobolev space, pp. 24, 41
$W^{k,p}(\nu, \Omega)$	a special weighted Sobolev space, pp. 26, 94
$W_L^{1,p}((a,b), w_1)$	a special weighted Sobolev space, p. 28

List of theorems

List of definitions

List of assumptions

List of examples

Chapter 0
Introduction

Let us start with some well-known facts. We will consider a (partial) differential operator A of order $2k$, $k \in \mathbb{N}$, in the divergence form:

$$(Au)(x) = \sum_{\substack{|\alpha|\le k\\ |\beta|\le k}} (-1)^{|\alpha|} D^{\alpha}(a_{\alpha\beta}(x) D^{\beta} u(x)) \tag{0.1}$$

for $x \in \Omega$, where Ω is a domain in $\mathbb{R}^N$. The coefficients $a_{\alpha\beta}$ of this *linear* operator are functions defined a.e. in Ω.

In the theory of boundary value problems for the equation

$$Au = f \quad \text{in} \quad \Omega\,, \tag{0.2}$$

in particular in the theory of *weak solutions*, an important role is played by the bilinear form $a(u, v)$ associated with the differential operator A which is defined by the formula

$$a(u, v) = \sum_{\substack{|\alpha|\le k\\ |\beta|\le k}} \int_{\Omega} a_{\alpha\beta}(x) D^{\beta} u(x) D^{\alpha} v(x)\, dx\,. \tag{0.3}$$

The weak solution u of the equation (0.2) is sought in the (classical) *Sobolev space*

$$W^{k,2}(\Omega) \tag{0.4}$$

defined as the set of all functions $u = u(x)$ whose (distributional) derivatives $D^{\alpha} u$ of orders $|\alpha|$ with $|\alpha| \le k$ are in $L^2(\Omega)$; this space is normed by

$$\|u\|_{k,2} = \left(\sum_{|\alpha|\le k} \int_{\Omega} |D^{\alpha} u(x)|^2\, dx \right)^{\frac{1}{2}}. \tag{0.5}$$

Let us summarize some assumptions concerning the coefficients $a_{\alpha\beta}$:

(i) If we assume that

$$a_{\alpha\beta} \in L^{\infty}(\Omega) \quad \text{for} \quad |\alpha|, |\beta| \le k \tag{0.6}$$

then, obviously, the form $a(u, v)$ becomes *bounded*, i.e. the *upper estimate*

$$|a(u, v)| \le c_0 \|u\|_{k,2} \|v\|_{k,2} \tag{0.7}$$

holds for every $u, v \in W^{k,2}(\Omega)$ with a constant $c_0 > 0$ independent of u, v.

(ii) In order to obtain assertions about the existence of a weak solution, we have to suppose some kind of *ellipticity* of the operator A, which can be expressed (in a rather special form, of course, with the possibility of some weakening) by the claim that a *lower estimate*

$$a(u,u) \geq c_1 \|u\|_{k,2}^2 \tag{0.8}$$

holds for every $u \in V$ with the so called *ellipticity constant* $c_1 > 0$ independent of u. Here V is a certain subspace of $W^{k,2}(\Omega)$ determined by the boundary conditions considered. Let us point out that the (integral) condition (0.8) is guaranteed if, e.g., the *algebraic* ellipticity condition

$$\sum_{\substack{|\alpha|\leq k\\|\beta|\leq k}} a_{\alpha\beta}(x)\xi_\alpha\xi_\beta \geq c_1 \sum_{|\alpha|\leq k} \xi_\alpha^2 \tag{0.9}$$

is satisfied for a.e. $x \in \Omega$ with a constant $c_1 > 0$ for every (real) vector $\xi = \{\xi_\alpha\,;\ |\alpha| \leq k\} \in \mathbb{R}^m$.

The conditions (0.6) and (0.9) can be considered to be the basic assumptions on the coefficients $a_{\alpha\beta}$ of the operator A, which allow to develop a rather general theory of weak solutions of the elliptic linear differential equation (0.2). This theory is now very well-known and has become part of university courses in mathematics — see, e.g., S. Agmon [2], J.-L. Lions, E. Magenes [49], C. Miranda [50], J. Nečas [52] and others.

The question arises, what happens if some of the fundamental assumptions (0.6) and (0.8) (i.e.,(0.9)) are *violated.* If, e.g., the conditions (0.6) are not fulfilled, the coefficients $a_{\alpha\beta}$ can become *singular* and we are not able to guarantee the boundedness of the bilinear form $a(u,v)$ — see the estimate (0.7). On the other hand, if the *positive definiteness* of the quadratic form

$$\sum_{\substack{|\alpha|\leq k\\|\beta|\leq k}} a_{\alpha\beta}(x)\xi_\alpha\xi_\beta$$

is not guaranteed but, nevertheless, the form is still *positive* — i.e., if, for example, condition (0.9) is replaced by the condition

$$\sum_{\substack{|\alpha|\leq k\\|\beta|\leq k}} a_{\alpha\beta}(x)\xi_\alpha\xi_\beta \geq \sum_{|\alpha|\leq k} w_\alpha(x)\xi_\alpha^2 \tag{0.10}$$

where now $w_\alpha(x)$ are given functions on Ω which are *positive* a.e. in Ω but are not "separated from zero" (so-called *weight* functions) — then we have to do with *degenerated* differential operators and we are not able to guarantee the (uniform) ellipticity (0.9) or (0.8). In these cases, we have to change our approach introducing a modified version of the Sobolev space, namely the so-called *weighted Sobolev space*

$$W^{k,2}(\Omega, w) \tag{0.11}$$

defined (roughly speaking) as the set of all functions $u = u(x)$ for which the distributional derivatives $D^\alpha u$ satisfy

$$\int_\Omega |D^\alpha u(x)|^2 w_\alpha(x)\, dx < \infty \quad \text{for} \quad |\alpha| \le k \tag{0.12}$$

with $w = \{w_\alpha = w_\alpha(x)\,;\ |\alpha| \le k\}$ a given *family of weight functions* w_α on Ω (i.e. w_α are measurable and positive a.e. in Ω). The space (0.11) can then be normed by

$$\|u\|_{k,2,w} = \left(\sum_{|\alpha|\le k} \int_\Omega |D^\alpha u(x)|^2 w_\alpha(x)\, dx \right)^{\frac{1}{2}} . \tag{0.13}$$

Let us illustrate the application of weighted Sobolev spaces in the case of singular coefficients or a degenerated operator by a very simple example: We consider the special *second order* differential operator

$$(Au)(x) = -\sum_{i=1}^{N} \frac{\partial}{\partial x_i} \left(a_i(x) \frac{\partial u(x)}{\partial x_i} \right) + a_0(x) u(x) \tag{0.14}$$

in $\Omega \subset \mathbb{R}^N$. The corresponding bilinear form (0.3) is now

$$a(u, v) = \sum_{i=1}^{N} \int_\Omega a_i(x) \frac{\partial u}{\partial x_i} \frac{\partial v}{\partial x_i}\, dx + \int_\Omega a_0(x) u v\, dx\,. \tag{0.15}$$

If the coefficients $a_i = a_i(x)$, $i = 0, 1, \ldots, N$, are *bounded*,

$$a_i \in L^\infty(\Omega)\,, \quad i = 0, 1, \ldots, N\,, \tag{0.16}$$

and *separated from zero*,

$$a_i(x) \ge c_1 > 0 \quad \text{for a.e.} \quad x \in \Omega \quad \text{and} \quad i = 0, 1, \ldots, N\,, \tag{0.17}$$

then we can use the classical theory, since conditions (0.6) and (0.9) are obviously satisfied and the corresponding space under consideration is the Sobolev space $W^{1,2}(\Omega)$.

If some of conditions (0.16) and (0.17) are violated, i.e. if some of the functions $a_i(x)$ become *singular* or are *only positive a.e. in* Ω but not separated from zero, then we cannot use the space $W^{1,2}(\Omega)$. Nonetheless, we can use the *weighted* Sobolev space

$$W^{1,2}(\Omega, w) \tag{0.18}$$

where now the family w is determined by the coefficients:

$$w = \{a_i(x)\,;\ i = 0, 1, \ldots, N\}\,. \tag{0.19}$$

Indeed, since due to (0.15)

$$a(u,u)=\sum_{i=1}^{N}\int_\Omega\left|\frac{\partial u}{\partial x_i}\right|^2 a_i(x)\,dx+\int_\Omega |u|^2 a_0(x)\,dx=\|u\|_{1,2,w}^2$$

with w given by (0.19), we immediately obtain the ellipticity condition of the type (0.8) with the norm of the Sobolev space $W^{1,2}(\Omega, w)$. Similarly, also the estimate

$$|a(u,v)|\le c_0\|u\|_{1,2,w}\|v\|_{1,2,w}$$

can be easily derived from (0.15) using simply the Hölder inequality:

$$\begin{aligned}|a(u,v)| &\le \sum_{i=1}^{N}\int_\Omega |a_i(x)|^{\frac12}\left|\frac{\partial u}{\partial x_i}\right| |a_i(x)|^{\frac12}\left|\frac{\partial v}{\partial x_i}\right| dx+\int_\Omega |a_0(x)|^{\frac12}|u||a_0(x)|^{\frac12}|v|\,dx\\ &\le \sum_{i=1}^{N}\left(\int_\Omega\left|\frac{\partial u}{\partial x_i}\right|^2 a_i(x)\,dx\right)^{\frac12}\left(\int_\Omega\left|\frac{\partial v}{\partial x_i}\right|^2 a_i(x)\,dx\right)^{\frac12}\\ &\quad+\left(\int_\Omega |u|^2 a_0(x)\,dx\right)^{\frac12}\left(\int_\Omega |v|^2 a_0(x)\,dx\right)^{\frac12}\\ &\le (N+1)\|u\|_{1,2,w}\|v\|_{1,2,w}.\end{aligned}$$

Consequently, for our special example, we have obtained again estimates of the type (0.7) and (0.8), but now *with the norm of the weighted space* $W^{1,2}(\Omega, w)$ *instead of the norm of the classical space* $W^{1,2}(\Omega)$.

So it seems that the application of weighted spaces can save the situation even in the case when the fundamental assumptions (0.6), (0.9) are violated, and this is true not only for our special operator (0.14), but sometimes also for the general operator A from (0.1) (where, of course, the weight functions $w_\alpha(x)$ should be chosen in a more sophisticated way). In any case, the weight functions $w_\alpha(x)$, which then appear in the corresponding weighted space $W^{k,2}(\Omega, w)$, *are closely related to the coefficients* $a_{\alpha\beta}$, i.e. to the singularity and/or degeneration of the operator A.

Let us mention that the approach to degenerated (second order) linear differential operators was due basically to M. K. V. Murthy, G. Stampacchia [51], being then extended to higher order linear degenerated elliptic operators by several authors, e.g. V. P. Glushko [29], F. Guglielmino, F. Nicolosi [30, 31], B. Hanouzet [32], I. A. Kiprijanov [34], S. M. Nikolskij [53]; see also A. Kufner, B. Opic [41].

Up to now, we have treated only the *linear case*. In the sixties, a theory of nonlinear (quasilinear) elliptic equations was developed, with nonlinearities of a "polynomial" type, based on the theory of monotone operators and using as a fundamental tool the Sobolev space

$$W^{k,p}(\Omega) \tag{0.20}$$

with arbitrary $p > 1$ (i.e., not only $p = 2$, the Hilbert space case), which is defined as the set of all functions $u = u(x)$ whose (distributional) derivatives $D^\alpha u$ of order $|\alpha| \le k$ are in $L^p(\Omega)$; this space is normed by

$$\|u\|_{k,p} = \left(\sum_{|\alpha| \le k} \int_\Omega |D^\alpha u(x)|^p \, dx \right)^{\frac{1}{p}} . \tag{0.21}$$

To be more specific, the *linear* operator A from (0.1) is now replaced by the *nonlinear* differential operator of order $2k$ in the divergence form,

$$(Au)(x) = \sum_{|\alpha| \le k} (-1)^{|\alpha|} D^\alpha a_\alpha(x; u, \nabla u, \dots, \nabla^k u) \tag{0.22}$$

for $x \in \Omega \subset \mathbb{R}^N$, where now the "coefficients"

$$a_\alpha = a_\alpha(x; \xi)$$

are defined on $\Omega \times \mathbb{R}^m$, and where

$$\nabla^j u = \{D^\gamma u ;\ |\gamma| = j\} \tag{0.23}$$

for $j = 0, 1, \dots, k$ is the "gradient of the j-th order".

Here we suppose that

(i) $a_\alpha(x; \xi)$ satisfy the *Carathéodory condition*, i.e., $a_\alpha(\cdot; \xi)$ is measurable in Ω for every $\xi \in \mathbb{R}^m$ and $a_\alpha(x; \cdot)$ is continuous in $\mathbb{R}^m$ for a.e. $x \in \Omega$;

(ii) $a_\alpha(x; \xi)$ satisfy the *growth condition*

$$|a_\alpha(x; \xi)| \le c_\alpha \left(g_\alpha(x) + \sum_{|\beta| \le k} |\xi_\beta|^{p-1} \right), \quad |\alpha| \le k , \tag{0.24}$$

for a.e. $x \in \Omega$ and every $\xi \in \mathbb{R}^m$; here c_α is a given positive constant and g_α is a given function from $L^{p'}(\Omega)$, $p' = \dfrac{p}{p-1}$;

(iii) $a_\alpha(x; \xi)$ satisfy the *monotonicity condition*

$$\sum_{|\alpha| \le k} [a_\alpha(x; \xi) - a_\alpha(x; \eta)](\xi_\alpha - \eta_\alpha) > 0 \tag{0.25}$$

for every $\xi, \eta \in \mathbb{R}^m$, $\xi \ne \eta$;

(iv) $a_\alpha(x;\xi)$ satisfy the *coercivity (ellipticity) condition*

$$\sum_{|\alpha|\le k} a_\alpha(x;\xi)\xi_\alpha \ge c_1 \sum_{|\alpha|\le k} |\xi_\alpha|^p \tag{0.26}$$

for every $\xi \in \mathbb{R}^m$ with a constant $c_1 > 0$ independent of ξ.

These conditions (which, of course, again can be weakened) guarantee that the form $a(u, v)$ given by

$$a(u,v) = \sum_{|\alpha|\le k} \int_\Omega a_\alpha(x; u(x), \nabla u(x), \dots, \nabla^k u(x)) D^\alpha v(x)\, dx \tag{0.27}$$

(which in fact is the counterpart of the bilinear form from (0.3)) is well-defined for $u, v \in W^{k,p}(\Omega)$. It determines an operator $T\colon W^{k,p}(\Omega) \to (W^{k,p}(\Omega))^*$ by the formula

$$a(u,v) = \langle Tu, v\rangle\,, \tag{0.28}$$

where $\langle\cdot,\cdot\rangle$ means the duality between $W^{k,p}(\Omega)$ and $(W^{k,p}(\Omega))^*$ (or, more precisely, between V and V^* where V is a subspace of $W^{k,p}(\Omega)$ determined by the *boundary conditions* considered). The operator

$$T\colon V \to V^* \tag{0.29}$$

is, in general, *nonlinear*, but in a certain sense *continuous* (as a consequence of the growth conditions (0.24)), *monotone*, i.e., satisfies

$$\langle Tu - Tv, u - v\rangle > 0 \quad \text{for} \quad u, v \in V\,,\ u \ne v \tag{0.30}$$

(as a consequence of the monotonicity condition (0.25)) and *coercive*, i.e., satisfies

$$\lim_{\|u\|_{k,p}\to\infty} \frac{\langle Tu, u\rangle}{\|u\|_{k,p}} = \infty \tag{0.31}$$

(due to condition (0.26)). Consequently, the boundary value problem for the *differential* equation

$$Au = f \quad \text{in} \quad \Omega \tag{0.32}$$

with A from (0.22) can be transformed into an *operator* equation

$$Tu = F\,, \quad F \in V^*\,, \tag{0.33}$$

whose solvability is guaranteed, e.g., by the theory of monotone operators.

A typical (and simple) example of a differential operator A satisfying all the foregoing conditions is the so-called *p-Laplacian* Δ_p defined for $p > 1$ by

$$\begin{aligned}\Delta_p u &= \operatorname{div}(|\nabla u|^{p-2}\nabla u)\\ &= \sum_{i=1}^N \frac{\partial}{\partial x_i}\left(\left[\left(\frac{\partial u}{\partial x_1}\right)^2 + \dots + \left(\frac{\partial u}{\partial x_N}\right)^2\right]^{\frac{p-2}{2}} \frac{\partial u}{\partial x_i}\right),\end{aligned} \tag{0.34}$$

or its modification, the operator $\tilde{\Delta}_p$ defined by

$$\tilde{\Delta}_p u = \sum_{i=1}^{N} \frac{\partial}{\partial x_i} \left(\left| \frac{\partial u}{\partial x_i} \right|^{p-2} \frac{\partial u}{\partial x_i} \right) \tag{0.35}$$

(notice that we define $|t|^{p-2}t = |t|^{p-1} \operatorname{sgn} t$, and consequently, as *zero* if $t = 0$), or the little more complicated, again *second order*, operator

$$Au = -\tilde{\Delta}_p u + |u|^{p-2}u \,. \tag{0.36}$$

Let us slightly change the foregoing operator into

$$(Au)(x) = -\sum_{i=1}^{N} \frac{\partial}{\partial x_i} \left(a_i(x) \left| \frac{\partial u}{\partial x_i} \right|^{p-2} \frac{\partial u}{\partial x_i} \right) + a_0(x)|u|^{p-2}u \tag{0.37}$$

with given functions (coefficients) $a_i(x)$, $i = 0, 1, \ldots, N$, satisfying

$$a_i \in L^\infty(\Omega) \quad \text{for} \quad i = 0, 1, \ldots, N \tag{0.38}$$

and

$$a_i(x) \geq c_1 > 0 \quad \text{for a.e. } x \in \Omega \quad \text{and} \quad i = 0, 1, \ldots, N \,. \tag{0.39}$$

Then, similarly as in the linear case, all conditions (0.24)–(0.26) remain satisfied and we can look for a weak solution in the same space as for the operator (0.36), namely in the Sobolev space

$$W^{1,p}(\Omega) \,.$$

However, again as in the linear case, the situation changes dramatically if some of the coefficients $a_i(x)$ *violate* condition (0.38) and/or condition (0.39). Since the form (0.27) can be written as

$$a(u, v) = \sum_{i=1}^{N} \int_\Omega a_i(x) \left| \frac{\partial u}{\partial x_i} \right|^{p-2} \frac{\partial u}{\partial x_i} \frac{\partial v}{\partial x_i} \, dx + \int_\Omega a_0(x)|u|^{p-2}uv \, dx \tag{0.40}$$

(notice that in our particular example, the "coefficients" $a_\alpha(x; \xi)$ have the form

$$a_i(x; \xi_0, \xi_1, \ldots, \xi_N) = a_i(x)|\xi_i|^{p-2}\xi_i \,, \quad i = 0, 1, \ldots, N) \,, \tag{0.41}$$

we can immediately see that

$$a(u, u) = \sum_{i=1}^{N} \int_\Omega a_i(x) \left| \frac{\partial u}{\partial x_i} \right|^{p} dx + \int_\Omega a_0(x)|u|^p \, dx$$

is nothing else than the p-th power of the norm in the *weighted Sobolev space*

$$W^{1,p}(\Omega, w)$$

with the family of weights

$$w = \{a_i(x)\,;\ i = 0, 1, \dots, N\} \tag{0.42}$$

and the norm

$$\|u\|_{1,p,w} = \left(\int_\Omega |u|^p a_0(x)\,dx + \sum_{i=1}^N \int_\Omega \left| \frac{\partial u}{\partial x_i} \right|^p a_i(x)\,dx \right)^{\frac{1}{p}}. \tag{0.43}$$

Since the coefficients $a_i(x)$ of the operator A are weight functions, we have

$$a(u, u) = \|u\|_{1,p,w}^p \tag{0.44}$$

and we can derive the estimate

$$|a(u, v)| \le (N + 1) \|u\|_{1,p,w}^{p-1} \|v\|_{1,p,w} \tag{0.45}$$

using the Hölder inequality. Indeed, with $p' = \dfrac{p}{p-1}$ we have from (0.40)

$$\begin{aligned} |a(u, v)| &\le \sum_{i=1}^N \int_\Omega |a_i(x)|^{\frac{1}{p'}} \left| \frac{\partial u}{\partial x_i} \right|^{p-1} |a_i(x)|^{\frac{1}{p}} \left| \frac{\partial v}{\partial x_i} \right| dx \\ &\quad + \int_\Omega |a_0(x)|^{\frac{1}{p'}} |u|^{p-1} |a_0(x)|^{\frac{1}{p}} |v|\,dx \\ &\le \sum_{i=1}^N \left(\int_\Omega |a_i(x)| \left| \frac{\partial u}{\partial x_i} \right|^p dx \right)^{\frac{1}{p'}} \left(\int_\Omega |a_i(x)| \left| \frac{\partial v}{\partial x_i} \right|^p dx \right)^{\frac{1}{p}} \\ &\quad + \left(\int_\Omega |a_0(x)| |u|^p\,dx \right)^{\frac{1}{p'}} \left(\int_\Omega |a_0(x)| |v|^p\,dx \right)^{\frac{1}{p}} \end{aligned}$$

and hence (0.45) follows immediately. Consequently, under the *only* assumption that the coefficients $a_i(x)$ are weight functions, i.e. measurable and positive a.e. in Ω, we can, not having assumptions (0.38) and (0.39), use the form

$$a(u, v) = \langle Tu, v \rangle$$

to introduce a (nonlinear) operator T, acting from $W^{1,p}(\Omega, w)$ (with w given by (0.42)) into its dual space $(W^{1,p}(\Omega, w))^*$ which is continuous due to (0.45), coercive due to (0.44) and monotone due to (0.41). Indeed, as far as monotonicity is concerned, we have

$$\begin{aligned} \langle Tu - Tv, u - v \rangle &= a(u, u - v) - a(v, u - v) \\ &= \sum_{i=1}^N \int_\Omega a_i(x) \left(\left| \frac{\partial u}{\partial x_i} \right|^{p-2} \frac{\partial u}{\partial x_i} - \left| \frac{\partial v}{\partial x_i} \right|^{p-2} \frac{\partial v}{\partial x_i} \right) \left(\frac{\partial u}{\partial x_i} - \frac{\partial v}{\partial x_i} \right) dx \\ &\quad + \int_\Omega a_0(x) \left(|u|^{p-2} u - |v|^{p-2} v \right) (u - v)\,dx > 0 \end{aligned}$$

for $u, v \in W^{1,p}(\Omega, w)$, $u \neq v$ due to the monotonicity of the function $s(t) = |t|^{p-1}\operatorname{sgn} t$ and the positivity of a_i and a_0. Now, the assertion about the existence of a *weak* solution u of a boundary value problem for the equation $Au = f$ in Ω with A from (0.37) can again be derived as in the "classical" case using the theory of monotone operators, only replacing the space $W^{1,p}(\Omega)$ by the *weighted* space $W^{1,p}(\Omega, w)$.

So, analogously to the linear case, also for our *nonlinear* operator A from (0.37) with coefficients which are *singular* ($a_i(x)$ are unbounded) and/or *degenerated* ($a_i(x)$ are only positive a.e.), the situation can be saved using the weighted Sobolev space $W^{1,p}(\Omega, w)$ with w from (0.42) instead of the classical Sobolev space $W^{1,p}(\Omega)$.

It is the aim of this book to show that this situation occurs not only for our nonlinear second order differential operator from (0.37), but for a large class of even higher order nonlinear operators of the form (0.22), where, of course, the growth conditions (0.24), the monotonicity condition (0.25) and the coerciveness conditon (0.26) have to be appropriately modified, if we use the *weighted Sobolev space*

$$W^{k,p}(\Omega, w)\,.$$

Here $k \in \mathbb{N}$, $p > 1$, and

$$w = \{w_\alpha(x)\,;\ |\alpha| \le k\}$$

describes the family of weight functions w_α. The space $W^{k,p}(\Omega, w)$ is defined as the set of all functions $u = u(x)$ whose (distributional) derivatives $D^\alpha u$ satisfy

$$\int_\Omega |D^\alpha u(x)|^p w_\alpha(x)\,dx < \infty \quad \text{for} \quad |\alpha| \le k\,.$$

This space is normed by

$$\|u\|_{k,p,w} = \left(\sum_{|\alpha|\le k} \int_\Omega |D^\alpha u(x)|^p w_\alpha(x)\,dx\right)^{\frac{1}{p}}. \tag{0.46}$$

It allows to describe various approaches which enable us to obtain existence theorems for weak solutions of boundary value problems for the differential equation $Au = f$ in Ω with A given by (0.22), where the operator A becomes in a certain sense singular or degenerated. This behaviour is described by certain conditions on the "coefficients" $a_\alpha(x;\xi)$ and the solution u is sought in a special weighted space $W^{k,p}(\Omega, w)$ with the parameter p and the weight w connected with the growth and singularity/degeneration of A, i.e., of $a_\alpha(x;\xi)$.

We will show how the interplay between the growth (described by p) and the singularity/degeneration (described by w) makes it possible to consider various classes of differential operators and to find the most appropriate space in which the solution u is looked for.

This, of course, requires also some knowledge of the structure of weighted spaces considered. Therefore, in Chapter 1 we will give a detailed description of the properties of these spaces which will be useful in the sequel. At this moment, let us point out that the most transparent and instructive examples of operators A of the form (0.22) are operators with the coefficients $a_\alpha = a_\alpha(x;\xi)$ looking like

$$w_\alpha(x)|\xi_\alpha|^{p-2}\xi_\alpha\,, \quad x \in \Omega\,, \quad p > 1\,, \tag{0.47}$$

where the singularity and/or degeneration appears immediately via the factor $w_\alpha(x)$. Note that most degenerations or singularities, i.e., most weights are of the type

$$w_a(x) = [\operatorname{dist}(x, M)]^\lambda\,, \tag{0.48}$$

where M is a part of $\overline{\Omega} = \Omega \cup \partial\Omega$, and λ is a real number. So the singularity ($\lambda < 0$) or degeneration ($\lambda > 0$) can appear on the boundary $\partial\Omega$ of Ω as well as in the interior of the domain. We will give several examples to show the variety of possibilities which arise.

Before starting the description of the properties of the spaces $W^{k,p}(\Omega, w)$, let us mention one approach which is a less general one, but needs no such information about the space and offers some insight into the structure of conditions on the operator A — i.e., on its "coefficients" $a_\alpha(x;\xi)$ — needed for the investigation of A from the point of view of the theory of monotone operators in the singular/degenerated case.

Let us suppose that we have a weighted Sobolev space $W^{k,p}(\Omega, w)$ with a family $w = \{w_\alpha(x)\,;\ |\alpha| \le k\}$ of weight functions $w_\alpha(x)$, $x \in \Omega$. Suppose that the weight functions satisfy

$$w_\alpha^{-\frac{1}{p-1}} \in L^1_{\mathrm{loc}}(\Omega)\,, \quad |\alpha| \le k \tag{0.49}$$

(this condition guarantees that the space $W^{k,p}(\Omega, w)$ normed by (0.46) is a reflexive *Banach* space). Taking into account the operator A from (0.22), we suppose that the "coefficients" satisfy again the Carathéodory condition and the (*modified*) *growth condition*

$$|a_\alpha(x;\xi)| \le c_\alpha w_\alpha^{\frac{1}{p}}(x)\left(g_\alpha(x) + \sum_{|\beta|\le k} w_\beta^{\frac{1}{p'}}(x)|\xi_\beta|^{p-1}\right), \quad |\alpha| \le k\,, \tag{0.50}$$

for a.e. $x \in \Omega$ and every $\xi \in \mathbb{R}^m$; here c_α is a given positive constant, g_α is a given function from $L^{p'}(\Omega)$ and $p' = \dfrac{p}{p-1}$.

(**Exercise:** Compare condition (0.50) with the (classical) condition (0.24). Obviously, the two conditions coincide if $w_\gamma(x) \equiv 1$ for every γ, $|\gamma| \le k$, and moreover, they coincide also if $0 < c_\gamma \le w_\gamma(x) \le \tilde{c}_\gamma < \infty$, i.e., if *no* singularity and *no* degeneration appear. — Verify that the "coefficients" a_α from (0.47) satisfy the growth condition (0.50).)

Further, keep the monotonicity condition *unchanged*, i.e. assume again that (0.25) holds, and modify the coerciveness condition (0.26) into the form

$$\sum_{|\alpha|\le k} a_\alpha(x;\xi)\xi_\alpha \ge c_1 \sum_{|\alpha|\le k} w_\alpha(x)|\xi_\alpha|^p\,. \tag{0.51}$$

Conditions (0.50), (0.25) and (0.51) are the simplest ones which allow to use a (modified) theory of monotone operators. Indeed, it follows from (0.50) due to (0.27) that (using the Hölder inequality and the Minkowski inequality)

$$
\begin{aligned}
|a(u,v)| &\le \sum_{|\alpha|\le k} \int_\Omega |a_\alpha(x;u,\nabla u,\dots,\nabla^k u)||D^\alpha v(x)|\,dx \\
&\le \sum_{|\alpha|\le k} c_\alpha \int_\Omega w_\alpha^{\frac{1}{p}}(x)|D^\alpha v(x)| \left(g_\alpha(x) + \sum_{|\beta|\le k} w_\beta^{\frac{1}{p'}}(x)|D^\beta u(x)|^{p-1}\right) dx \\
&\le \sum_{|\alpha|\le k} c_\alpha \left(\int_\Omega w_\alpha(x)|D^\alpha v(x)|^p\,dx\right)^{\frac{1}{p}} \left[\left(\int_\Omega |g_\alpha(x)|^{p'}\,dx\right)^{\frac{1}{p'}} \right.\\
&\quad \left. + \sum_{|\beta|\le k}\left(\int_\Omega w_\beta(x)|D^\beta u(x)|^p\right)^{\frac{1}{p'}}\right] \\
&= \sum_{|\alpha|\le k} c_\alpha \|D^\alpha v\|_{p,w_\alpha} \left(\|g_\alpha\|_{p'} + \sum_{|\beta|\le k} \|D^\beta u\|_{p,w_\beta}^{p-1}\right) \\
&\le c_0 \|v\|_{k,p,w} \left(\sum_{|\alpha|\le k} \|g_\alpha\|_{p'} + \|u\|_{k,p,w}^{p-1}\right),
\end{aligned}
$$

where

$$
\|g_\alpha\|_{p'} = \left(\int_\Omega |g_\alpha(x)|^{p'}\,dx\right)^{\frac{1}{p'}}, \qquad \|h\|_{p,w_\alpha} = \left(\int_\Omega |h(x)|^p w_\alpha(x)\,dx\right)^{\frac{1}{p}}.
$$

Hence, for $u, v \in W^{k,p}(\Omega, w)$ we have

$$
|a(u,v)| \le c_0 \|v\|_{k,p,w}(c_2 + \|u\|_{k,p,w}^{p-1}).
$$

Consequently, the operator T defined by the duality formula

$$
a(u,v) = \langle Tu, v\rangle
$$

is meaningful and maps $W^{k,p}(\Omega, w)$ into $(W^{k,p}(\Omega, w))^*$. Moreover, T is monotone and, due to (0.51), also coercive:

$$
\langle Tu, u\rangle = a(u,u) \ge c_1 \sum_{|\alpha|\le k} \int_\Omega w_\alpha(x)|D^\alpha u(x)|^p\,dx = c_1 \|u\|_{k,p,w}^p.
$$

Consequently, using conditions (0.50), (0.25) and (0.51), we can handle the boundary value problem for the singular/degenerated differential operator A in the same way in terms of the weighted space $W^{k,p}(\Omega, w)$ as it was done before, when using the conditions (0.24), (0.25) and (0.26) and the (unweighted!) space $W^{k,p}(\Omega)$.

It should be pointed out that the foregoing approach is very simple and serves mainly for illustration. More sophisticated approaches allow to consider more general operators.

Finally, let us add one important **remark**: Usually, we are investigating a (homogeneous) *boundary value problem*, i.e., we are looking for a function u from a certain space X, satisfying

(i) the equation $Au = f$ in Ω and

(ii) the (homogeneous) boundary conditions $B_i u = 0$ on $\partial\Omega$,

where A is a differential operator of order $2k$ (see (0.22)) and B_i $(i = 1, 2, \dots, k)$ are also some (linear) differential operators on $\partial\Omega$.

Here, we are solving the *operator equation* $Tu = F$ with T an operator from the space V into V^* determined by the differential operator A through formula (0.28), and with a space V which "covers" the boundary conditions.

Moreover, we are considering *homogeneous* boundary conditions which is made possible due to the *linearity* of the operators B_i: the step from nonhomogeneous boundary conditions (formally: $B_i u = g_i$ on $\partial\Omega$, $i = 1, 2, \dots, k$) can be made with help of a function $u_0 \in X$ (if it exists!!) such that $B_i u_0 = g_i$. The "translation" $w = u - u_0$ then reduces the nonhomogeneous problem (with unknown function u) to a homogeneous one (with unknown function w). (See, e.g., S. Fučík, A. Kufner [27].)

Thus, our main aim is the investigation of the operator equation $Tu = F$, i.e. the study of the properties of the *operator* T, and since T is defined by formula

$$a(u, v) = \langle Tu, v \rangle ,$$

the book deals mainly with the form $a(u, v)$.

This approach has — among other — the advantage that we need not introduce and investigate the rather complicated concept of the *trace* of a function (defined on Ω) on the boundary $\partial\Omega$ and the concept of "normal derivative" $\frac{\partial}{\partial n}$ which needs a detailed and tedious description of the geometric properties of the boundary of the domain Ω.

Finally, let us emphasize that we study various types of equations and boundary value problems in this book. Since each situation requires its own functional setting we do not give a general definition of the weak solution. We prefer to define the concept of the weak solution in every particular situation separately but — roughly speaking — to look for a weak solution means to find a function $u \in V$ satisfying the identity $a(u, v) = \langle F, v \rangle$ for every $v \in V$. Here, of course, also the appropriate choice of the space V is important.

Chapter 1
Preliminaries

Here, we intend to recall some concepts and results which will play an important — even if auxiliary — role in our further considerations. We will not go into details, refering the reader to the corresponding literature. The standard references are: R. A. Adams [1] and A. Kufner, O. John, S. Fučík [38] for function spaces, B. Opic, A. Kufner [55] and A. Kufner [37] for weighted Sobolev spaces, I.V. Skrypnik [61] for degree theory, and J.-L. Lions [47] for Leray–Lions theorem.

1.1 The domain Ω

We will deal with *real* functions $u = u(x)$ defined *almost everywhere* (a.e.) on an N-dimensional *domain* Ω,

$$\Omega \subset \mathbb{R}^N ,$$

i.e., Ω is an *open, connected* set of points $x = (x_1, x_2, \dots, x_N)$ of the N-dimensional Euclidean space $\mathbb{R}^N$, bounded or unbounded, possibly even equal to the whole $\mathbb{R}^N$.

The *boundary* of Ω will be denoted by

$$\partial\Omega .$$

We will suppose that $\partial\Omega$ is "sufficiently smooth" which — roughly speaking — means that $\partial\Omega$ is *locally Lipschitzian.* This property, whose exact description can be found in the literature (see, e.g. [1], [38] etc.) will guarantee that the results mentioned in the sequel (mainly, imbedding theorems for some function spaces) are valid.

Let us also point out that the property of $\partial\Omega$ mentioned above appears also under the assumption that Ω has the *cone property.*

1.2 Function spaces

We will consider functions $u = u(x)$ defined (and measurable) a.e. in Ω. The following spaces will be considered:

(i) *Lebesgue spaces* $L^p(\Omega)$, $1 \le p \le \infty$, where

$$L^p(\Omega) = \left\{ u = u(x)\,;\ \|u\|_p = \left(\int_\Omega |u(x)|^p \, dx \right)^{\frac{1}{p}} < \infty \right\} \quad \text{for} \quad 1 \le p < \infty\,,$$

$$L^\infty(\Omega) = \left\{ u = u(x)\,;\ \|u\|_\infty = \operatorname*{supess}_{x\in\Omega} |u(x)| < \infty \right\}.$$

The space $L^p(\Omega)$ equipped with the norm $\|u\|_p$ is a *Banach space*, which is for $1 < p < \infty$ uniformly convex and hence reflexive.

By

$$L^p_{\mathrm{loc}}(\Omega)$$

we will denote the set of all functions $u = u(x)$ defined a.e. on Ω, for which

$$u \in L^p(Q) \quad \text{for every } \textit{compact} \text{ set} \quad Q \subset \Omega\,.$$

(ii) *Weighted Lebesgue spaces* $L^p(\Omega, w)$, $1 \le p < \infty$, where $w = w(x)$ is a *weight function*, i.e., a function measurable and *positive a.e.* in Ω:

$$L^p(\Omega, w) = \left\{ u = u(x)\,;\ u w^{\frac{1}{p}} \in L^p(\Omega) \right\}.$$

It is again a *Banach space* (uniformly convex and hence reflexive if $p > 1$) equipped with the norm

$$\|u\|_{p,w} = \left(\int_\Omega |u(x)|^p w(x) \, dx \right)^{\frac{1}{p}}.$$

If the domain Ω is fixed or if it is clear what domain we have in mind, we will omit it and use — instead of $L^p(\Omega)$ and $L^p(\Omega, w)$ — the *short notation*

$$L^p \quad \text{and} \quad L^p(w)\,.$$

Lemma 1.1 *Let v be a weight function, M_v be the operator of pointwise multiplication,*

$$(M_v u)(x) := u(x)v(x)\,, \qquad x \in \Omega\,.$$

If w is a weight function and $v = w^{\frac{1}{p}}$, $1 \le p < \infty$, then $M_v : L^p(w) \to L^p$ is an isometric isomorphism.

This lemma, whose proof is straightforward, allows us to reduce all considerations connected with *weighted* Lebesgue spaces to the *nonweighted* ones.

1.3 Carathéodory functions, Nemytskij (superposition) operators

We have already used the concept of a Carathéodory function in Introduction. Therefore, let us only recall that a function $g = g(x, s)$, defined for $x \in \Omega$ and $s \in \mathbb{R}^m$, is called a *Carathéodory function*, shortly denoted as

$$g \in \text{CAR},$$

if
(i) the function $g(x, \cdot)$ is *continuous* on $\mathbb{R}^m$ for a.e. $x \in \Omega$, and

(ii) the function $g(\cdot, s)$ is *measurable* on Ω for every $s \in \mathbb{R}^m$.

For $g \in$ CAR, we define the *Nemytskij* (or *superposition) operator* G generated by g and acting on vector-valued functions $u = u(x)$, $u: \Omega \to \mathbb{R}^m$, by the formula

$$(Gu)(x) = g(x, u(x)), \qquad x \in \Omega.$$

It is well known (see, e.g., M. M. Vajnberg [64] or J. Appell, P. P. Zabreiko [6]) that the operator G maps the space

$$\prod_{i=1}^{m} L^{p(i)}, \quad 1 \le p(i) < \infty,$$

continuously into L^p, $1 \le p < \infty$, if and only if the following estimate holds:

$$|g(x, s)| \le a(x) + c \sum_{i=1}^{m} |s_i|^{\frac{p(i)}{p}} \tag{1.1}$$

for a.e. $x \in \Omega$ and every $s = (s_1, s_2, \dots, s_m) \in \mathbb{R}^m$ with a fixed (nonnegative) function $a \in L^p$ and a fixed nonnegative constant c.

The following assertion is an easy consequence of this result and Lemma 1.1.

Theorem 1.1 *Let $g \in$ CAR, $g: \Omega \times \mathbb{R}^m \to \mathbb{R}$. Let $w_0, w_1, \dots, w_m$ be weight functions on Ω. Then the corresponding Nemytskij operator G maps continuously*

$$\prod_{i=1}^{m} L^{p(i)}(w_i) \quad \textit{into} \quad L^p(w_0)$$

if and only if g satisfies

$$|g(x, s)| \le a(x) w_0^{-\frac{1}{p}}(x) + c w_0^{-\frac{1}{p}}(x) \sum_{i=1}^{m} |s_i|^{\frac{p(i)}{p}} w_i^{\frac{1}{p}}(x) \tag{1.2}$$

for a.e. $x \in \Omega$ and every $s \in \mathbb{R}^m$ with a fixed (nonnegative) function $a \in L^p$ and a fixed nonnegative constant c.

1.4 Function spaces (continued)

(i) For m a nonnegative integer and $\lambda \in (0, 1]$, we denote by

$$C^{m,\lambda}(\overline{\Omega})$$

the set of all functions $u = u(x)$ whose derivatives $D^\alpha u$ of order $|\alpha| \leq m$ are *continuous* and *bounded* on Ω and whose m-th order derivatives satisfy the *Hölder condition* with exponent λ. The norm on $C^{m,\lambda}(\overline{\Omega})$ is given by

$$\|u\|_{(m,\lambda)} = \sum_{|\alpha|\leq m} \sup_{x\in\Omega} |D^\alpha u(x)| + \sum_{|\alpha|=m} \sup_{\substack{x,y\in\Omega \\ x\neq y}} \frac{|D^\alpha u(x) - D^\alpha u(y)|}{|x-y|^\lambda} . \tag{1.3}$$

(ii) By

$$C_0^\infty(\Omega)$$

we denote the set of all functions $u = u(x)$ defined and infinitely differentiable on $\mathbb{R}^N$ and such that their *support* $\operatorname{supp} u$ is bounded and satisfies

$$\operatorname{supp} u \subset \Omega .$$

Recall that $\operatorname{supp} u$ is the closure (in the Euclidean norm of $\mathbb{R}^N$) of the set

$$\{x \in \mathbb{R}^N ;\ u(x) \neq 0\} .$$

(iii) For α a multiindex, $|\alpha| \geq 1$, the function v_α is called a *weak* (or *distributional*) derivative of u (of order α) if the identity

$$\int_\Omega v_\alpha(x)\varphi(x)\,dx = (-1)^{|\alpha|} \int_\Omega u(x) D^\alpha \varphi(x)\,dx$$

holds for every $\varphi \in C_0^\infty(\Omega)$. Then v_α is denoted by $D^\alpha u$.

(iv) *Sobolev spaces.* For $k \in \mathbb{N}$ and $1 \leq p < \infty$ we denote by

$$W^{k,p}(\Omega)$$

the set of all functions $u \in L^p(\Omega)$ for which the weak derivatives $D^\alpha u$ with $|\alpha| \leq k$ exist from $L^p(\Omega)$ as well. The *Sobolev space* $W^{k,p}(\Omega)$ is a Banach space (uniformly convex and hence reflexive if $1 < p < \infty$) if equipped with the norm

$$\|u\|_{k,p} = \left(\sum_{|\alpha|\leq k} \|D^\alpha u\|_p^p \right)^{\frac{1}{p}} . \tag{1.4}$$

We denote by

$$W_{\text{loc}}^{k,p}(\Omega)$$

the set of all functions u defined on Ω, for which $u \in W^{k,p}(Q)$ for every compact set $Q \subset \Omega$.

Further, the space

$$W_0^{k,p}(\Omega)$$

is defined as the closure of $C_0^\infty(\Omega)$ with respect to the norm $\|\cdot\|_{k,p}$. For Ω bounded, the expression

$$|||u|||_{k,p} = \left(\sum_{|\alpha|=k} \|D^\alpha u\|_p^p \right)^{\frac{1}{p}} \tag{1.5}$$

is a norm on $W_0^{k,p}(\Omega)$ *equivalent to* $\|\cdot\|_{k,p}$. This is a consequence of the famous *Friedrichs inequality* which claims that for $u \in C_0^\infty(\Omega)$, the estimate

$$\int_\Omega |u(x)|^p \, dx \leq c \int_\Omega |\nabla u(x)|^p \, dx \tag{1.6}$$

holds with $c > 0$ independent of u (but depending on Ω).

Let us mention that throughout the book, we will use the notation

$$Y \hookrightarrow X$$

for the *continuous* imbedding $Y \subset X$, i.e., for the estimate

$$\|u\|_X \leq C\|u\|_Y$$

for every $u \in Y$ with a constant $C > 0$ independent of u. The imbedding is thus realized by the identity operator $I: Y \to X$, and if this operator is, moreover, compact, we will speak about a *compact imbedding* which will be denoted by

$$Y \hookrightarrow\hookrightarrow X\,.$$

The most famous result concerning Sobolev spaces is the following *imbedding theorem*:

Theorem 1.2 (i) *The Sobolev space $W^{k,p}(\Omega)$ is continuously imbedded into the Banach space X, i.e., the estimate*

$$\|u\|_X \leq c\|u\|_{k,p} \tag{1.7}$$

holds for every function $u \in W^{k,p}(\Omega)$ with a constant $c > 0$ independent of u, where

(a) $X = L^q(\Omega)$ *with*

$$\frac{1}{q} \geq \frac{1}{p} - \frac{k}{N} \tag{1.8}$$

provided $kp < N$;

(b) $X = L^r(\Omega)$ *with arbitrary* $r \geq 1$ *if* $kp = N$;

(c) $X = C^{m,\lambda}(\overline{\Omega})$ *provided* $kp > N$,

where the nonnegative integer m *and the number* λ *are chosen in such a way that either*

$$(k - m - 1)p < N < (k - m)p \quad \textit{and} \quad 0 < \lambda \leq \frac{(k-m)p - N}{p} \tag{1.9}$$

or

$$(k - m - 1)p = N \quad \textit{and} \quad \lambda \in (0, 1) \quad \textit{is arbitrary.} \tag{1.10}$$

(ii) *Moreover, the imbedding* (b) *is compact, and the imbeddings* (a) *and* (c) *are compact if the inequalities in* (1.8) *and* (1.9) *are sharp.*

Example 1.1 (Sobolev inequality) Let us take $k = 1$ and $\Omega = \mathbb{R}^N$. Then the spaces $W^{1,p}(\mathbb{R}^N)$ and $W_0^{1,p}(\mathbb{R}^N)$ *coincide* (see, e.g., [1] or [38]) and the expression

$$\left(\int_{\mathbb{R}^N} |\nabla u|^p \, dx\right)^{\frac{1}{p}}$$

is also a *norm* in $W^{1,p}(\mathbb{R}^N)$. If we denote by p^* the value

$$p^* = \frac{Np}{N - p} \qquad (p < N) \tag{1.11}$$

(notice that this is the limit value in (1.8), called also the *critical Sobolev exponent*) then the following estimate holds for every $u \in W^{1,p}(\mathbb{R}^N)$:

$$\left(\int_{\mathbb{R}^N} |u(x)|^{p^*} \, dx\right)^{\frac{1}{p^*}} \leq c \left(\int_{\mathbb{R}^N} |\nabla u(x)|^p \, dx\right)^{\frac{1}{p}}. \tag{1.12}$$

This inequality, called also *Sobolev inequality*, will be used in the sequel. ◇

1.5 Weighted Sobolev spaces

These spaces are only a slight modification of the (classical) Sobolev spaces $W^{k,p}(\Omega)$, but their properties (mainly the imbeddings) are much more complicated and less transparent. These spaces will play an important role in our further considerations.

(i) Let $k \in \mathbb{N}$ and $1 \leq p < \infty$. Further, let w be a given *family of weight functions* w_α, $|\alpha| \leq k$:

$$w = \{w_\alpha(x), \; x \in \Omega; \; |\alpha| \leq k\}.$$

We denote by

$$W^{k,p}(\Omega, w)$$

the set of all functions $u \in L^p(\Omega, w_\Theta)$ (where Θ is the "zero"-multiindex, $\Theta = (0, 0, \dots, 0)$) for which the weak derivatives $D^\alpha u$ with $|\alpha| \le k$ belong to $L^p(\Omega, w_\alpha)$. The *weighted Sobolev space* $W^{k,p}(\Omega, w)$ is a normed linear space if equipped with the norm

$$\|u\|_{k,p,w} = \left(\sum_{|\alpha| \le k} \|D^\alpha u\|^p_{p,w_\alpha} \right)^{\frac{1}{p}} .$$

Theorem 1.3 *Let* $1 < p < \infty$ *and suppose that the weight functions* w_α *satisfy*

$$w_\alpha^{-\frac{1}{p-1}} \in L^1_{\mathrm{loc}}(\Omega), \quad |\alpha| \le k . \tag{1.13}$$

Then $W^{k,p}(\Omega, w)$ *is a uniformly convex (and hence reflexive) Banach space.*

(ii) If we additionally suppose that also

$$w_\alpha \in L^1_{\mathrm{loc}}(\Omega), \quad |\alpha| \le k , \tag{1.14}$$

then $C_0^\infty(\Omega)$ is a subset of $W^{k,p}(\Omega, w)$, and we can introduce the space

$$W_0^{k,p}(\Omega, w)$$

as the closure of $C_0^\infty(\Omega)$ with respect to the norm $\|\cdot\|_{k,p,w}$.

Remark 1.1 In all our further considerations, we will assume that the weight functions w_α from the family w *satisfy both conditions* (1.13) and (1.14), since only for such weight functions, the weighted Sobolev spaces $W^{k,p}(\Omega, w)$ and $W_0^{k,p}(\Omega, w)$ are *reasonably defined.* (For details and arguments, see [40], [55].)

Example 1.2 Let M be a nonempty subset of $\overline{\Omega} = \Omega \cup \partial\Omega$ and denote

$$d_M(x) = \mathrm{dist}(x, M), \quad x \in \mathbb{R}^N . \tag{1.15}$$

The most important — and in applications frequently appearing — case are weighted Sobolev spaces where the weight functions w_α are *functions of the distance* d_M, in particular, power functions

$$w_\alpha(x) = [d_M(x)]^{\varepsilon(\alpha)}, \quad \varepsilon(\alpha) \in \mathbb{R} . \tag{1.16}$$

The properties of spaces with these weights are very thoroughly described (see, e.g., [37]) and we will use them to illustrate our general results.

The set M is very often a *closed part of the boundary* $\partial\Omega$,

$$M \subset \partial\Omega .$$

In this case, the conditions (1.13) and (1.14) are obviously satisfied and the corresponding weighted Sobolev spaces $W^{k,p}(\Omega, w)$ and $W_0^{k,p}(\Omega, w)$ are reasonably defined Banach spaces. Moreover, it is also possible (and useful) to introduce another weighted Sobolev space denoted by

$$W_M^{k,p}(\Omega, w)$$

and defined as the closure — with respect to the norm $\|\cdot\|_{k,p,w}$ where w is the family of weight functions w_α given by (1.16), $w = \{w_\alpha;\ |\alpha| \leq k\}$ — of the set of all functions u infinitely differentiable on $\overline{\Omega}$ and such that $\operatorname{supp} u \cap M = \emptyset$. Notice that then we have

$$W_0^{k,p}(\Omega, w) = W_{\partial\Omega}^{k,p}(\Omega, w)\,.$$

◇

(iii) Also for weighted Sobolev spaces we can derive imbedding theorems, but the description of conditions under which they hold is much more complicated than in the case of nonweighted spaces $W^{k,p}(\Omega)$.

For simplicity, we will deal with the case $k = 1$, i.e., with spaces $W^{1,p}(\Omega, w)$ (and their subspaces) where the family w is in fact an $(N+1)$-tuple

$$w = \{w_0, w_1, \ldots, w_N\} \tag{1.17}$$

and where the norm $\|\cdot\|_{1,p,w}$ can be expressed in the form

$$\|u\|_{1,p,w} = \left(\int_\Omega |u(x)|^p w_0(x)\,dx + \sum_{i=1}^N \int_\Omega \left|\frac{\partial u}{\partial x_i}(x)\right|^p w_i(x)\,dx\right)^{\frac{1}{p}}. \tag{1.18}$$

The higher order case ($k > 1$) can be investigated using repeatedly the results for order one.

The easiest way how to derive imbeddings for *weighted* Sobolev spaces is to reduce them to *unweighted* spaces. This can be done, e.g., by the following procedure:

Example 1.3 (a) Let us consider, for simplicity, the weighted Sobolev space $W^{1,p}(\Omega, w)$ with a special choice of the family w:

$$w_0(x) \equiv 1\,, \quad w_1(x) = w_2(x) = \cdots = w_N(x) = \nu(x)\,.$$

In this case, we will use for the space $W^{1,p}(\Omega, w)$ the special notation $W^{1,p}(\nu, \Omega)$. This space is normed by

$$\|u\|_{1,p,\nu} = \left(\int_\Omega |u(x)|^p\,dx + \int_\Omega |\nabla u(x)|^p \nu(x)\,dx\right)^{\frac{1}{p}}. \tag{1.19}$$

Let us suppose that the weight function ν satisfies — besides (1.13), (1.14) — also the condition

$$\nu^{-s} \in L^1(\Omega) \tag{1.20}$$

with a certain $s > 0$ which will be specified later. (Notice that condition (1.20) is *stronger* than (1.13) since here we claim the integrability over the *whole* domain Ω while in (1.13) we only need the *local* integrability.)

Introducing the parameter p_s by

$$p_s = \frac{ps}{s+1} < p$$

and using the Hölder inequality with the parameter $r = \dfrac{s+1}{s} = \dfrac{p}{p_s} > 1$, we obtain

$$\begin{aligned} \int_\Omega |v(x)|^{p_s}\,dx &= \int_\Omega |v(x)|^{p_s} \nu^{\frac{p_s}{p}}(x)\nu^{-\frac{p_s}{p}}(x)\,dx \\ &\le \left(\int_\Omega |v(x)|^p \nu(x)\,dx\right)^{\frac{p_s}{p}} \left(\int_\Omega \nu^{-s}(x)\,dx\right)^{\frac{1}{s+1}} . \end{aligned} \tag{1.21}$$

Taking here $v = \dfrac{\partial u}{\partial x_i}$, $i = 1, 2, \dots, N$, and considering, moreover, a *bounded* domain, we see that a function $u \in W^{1,p}(\nu, \Omega)$ belongs to the nonweighted space $W^{1,p_s}(\Omega)$:

$$\|u\|_{1,p_s} \le c\|u\|_{1,p,\nu}$$

i.e.

$$W^{1,p}(\nu, \Omega) \hookrightarrow W^{1,p_s}(\Omega)\,, \tag{1.22}$$

of course, with a parameter p_s which is *less than* p.

Now, we can use Theorem 1.2 and obtain that

$$W^{1,p}(\nu, \Omega) \hookrightarrow L^r(\Omega) \tag{1.23}$$

where

$$1 \le r \le p_s^* = \frac{Np_s}{N - p_s} = \frac{Nps}{N(s+1) - ps} \quad \text{for} \quad ps < N(s+1)\,,$$

and $r \ge 1$ is arbitrary for $ps \ge N(s+1)$ (cf. Theorem 1.2 (a), (b)). Moreover, we have the *compact* imbedding

$$W^{1,p}(\nu, \Omega) \hookrightarrow\hookrightarrow L^r(\Omega) \tag{1.24}$$

provided $1 \le r < p_s^*$.

In particular, we have $p_s^* > p$ if $s > \dfrac{N}{p}$, and consequently,

$$W^{1,p}(\nu,\Omega) \hookrightarrow\hookrightarrow L^p(\Omega) \quad \text{for} \quad s > \frac{N}{p}\,. \tag{1.25}$$

We will use this compact imbedding later. Let us point out that (1.25) holds if

$$\nu^{-s} \in L^1(\Omega) \quad \text{with} \quad s \in \left(\frac{N}{p}, \infty\right) \cap \left[\frac{1}{p-1}, \infty\right) \tag{1.26}$$

since, to be in accordance with (1.13), we have to suppose that also $s \geq \dfrac{1}{p-1}$.

(b) Let us consider a higher order weighted Sobolev space $W^{k,p}(\Omega, w)$ with the special choice of w, namely

$$w_\alpha(x) \equiv 1 \quad \text{for} \quad |\alpha| < k\,, \qquad w_\alpha(x) = \nu(x) \quad \text{for} \quad |\alpha| = k\,.$$

This special space will be denoted by $W^{k,p}(\nu,\Omega)$ and normed by

$$\|u\|_{k,p,\nu} = \left(\sum_{|\alpha|\leq k-1} \int_\Omega |D^\alpha u(x)|^p\, dx + \sum_{|\alpha|=k} \int_\Omega |D^\alpha u(x)|^p \nu(x)\, dx\right)^{\frac{1}{p}}\,. \tag{1.27}$$

If we use the results of part (a) for the $(k-1)$-st derivatives of u, we immediately obtain from (1.25) that

$$W^{k,p}(\nu,\Omega) \hookrightarrow\hookrightarrow W^{k-1,p}(\Omega)$$

provided (1.26) holds.

(c) Let us consider the space $W^{1,p}(\nu,\Omega)$ from part (a) and its subspace $W_0^{1,p}(\nu,\Omega)$ with Ω a bounded domain. The imbeddings derived for $W^{1,p}(\nu,\Omega)$ hold also for $W_0^{1,p}(\nu,\Omega)$.

For $p < p_s^*$ we have, in virtue of the imbedding $W_0^{1,p_s}(\Omega) \hookrightarrow L^{p_s^*}(\Omega)$, that

$$\begin{aligned}\left(\int_\Omega |u(x)|^p\, dx\right)^{\frac{1}{p}} &\leq c_1 \left(\int_\Omega |u(x)|^{p_s^*}\, dx\right)^{\frac{1}{p_s^*}} \\ &\leq c_2 \left(\int_\Omega \left(|u(x)|^{p_s} + |\nabla u(x)|^{p_s}\right) dx\right)^{\frac{1}{p_s}}.\end{aligned}$$

The Friedrichs inequality in $W_0^{1,p_s}(\Omega)$ — see (1.6) — yields

$$\left(\int_\Omega \left(|u(x)|^{p_s} + |\nabla u(x)|^{p_s}\right) dx\right)^{\frac{1}{p_s}} \leq c_3 \left(\int_\Omega |\nabla u(x)|^{p_s}\, dx\right)^{\frac{1}{p_s}}.$$

Since in part (a) we have shown that

$$\left(\int_\Omega |\nabla u(x)|^{p_s}\,dx\right)^{\frac{1}{p_s}} \le \left(\int_\Omega \nu^{-s}(x)\,dx\right)^{\frac{1}{ps}} \left(\int_\Omega |\nabla u(x)|^p \nu(x)\,dx\right)^{\frac{1}{p}},$$

we immediately have that the estimate

$$\int_\Omega |u(x)|^p\,dx \le c_4 \int_\Omega |\nabla u(x)|^p \nu(x)\,dx \tag{1.28}$$

holds for every $u \in C_0^\infty(\Omega)$ with a constant $c_4 > 0$ independent of u provided (1.26) is satisfied.

Inequality (1.28) can be considered a *weighted Friedrichs inequality*. It implies that the expression

$$|||u|||_{1,p,\nu} = \left(\int_\Omega |\nabla u(x)|^p \nu(x)\,dx\right)^{\frac{1}{p}} \tag{1.29}$$

is a norm on the space $W_0^{1,p}(\nu, \Omega)$ equivalent with the norm (1.19). ◇

A counterpart of the imbedding (a) from Theorem 1.2 for *weighted* Sobolev spaces is the imbedding

$$W^{1,p}(\Omega, w) \hookrightarrow L^q(\Omega, \omega),$$

i.e., the estimate

$$\|u\|_{q,\omega} \le c\|u\|_{1,p,w}, \tag{1.30}$$

and what we need are conditions on the parameters p, q and the weights w (from (1.17)) and ω under which (1.30) holds.

Inequality (1.30) is a consequence of the so-called *Hardy-type inequality*

$$\left(\int_\Omega |u(x)|^q \omega(x)\,dx\right)^{\frac{1}{q}} \le c \left(\sum_{i=1}^N \int_\Omega \left|\frac{\partial u}{\partial x_i}(x)\right|^p w_i(x)\,dx\right)^{\frac{1}{p}}. \tag{1.31}$$

Of course, here we need some additional conditions on u since the expression on the right hand side of (1.31) is in general only a *seminorm*. But similarly to the case of classical Sobolev spaces, we can consider inequality (1.31) for $u \in C_0^\infty(\Omega)$, and then it describes in fact the imbedding

$$W_0^{1,p}(\Omega, w) \hookrightarrow L^q(\Omega, \omega). \tag{1.32}$$

Sometimes we can replace the space $W_0^{1,p}(\Omega, w)$ by the space $W_M^{1,p}(\Omega, w)$ introduced in Example 1.2.

We will not give here details about the connections between p and q and between ω and $w_0, w_1, \dots, w_N$ which guarantee that the inequality (1.30) or inequality (1.31) holds, since the precise assumptions depend on the particular situation. We will only mention some special cases; the general situation is described in [55] for rather broad classes of weights and in [37] for the special weights mentioned in Example 1.2.

Remark 1.2 It should be emphasized that in (1.30) and (1.31) *both* possibilities,

$$p \le q \quad \text{and} \quad p > q\,,$$

may occur. In the latter case, the corresponding imbedding is not only continuous but, moreover, *compact*, while in the former case, some additional assumptions are needed to guarantee compactness.

Example 1.4 (the one-dimensional case) For $N = 1$ and $\Omega = (a, b)$ with $-\infty \le a < b \le \infty$, the conditions which guarantee the validity of (1.30) have been described completely (see, e.g., [55]). Here, let us consider the case of functions u satisfying the additional condition

$$u(a) = 0\,; \tag{1.33}$$

if we denote by $W_L^{1,p}((a,b), w_1)$ the set of all functions u for which the right-hand side in the *Hardy inequality*

$$\left(\int_a^b |u(t)|^q \omega(t)\,dt\right)^{\frac{1}{q}} \le c\left(\int_a^b |u'(t)|^p w_1(t)\,dt\right)^{\frac{1}{p}} \tag{1.34}$$

is finite and which satisfy (1.33), then the right-hand side in (1.34) is a *norm* in the space $W_L^{1,p}((a,b), w_1)$ and (1.34) describes the imbedding

$$W_L^{1,p}((a,b), w_1) \hookrightarrow L^q((a,b), \omega)\,. \tag{1.35}$$

Now, inequality (1.34) holds — and the imbedding (1.35) is continuous — for $1 < p \le q < \infty$ if and only if the following condition is satisfied:

$$\sup_{a<x<b} B(x) < \infty \tag{1.36}$$

where

$$B(x) = \left(\int_x^b \omega(t)\,dt\right)^{\frac{1}{q}} \left(\int_a^x w_1^{1-p'}(t)\,dt\right)^{\frac{1}{p'}}, \quad p' = \frac{p}{p-1}\,. \tag{1.37}$$

Moreover, the imbedding (1.35) is *compact* if and only if

$$\lim_{x\to a+} B(x) = \lim_{x\to b-} B(x) = 0\,. \tag{1.38}$$

For the case $1 < q < p < \infty$ the necessary and sufficient condition for the continuity as well as for the compactness of the imbedding (1.35) reads as follows:

$$A = \left(\int_a^b \left(\int_x^b \omega(t)\,dt\right)^{\frac{r}{q}} \left(\int_a^x w_1^{1-p'}(t)\,dt\right)^{\frac{r}{q'}} w_1^{1-p'}(x)\,dx\right)^{\frac{1}{r}} < \infty \tag{1.39}$$

where $\dfrac{1}{r} = \dfrac{1}{q} - \dfrac{1}{p}$. ◇

Example 1.5 Let us take $w_0(x) = w_1(x) = \cdots = w_N(x) = [\operatorname{dist}(x, \partial\Omega)]^\lambda$ and $\omega(x) = [\operatorname{dist}(x, \partial\Omega)]^\kappa$, $\Omega \subset \mathbb{R}^N$ bounded, $\lambda, \kappa \in \mathbb{R}$, $\lambda < p - 1$. Then the inequality (1.31) holds for every $u \in C_0^\infty(\Omega)$ — and consequently, the imbedding (1.32) is continuous — if and only if either

$$1 < p \le q < \infty, \quad \frac{N}{q} - \frac{N}{p} + 1 \ge 0, \quad \kappa \ge \lambda\frac{q}{p} - N + N\frac{q}{p} - q \tag{1.40}$$

or

$$1 \le q < p < \infty, \quad \kappa > \lambda\frac{q}{p} - 1 + \frac{q}{p} - q\,.$$

Moreover, the imbedding (1.32) is *compact* not only for $p > q$ (as mentioned in Remark 1.2) but also for $p \le q$ if the last inequality in (1.40) is sharp. ◇

Example 1.6 Let us take $p = q$ and consider the inequality

$$\int_\Omega |u(x)|^p d_M^\kappa(x)\,dx \le c \int_\Omega |\nabla u(x)|^p d_M^\lambda(x)\,dx \tag{1.41}$$

where $M \subset \overline{\Omega}$ and d_M is the weight from Example 1.2. Under some additional conditions on u, this inequality holds — roughly speaking — for

$$\kappa = \lambda - p$$

where the set of admissible values of λ depends on the dimension m of the set M. E.g., if $u \in C_0^\infty(\Omega)$, $M \subset \partial\Omega$ and $\dim M = m$, then (1.41) holds for

$$\lambda \ne p - N + m\,.$$

◇

Example 1.7 For *unbounded* domains, weight functions of the type

$$(1 + |x|)^{-\varepsilon}, \tag{1.42}$$

mostly with $\varepsilon > 0$, appear in various circumstances. E.g., the inequality

$$\int_{\mathbb{R}^N} \frac{|u(x)|^p}{(1 + |x|)^\kappa}\,dx \le \left(\frac{p}{N - \kappa}\right)^p \int_{\mathbb{R}^N} \frac{|\nabla u(x)|^p}{(1 + |x|)^\lambda}\,dx \tag{1.43}$$

(which is in fact inequality (1.31) for $\Omega = \mathbb{R}^N$, $p = q$, $\omega(x) = (1 + |x|)^{-\kappa}$, $w_1(x) = \cdots = w_N(x) = (1 + |x|)^{-\lambda}$) holds for $u \in C_0^\infty(\mathbb{R}^N)$ provided

$$\kappa < N\,, \quad \kappa = p + \lambda\,. \tag{1.44}$$

Indeed, (1.43) follows from the following calculations where we use Green's formula and Hölder's inequality:

$$\int_{\mathbb{R}^N} \frac{|u(x)|^p}{(1+|x|)^\kappa}\,dx = \frac{1}{N}\int_{\mathbb{R}^N} (\operatorname{div} x)\frac{|u(x)|^p}{(1+|x|)^\kappa}\,dx$$

$$= -\frac{1}{N}\left(\int_{\mathbb{R}^N} px\cdot\nabla u(x)\frac{|u(x)|^{p-2}u(x)}{(1+|x|)^\kappa}\,dx - \kappa\int_{\mathbb{R}^N}\frac{|u(x)|^p}{(1+|x|)^{\kappa+1}}|x|\,dx\right)$$

$$\leq \frac{1}{N}\left(p\int_{\mathbb{R}^N}\frac{|u(x)|^{p-1}}{(1+|x|)^{\frac{\kappa(p-1)}{p}}}\frac{|\nabla u(x)|}{(1+|x|)^{-1+\frac{\kappa}{p}}}\,dx + \kappa\int_{\mathbb{R}^N}\frac{|u(x)|^p}{(1+|x|)^\kappa}\,dx\right)$$

$$\leq \frac{p}{N}\left(\int_{\mathbb{R}^N}\frac{|\nabla u(x)|^p}{(1+|x|)^{\kappa-p}}\,dx\right)^{\frac{1}{p}}\left(\int_{\mathbb{R}^N}\frac{|u(x)|^p}{(1+|x|)^\kappa}\,dx\right)^{1-\frac{1}{p}}$$

$$+\frac{\kappa}{N}\int_{\mathbb{R}^N}\frac{|u(x)|^p}{(1+|x|)^\kappa}\,dx\,.$$

Let us note that for $\kappa = p$ i.e., for $\lambda = 0$, we obtain from (1.43) the inequality

$$\int_{\mathbb{R}^N}\frac{|u(x)|^p}{(1+|x|)^p}\,dx \leq \left(\frac{p}{N-p}\right)^p\int_{\mathbb{R}^N}|\nabla u(x)|^p\,dx\,. \tag{1.45}$$

◇

Example 1.8 (generalized Sobolev inequality) The Sobolev inequality (1.12) can be extended to the case of weights. Here, we will use weight functions of the form (1.42). Then we have that the inequality

$$\left(\int_{\mathbb{R}^N}\frac{|u(x)|^{p^*}}{(1+|x|)^\gamma}\,dx\right)^{\frac{1}{p^*}} \leq c\left(\int_{\mathbb{R}^N}\frac{|\nabla u(x)|^p}{(1+|x|)^\lambda}\,dx\right)^{\frac{1}{p}} \tag{1.46}$$

holds for $u \in C_0^\infty(\mathbb{R}^N)$ if $1 < p < N$, provided

$$p^* = \frac{Np}{N-p}\,,\quad \lambda < N-p\,,\quad \gamma = \lambda\frac{p^*}{p} = \lambda\frac{N}{N-p}\,. \tag{1.47}$$

Indeed, for any $u \in C_0^\infty(\mathbb{R}^N)$ and any differentiable weight function ω on $\mathbb{R}^N$ we have from (1.12) that

$$\left(\int_{\mathbb{R}^N}\omega(x)|u(x)|^{p^*}\,dx\right)^{\frac{1}{p^*}} \leq c_1\left(\int_{\mathbb{R}^N}|\nabla(\omega^{\frac{1}{p^*}}(x)u(x))|^p\,dx\right)^{\frac{1}{p}}.$$

Taking now $\omega(x) = (1+|x|)^{-\gamma}$ and using Minkowski's inequality we obtain that

$$\left(\int_{\mathbb{R}^N}\frac{|u(x)|^{p^*}}{(1+|x|)^\gamma}\,dx\right)^{\frac{1}{p^*}} \leq c_1\left(\int_{\mathbb{R}^N}\left|\frac{\nabla u(x)}{(1+|x|)^{\frac{\gamma}{p^*}}} - \frac{\gamma}{p^*}\frac{u(x)}{(1+|x|)^{1+\frac{\gamma}{p^*}}}\frac{x}{|x|}\right|^p dx\right)^{\frac{1}{p}} \leq$$

$$\le c_2 \left(\int_{\mathbb{R}^N} \left(\frac{|\nabla u(x)|^p}{(1+|x|)^{\gamma \frac{p}{p^*}}} + \frac{|u(x)|^p}{(1+|x|)^{p+\gamma \frac{p}{p^*}}} \right) dx \right)^{\frac{1}{p}}$$

$$\le c_2 \left[\left(\int_{\mathbb{R}^N} \frac{|\nabla u(x)|^p}{(1+|x|)^{\gamma \frac{p}{p^*}}} dx \right)^{\frac{1}{p}} + \left(\int_{\mathbb{R}^N} \frac{|u(x)|^p}{(1+|x|)^{p+\gamma \frac{p}{p^*}}} dx \right)^{\frac{1}{p}} \right]$$

and (1.46) follows if we estimate the last integral by (1.43) since due to (1.47), we have $\gamma \dfrac{p}{p^*} = \lambda$. ◇

Notice that for $\lambda = 0$, we obtain from (1.46) the Sobolev inequality (1.12).

(iv) As a counterpart of the imbedding (c) from Theorem 1.2, there are also imbeddings of weighted Sobolev spaces into (weighted) spaces of continuous or Hölder-continuous functions, i.e., estimates of the type

$$\operatorname*{supess}_{x \in \Omega} |u(x)\omega(x)| \le c\|u\|_{1,p,w} \tag{1.48}$$

for $p > N$ with $u = u(x)$ continuous on Ω. E.g., for the weights ω and w_i, $i = 0, 1, \ldots, N$, from Example 1.2, inequality (1.48) holds (with a constant $c > 0$ independent of $u \in W_0^{1,p}(\Omega, w)$) if and only if

$$\kappa \ge \frac{\lambda}{p} - 1 + \frac{N}{p}$$

provided $p > N$, $\lambda \ne p - 1$. For details, see R. C. Brown, B. Opic [12].

1.6 Leray–Lions theorem

The existence result for the second order equations in Chapter 2 is based on the *Leray–Lions theorem.* In this section we will formulate this assertion for the reader's convenience.

Theorem 1.4 *Let X be a reflexive Banach space. Let T be an operator defined on X with values in the dual space X^*, and let the following conditions* (i)–(viii) *be satisfied:*

(i) *the operator T is bounded;*

(ii) *the operator T is demicontinuous (i.e. T maps strongly convergent sequences in X to weakly convergent sequences in X^*);*

(iii) *the operator T is coercive (i.e.* $\lim_{\|u\|\to\infty} \frac{\langle Tu, u\rangle}{\|u\|} = \infty$).

Let there exist a bounded mapping Φ from the space $X \times X$ into the space X^* such that

(iv) $\Phi(u, u) = Tu$ *for every* $u \in X$*;*

(v) *for all* $u, w, h \in X$ *and any sequence* $\{t_n\}_{n=1}^{\infty}$ *of real numbers such that* $t_n \to 0$*, it is*

$$\Phi(u + t_n h, w) \rightharpoonup \Phi(u, w)\,;$$

(vi) *for all* $u, w \in X$*, it is*

$$\langle \Phi(u, u) - \Phi(w, u), u - w\rangle \geq 0$$

(*the so-called* condition of monotonicity in the principal part)*;*

(vii) *if* $u_n \rightharpoonup u$ *and* $\langle \Phi(u_n, u_n) - \Phi(u, u_n), u_n - u\rangle \to 0$ *then for arbitrary* $w \in X$ *it is*

$$\Phi(w, u_n) \rightharpoonup \Phi(w, u)\,;$$

(viii) *if* $w \in X$*,* $u_n \rightharpoonup u$*,* $\Phi(w, u_n) \rightharpoonup z$*, then*

$$\langle \Phi(w, u_n), u_n\rangle \to \langle z, u\rangle\,.$$

Then the equation

$$Tu = f$$

has at least one solution $u \in X$ *for every* $f \in X^*$.

We refrain from giving the proof of the Leray–Lions theorem here. The reader is refered to [47].

1.7 Degree of mappings of monotone type

In this section we define the degree for generalized monotone mappings and summarize its properties which we will use in the forthcoming chapters to prove our existence results.

Definition 1.1 Let X be a reflexive Banach space and X^* its dual. The operator $T\colon X \to X^*$ is said to satisfy *condition* $\alpha(X)$ if the assumptions

$$u_n \rightharpoonup u_0 \quad \text{in} \quad X \quad \text{and} \quad \limsup_{n\to\infty} \langle Tu_n, u_n - u_0\rangle \leq 0$$

imply

$$u_n \to u_0 \quad \text{in} \quad X\,.$$

Then it follows from the results of [61] and F. E. Browder [10] that the notion of the degree can be defined for a class of mappings from X into X^*. Namely, we have the following basic assertion.

Theorem 1.5 *Let $T: X \to X^*$ be a bounded and demicontinuous operator satisfying condition $\alpha(X)$. Let $D \subset X$ be an open, bounded and nonempty set with the boundary ∂D such that $Tu \neq 0$ for $u \in \partial D$. Then there exists an integer*

$$\mathrm{Deg}[T;\, D, 0]$$

(*called the* degree of the mapping T with respect to the set D and the origin 0) *such that*

(i) $\mathrm{Deg}[T;\, D, 0] \neq 0$ *implies that there exists an element $u_0 \in D$ such that*

$$Tu_0 = 0\,.$$

(ii) *If D is symmetric with respect to the origin and T is an odd mapping (i.e. $Tu = -T(-u)$ for any $u \in D$) then*

$$\mathrm{Deg}[T;\, D, 0]$$

is an odd number (and thus different from zero).

(iii) (Invariance with respect to homotopy) *Let T_λ be a family of mappings depending continuously on a real parameter $\lambda \in [0, 1]$, and $T_\lambda(u) \neq 0$ for any $u \in \partial D$ and $\lambda \in [0, 1]$* (the map $(\lambda, u) \mapsto T_\lambda u$ from $\mathbb{R} \times X$ into X^* is called an *admissible* homotopy (with respect to D and 0)). *Then*

$$\mathrm{Deg}[T_\lambda;\, D, 0]$$

is constant with respect to $\lambda \in [0, 1]$. In particular, we have

$$\mathrm{Deg}[T_0;\, D, 0] = \mathrm{Deg}[T_1;\, D, 0]\,.$$

We will not give the proof here. The reader who is interested in the construction of the degree is refered to the book [61]. Note only that the main idea consists in a suitable approximation of T by operators with finite dimensional ranges. To show correctness of these approximations the validity of the condition $\alpha(X)$ is essential. This idea extends a similar approach working with the so-called Leray–Schauder degree and using compact perturbations of the identity instead of condition $\alpha(X)$ (see e.g. S. Fučík, A. Kufner [27]).

The following assertion is an immediate consequence of Definition 1.1 and the properties of compact operators (which map weakly convergent sequences in X to strongly convergent sequences in X^*).

Lemma 1.2 *Let $T: X \to X^*$ satisfy condition $\alpha(X)$ and let $K: X \to X^*$ be a compact operator. Then the sum $T + K: X \to X^*$ satisfies condition $\alpha(X)$.*

The following assertion combined with Theorem 1.5 (i) is a crucial tool in proving the existence of a solution.

Theorem 1.6 *Let $T: X \to X^*$ be a bounded, demicontinuous mapping satisfying condition $\alpha(X)$, $0 \in \overline{D} \setminus \partial D$, $Tu \neq 0$ for $u \in \partial D$, D being as in Theorem* 1.5, *and let for $u \in \partial D$ the inequality*

$$\langle Tu, u \rangle \geq 0$$

be valid. Then

$$\mathrm{Deg}[T; D, 0] = 1 .$$

The reader is refered to [61] for the proof.

Definition 1.2 Let $u_0 \in X$ and let $T: X \to X^*$. If $Tu_0 = 0$ then u_0 is called a *critical point of the mapping T*. A point u_0 is called *an isolated critical point* of T if there exists a ball $B_r(u_0) = \{u \in X;\ \|u - u_0\| < r\}$ which contains no critical points of T except u_0. If $F: X \to \mathbb{R}$ is a functional with the Fréchet derivative $F' = T$ then (isolated) critical point u_0 of T is called an (isolated) critical point of the functional F.

Definition 1.3 The number

$$\lim_{r \to 0} \mathrm{Deg}[T; B_r(u_0), 0]$$

will be called *the index of the isolated critical point u_0* of the mapping T and denoted by

$$\mathrm{Ind}(T, u_0) .$$

Theorem 1.7 (Additivity property of the degree) *Assume that the mapping $T: X \to X^*$ is bounded, demicontinuous and satisfies condition $\alpha(X)$. Let T have only isolated critical points in D (D being as in Theorem* 1.5) *and let $Tu \neq 0$ for $u \in \partial D$. Then there is only a finite number of critical points u_i in D, $i = 1, \dots, n$, and the equality*

$$\mathrm{Deg}[T; D, 0] = \sum_{i=1}^{n} \mathrm{Ind}(T, u_i)$$

holds.

Theorem 1.8 *Assume that the functional $F: X \to \mathbb{R}$ has a local minimum at $u_0 \in X$ and its Fréchet derivative $F': X \to X^*$ is a bounded and demicontinuous mapping which satisfies condition $\alpha(X)$. Let, moreover, u_0 be an isolated critical point of F'. Then* $\mathrm{Ind}(F', u_0) = 1$.

The reader is refered to [61] for the proof of the above theorems. Note also that the finite dimensional analogue of Theorem 1.8 can be found in H. Amann [4].

1.8 Harnack-type inequality, decay of solution, local regularity and interpolation inequality

For the reader's convenience we will now formulate some useful results which are quite often refered to in the forthcoming chapters. Note that we do not formulate these results in full generality. For the sake of brevity we do it for simpler equations which cover all cases dealt with in Chapters 2, 3 and 4. Let us also emphasize that the assertions presented below are rather technical. To see the precise meaning of all constants which appear there the reader is invited to consult the refered literature.

Let us consider the equation

$$-\operatorname{div}(a(x,u)|\nabla u|^{p-2}\nabla u) = h(x,u) \tag{1.49}$$

in a domain $\Omega \subset \mathbb{R}^N$ (where possibly $\Omega = \mathbb{R}^N$), $p > 1$ and functions a and h satisfy the following assumptions:

(a) $a \in \mathrm{CAR}$, $a(x,s)$ is uniformly separated from zero by a positive constant and bounded for a.e. $x \in \Omega$ and all $s \in \mathbb{R}$;

(b) $h \in \mathrm{CAR}$ and for any $M > 0$ there exists a constant $c_M > 0$ such that

$$|h(x,s)| \le c_M |s|^{p-1} \tag{1.50}$$

for a.e. $x \in \Omega$ and for all $s \in (-M, M)$.

A function $u \in W^{1,p}_{\mathrm{loc}}(\Omega)$ is said to be a *distributional solution* of (1.49) in Ω if

$$\int_\Omega a(x,u)|\nabla u|^{p-2}\nabla u \nabla\varphi\, dx = \int_\Omega h(x,u)\varphi\, dx$$

holds for any $\varphi \in C_0^\infty(\Omega)$.

Let us denote by $K = K(\varrho)$ a cube with the edge of length equal to $\varrho > 0$. The next result follows from the more general Theorem 1.1 in N. S. Trudinger [63]. Note, for completeness, that the distributional solution is called the weak solution in [63]. The notion of the weak solution in our book has a different meaning but any weak solution of second order equations from Chapter 4 (where Theorem 1.9 is applied) will be simultaneously a distributional solution in the sense presented above.

Theorem 1.9 *Let $u = u(x)$ be a distributional solution of* (1.49) *in a cube $K = K(3\varrho) \subset \Omega$ with $0 \le u < M$ in K. Then the so-called Harnack-type inequality holds:*

$$\max_{x \in K(\varrho)} u(x) \le C \min_{x \in K(\varrho)} u(x),$$

where $C = C(p, N, a, h, \varrho c_M)$ and the centres of $K(\rho)$ and $K(3\rho)$ coincide.

Remark 1.3 If we allow $a(x, s)$ to be unbounded with respect to $|s| \to +\infty$, the nonlinearity h satisfies on $\Omega \times \mathbb{R}$ the growth condition

$$|h(x, s)| \leq c|s|^{\gamma-1} \tag{1.51}$$

with some $c > 0$, $p < \gamma < p^* = \dfrac{Np}{N-p}$ $(1 < p < N)$ and we have an *apriori* L^∞*-bound* for the distributional solution of (1.49), then the assertion of Theorem 1.9 remains true. Indeed, we can take $M > 0$ and $c_M > 0$ so large that (1.51) implies (1.50) on $\Omega \times (-M, M)$. Also $a(x, s)$ can be modified outside of $(-M, M)$ in order to satisfy (a).

In particular, if the distributional solution $u \not\equiv 0$ of (1.49) satisfies an apriori L^∞-bound and $u \geq 0$ in Ω then it follows from Theorem 1.9 that u is *stricly positive* in Ω: $u > 0$.

Then the result below is a consequence of the more general Theorem 1 in J. Serrin [60].

Theorem 1.10 *Let us assume the same as in Theorem* 1.9, *or Remark* 1.3, $1 < p < N$, *respectively, and let the distributional solution u of* (1.49) *be bounded in $L^\infty(\Omega)$. Assume, moreover, that $h(x, u) \in L^{\gamma_1}(\Omega)$ for some $\gamma_1 > \dfrac{N}{p}$. Then for any $x \in \Omega$ we have*

$$\|u\|_{L^\infty(B_1(x))} \leq \left(\|u\|_{L^{p^*}(B_2(x))} + \|h(x, u)\|_{L^{\gamma_1}(B_2(x))}\right), \tag{1.52}$$

where $c = c(p, N, \gamma_1) > 0$ and $B_1(x) \subset B_2(x) \subset \Omega$.

Remark 1.4 In particular, if we have

$$\lim_{|x|\to\infty} \|u\|_{L^{p^*}(B_2(x))} = 0 \quad \text{and} \quad \lim_{|x|\to\infty} \|h(x, u)\|_{L^{\gamma_1}(B_2(x))} = 0$$

then (1.52) implies the *decay* of $u = u(x)$ as $|x| \to \infty$.

We have also a *regularity result* which follows from Theorem 1 in P. Tolksdorf [62].

Theorem 1.11 *Let us assume the same as in Theorem* 1.9, *or Remark* 1.3, *respectively. Let, moreover, $a \in C^1(\Omega \times \mathbb{R})$ and let the distributional solution u of* (1.49) *satisfy $u \in L^\infty(\Omega)$. Then u is locally in $C^{1,\alpha}(\Omega)$ with $\alpha \in (0, 1)$.*

Remark 1.5 More precisely, the assertion of Theorem 1.11 means that given $x_0 \in \Omega$ and $B_{3r}(x_0) \subset \Omega$ there exists $\alpha \in (0, 1)$ such that $u \in C^{1,\alpha}(\overline{B_r(x_0)})$.

Theorem 1.12 (Lions' interpolation lemma — see J. L. Lions [47]) *Let X, Y, Z be Banach spaces satisfying the imbeddings*

$$X \hookrightarrow\hookrightarrow Y \hookrightarrow Z\,.$$

Then, for any $\varepsilon > 0$, there is a positive constant $c(\varepsilon)$ such that for every $u \in X$,

$$\|u\|_Y \leq \varepsilon\|u\|_X + c(\varepsilon)\|u\|_Z\,.$$

1.9 Some technical lemmas

This section is devoted to some technical lemmas which play an important role in the proofs of our existence results. The first result is based on *Clarkson's inequality* and its precise proof can be found in P. Lindqvist [46].

Lemma 1.3 *Let $p \geq 2$. Then for every $\chi_1, \chi_2 \in \mathbb{R}^N$ we have*

$$|\chi_2|^p - |\chi_1|^p \geq p|\chi_1|^{p-2}\chi_1(\chi_2 - \chi_1) + \frac{|\chi_2 - \chi_1|^p}{2^{p-1} - 1}. \tag{1.53}$$

Let $1 < p < 2$. Then for every $\chi_1, \chi_2 \in \mathbb{R}^N$ which are not simultaneously equal to zero we have

$$|\chi_2|^p - |\chi_1|^p \geq p|\chi_1|^{p-2}\chi_1(\chi_2 - \chi_1) + \frac{3p(p-1)}{16} \cdot \frac{|\chi_2 - \chi_1|^2}{(|\chi_1| + |\chi_2|)^{2-p}}. \tag{1.54}$$

Remark 1.6 It follows from (1.53) and (1.54) that the inequality

$$|\chi_2|^p - |\chi_1|^p > p|\chi_1|^{p-2}\chi_1(\chi_2 - \chi_1) \tag{1.55}$$

holds for any $\chi_1, \chi_2 \in \mathbb{R}^N$, $\chi_1 \neq \chi_2$ and for any $p > 1$. Note that the inequality (1.55) is just a restating of the strict convexity of the mapping $\chi \mapsto |\chi|^p$ and can be proved independently of (1.53) and (1.54).

The second result we want to present in this section is a special property of real functions of one real variable which plays an important role in L^∞-estimates of solutions. The assertion is due to Stampacchia, and in M. K. V. Murthy, G. Stampacchia [51] it was used to prove an L^∞-bound of solutions of certain second order degenerated elliptic equations. Since the proof of this assertion is not easily available, we prefer to give it here.

Lemma 1.4 *Let $\zeta = \zeta(t)$ be a nonnegative, nonincreasing function on the half line $t \geq K_0 \geq 0$ such that*

$$\zeta(h) \leq c\frac{(\zeta(K))^\delta}{(h-K)^\sigma} \tag{1.56}$$

for $h > K \geq K_0$ with some $c > 0$. Then $\sigma > 0$, $\delta > 1$ imply

$$\zeta(K_0 + \tau) = 0,$$

for any $\tau \geq d$, where $d = c^{\frac{1}{\sigma}}(\zeta(K_0))^{\frac{\delta-1}{\sigma}} 2^{\frac{\delta}{\delta-1}}$.

Proof. Let d be as above and define a sequence $\{k_n\}$ by

$$k_0 = K_0, \quad k_n = k_{n-1} + \frac{d}{2^n}, \quad n = 1, 2, \dots . \tag{1.57}$$

Substituting (1.57) into (1.56) we get by induction

$$\zeta(k_n) \le \zeta(K_0) 2^{\frac{n\sigma}{1-\delta}} \to 0$$

for $n \to \infty$. Since $\lim_{n\to\infty} k_n = K_0 + d$ and ζ is nonincreasing, we obtain $\zeta(K_0 + \tau) = 0$ for any $\tau \ge d$. □

Chapter 2

Solvability of nonlinear boundary value problems

2.1 Formulation of the problem

Let Ω be a domain in $\mathbb{R}^N$ with boundary $\partial\Omega$. We will consider the following (partial) differential operator of order $2k$ in the divergence form:

$$(Au)(x) = \sum_{|\alpha|\le k} (-1)^{|\alpha|} D^\alpha a_\alpha(x; u, \nabla u, \ldots, \nabla^k u) \tag{2.1}$$

(see Introduction, formula (0.22)) where $a_\alpha = a_\alpha(x; \xi)$ are functions defined on $\Omega \times \mathbb{R}^m$, $a_\alpha \in \mathrm{CAR}$.

Further, let $w = \{w_\alpha(x);\ |\alpha| \le k\}$ be a family of weight functions on Ω satisfying the conditions

$$w_\alpha \in L^1_{\mathrm{loc}}(\Omega)\,, \quad w_\alpha^{-\frac{1}{p-1}} \in L^1_{\mathrm{loc}}(\Omega)\,, \quad |\alpha| \le k\,, \tag{2.2}$$

with some parameter $p > 1$, and let us consider the weighted Sobolev space

$$W^{k,p}(\Omega, w) \tag{2.3}$$

as well as its subspace

$$W_0^{k,p}(\Omega, w)\,. \tag{2.4}$$

Due to (2.2), both spaces are well-defined reflexive Banach spaces (see Theorem 1.3).

We will consider a boundary value problem (BVP) for the equation

$$Au = f \quad \text{in} \quad \Omega \tag{2.5}$$

with f a given function on Ω. Without going into details, we will replace the description of the corresponding *boundary conditions* by the introduction of a certain subspace V of $W^{k,p}(\Omega, w)$ which has the property that the imbeddings

$$W_0^{k,p}(\Omega, w) \hookrightarrow V \hookrightarrow W^{k,p}(\Omega, w) \tag{2.6}$$

are continuous, and we will simply speak about a *solution of* (2.5) *with respect to* V or about a V*-solution* of (2.5) or about a *weak solution* of (2.5) (see Definition 2.1).

Remark 2.1 As already mentioned at the end of Introduction, we adopt here a concept which is commonly used, e.g., in the theory of weak solutions of linear as well as nonlinear elliptic problems. Here, let us only mention that for the case of the *Dirichlet problem*, characterized by the boundary conditions

$$\frac{\partial^i u}{\partial n^i} = g_i \quad \text{on} \quad \partial\Omega \quad \text{for} \quad i = 0, 1, \ldots, k-1\,, \tag{2.7}$$

or

$$D^\beta u = g_\beta \quad \text{on} \quad \partial\Omega \quad \text{for} \quad |\beta| \le k-1\,,$$

where $\dfrac{\partial^i}{\partial n^i}$ means the derivative of order i in the direction of the outer normal to $\partial\Omega$ (the so called *normal derivative*), we take for V the space $W_0^{k,p}(\Omega, w)$, while for the case of the *Neumann problem*, characterized by the boundary conditions

$$\frac{\partial^i u}{\partial n^i} = g_i \quad \text{on} \quad \partial\Omega \quad \text{for} \quad i = k, k+1, \ldots, 2k-1\,, \tag{2.8}$$

we take for V the space $W^{k,p}(\Omega, w)$.

The advantage of our approach consists in the fact that we avoid the concept of normal derivatives of $u \in W^{k,p}(\Omega, w)$ on $\partial\Omega$.

With the operator A from (2.1), we associate formally the form

$$a(u, v) = \sum_{|\alpha| \le k} \int_\Omega a_\alpha(x; u, \nabla u, \ldots, \nabla^k u) D^\alpha v(x)\, dx \tag{2.9}$$

which is linear in v, and we will suppose that

$$a(u, v) \quad \text{is defined on} \quad V \times V \tag{2.10}$$

without going into details for now.

Definition 2.1 We will say that a function $u \in W^{k,p}(\Omega, w)$ is a V-*solution* of the equation

$$Au = f^* \quad \text{in} \quad \Omega \tag{2.11}$$

with $f^* \in V^*$ (a given functional), if the identity

$$a(u, \varphi) = \langle f^*, \varphi \rangle \tag{2.12}$$

holds for every $\varphi \in V$ (here $\langle \cdot, \cdot \rangle$ denotes the duality between V^* and V).

2.2 Second order equations (bounded domains)

In this section, we will deal with a special *second order* operator A:

$$(Au)(x) = -\sum_{i=1}^{N} \frac{\partial}{\partial x_i}(a_i(x; u, \nabla u)) + c_0|u|^{p-2}u + a_0(x; u, \nabla u) \tag{2.13}$$

where $p > 1$, $c_0 > 0$ and we will also deal with the special weighted Sobolev space denoted by

$$W^{1,p}(\nu, \Omega) \tag{2.14}$$

and normed by

$$\|u\|_{1,p,\nu} = \left(\int_\Omega |u|^p\,dx + \int_\Omega |\nabla u|^p \nu\,dx\right)^{1/p} \tag{2.15}$$

where $\nu = \nu(x)$ is a weight function defined on Ω. (Notice that $W^{1,p}(\nu, \Omega)$ is a special case of the space $W^{1,p}(\Omega, w)$ with $w = \{w_0, w_1, \dots, w_N\}$ such that $w_0(x) \equiv 1$, $w_i(x) = \nu(x)$ for $i = 1, 2, \dots, N$; see Section 1.5, Example 1.3.)

The degeneracy of the operator A is expressed by the assumption that

$$\sum_{i=1}^{N} a_i(x; \xi)\xi_i \geq \frac{\nu(x)}{\lambda(|\xi_0|)} \sum_{i=1}^{N} |\xi_i|^p \tag{2.16}$$

(notice that $\xi = (\xi_0, \xi_1, \dots, \xi_N)$!) where ν is the weight function appearing above and $\lambda = \lambda(t)$ is a nondecreasing function defined on $[0, \infty)$ and such that $\lambda(t) \geq 1$.

Remark 2.2 Notice that condition (2.16) is in fact a modified (weaker) form of condition (0.51) in Introduction.

According to Section 2.1, we will suppose that

$$\nu \in L^1_{\mathrm{loc}}(\Omega)\,, \quad \nu^{-\frac{1}{p-1}} \in L^1_{\mathrm{loc}}(\Omega) \tag{2.17}$$

and we will introduce the space $W_0^{1,p}(\nu, \Omega)$ as the closure of $C_0^\infty(\Omega)$ in $W^{1,p}(\nu, \Omega)$ and the closed space V such that

$$W_0^{1,p}(\nu, \Omega) \hookrightarrow V \hookrightarrow W^{1,p}(\nu, \Omega) \tag{2.18}$$

with continuous imbeddings.

Moreover, we will look for *bounded* solutions, and therefore we denote by W the space

$$W = V \cap L^\infty(\Omega)\,. \tag{2.19}$$

Remark 2.3 The approach described in this section is due to F. Guglielmino, F. Nicolosi [30], who considered the special case $p = 2$, and to P. Drábek, F. Nicolosi [22] who extended the results to an arbitrary $p > 1$. The reduction to the *second order* differential operator is essential since we use here the *truncation technique* which works for Sobolev spaces $W^{k,p}(\Omega)$ with $k = 1$ only. This important restriction is expressed by the first two assumptions below.

Assumption A There exists a positive number K_0 such that for any $u \in V$, we have also

$$\min_{x\in\Omega}\{u(x), K\} \in V \tag{2.20}$$

for any $K \geq K_0$.

Assumption B For any $u \in W$ and for any $\gamma > 0$ we have

$$u(x)|u(x)|^{\gamma} \in V\,. \tag{2.21}$$

In addition to (2.17), we will suppose

Assumption C There is $s \in \left(\frac{N}{p}, +\infty\right) \cap \left[\frac{1}{p-1}, +\infty\right)$ such that

$$\nu^{-s} \in L^1(\Omega) \tag{2.22}$$

and Ω has the cone property (see, e.g., [1] or [38]).

Remark 2.4 Assumption C guarantees some important imbeddings of our space V into Lebesgue spaces. Indeed, as was shown in Example 1.3 (a), for $u \in V$ we have $|\nabla u| \in L^{p_s}(\Omega)$ with $p_s = \frac{ps}{s+1}$, $p_s \in [1, p)$, and with $p_s^* = \frac{Np_s}{N - p_s} = \frac{spN}{N(s+1) - sp}$, $p_s^* > p$, we have a *continuous* imbedding

$$V \hookrightarrow L^{p_s^*}(\Omega)\,. \tag{2.23}$$

If Ω is a bounded domain, we have even a *compact* imbedding

$$V \hookrightarrow\hookrightarrow L^r(\Omega)$$

for arbitrary $r \in [1, p_s^*)$. In particular, we have a *compact* imbedding

$$V \hookrightarrow\hookrightarrow L^p(\Omega)\,.$$

Now, let us formulate some assumptions about the growth of the functions a_i in (2.13).

Assumption D The functions $a_i(x;\xi) = a_i(x;\xi_0,\xi_1,\dots,\xi_N)$ satisfy the Carathéodory condition and, moreover,

(i) there exists a number σ such that

$$\max\left\{0, \frac{2-p}{2}\right\} < \sigma < 1, \tag{2.24}$$

and a function $g_0 \in L^1(\Omega) + L^{p'}(\Omega)$ such that the estimate

$$\begin{aligned} &|a_0(x;\xi_0,\xi_1,\dots,\xi_N)| \\ &\leq \lambda(|\xi_0|)\{g_0(x) + |\xi_0|^{p-1+\sigma} + \left(\nu^{\frac{1}{p}}(x)|\xi|\right)^{p-1+\sigma} + \nu(x)|\xi|^p\} \end{aligned} \tag{2.25}$$

holds for a.e. $x \in \Omega$ and for all real $\xi_0,\xi_1,\dots,\xi_N$; here $|\xi| = \left(\sum_{i=1}^N |\xi_i|^2\right)^{\frac{1}{2}}$ and $\lambda = \lambda(t)$ is the function from the condition (2.16);

(ii) there exists a function $g \in L^{p'}(\Omega)$, such that the estimates

$$\begin{aligned} &|a_i(x;\xi_0,\xi_1,\dots,\xi_N)| \\ &\leq \nu^{\frac{1}{p}}(x)\lambda(|\xi_0|)\{g(x) + |\xi_0|^{p-1} + \nu^{1-\frac{1}{p}}(x)|\xi|^{p-1}\} \end{aligned} \tag{2.26}$$

hold for a.e. $x \in \Omega$ and for all real $\xi_0,\xi_1,\dots,\xi_N$, $i = 1,2,\dots,N$.

Remark 2.5 (i) Compare conditions (2.25) and (2.26) with (a little simpler) growth conditions (0.50) in Introduction. The conditions appearing here are *more complicated* since the space $W^{1,p}(\nu,\Omega)$ used is *simpler*. We will meet this situation — the interrelation between the structure of the spaces used and the differential operator in question — also at several places later.

(ii) Recall that $L^1(\Omega) + L^{p'}(\Omega)$ is the set of all functions g of the form $g = g_1 + g_2$ with $g_1 \in L^1(\Omega)$ and $g_2 \in L^{p'}(\Omega)$. The norm $\|g\|_{L^1(\Omega)+L^{p'}(\Omega)}$ is defined as $\inf\{\|g_1\|_{L^1(\Omega)} + \|g_2\|_{L^{p'}(\Omega)}\}$, the infimum being taken over all decompositions of g into $g_1 + g_2$.

Also the monotonicity condition (see (0.27)) is slightly modified:

Assumption E For a.e. $x \in \Omega$ and for any real numbers $\xi_0,\xi_1,\dots,\xi_N,\eta_1,\dots,\eta_N$ the following inequality holds:

$$\sum_{i=1}^N (a_i(x;\xi_0,\xi_1,\dots,\xi_N) - a_i(x;\xi_0,\eta_1,\dots,\eta_N))(\xi_i - \eta_i) > 0 \tag{2.27}$$

whenever $(\xi_1,\dots,\xi_N) \neq (\eta_1,\dots,\eta_N)$.

Finally, we have to introduce some ellipticity (coercivity) conditions involving the weight function $\nu(x)$. In fact, we have met one of these conditions in (2.16).

Assumption F (i) There is a function $f \in L^1(\Omega) \cap L^\infty(\Omega)$ and a nonnegative number $c_3 < c_0$ (for c_0, see formula (2.13)) such that the inequality

$$\xi_0 a_0(x; \xi_0, \xi_1, \dots, \xi_N) + c_3|\xi_0|^p + \lambda(|\xi_0|)\nu(x)|\xi|^p + f(x) \geq 0 \tag{2.28}$$

holds for a.e. $x \in \Omega$ and for all real $\xi_0, \xi_1, \dots, \xi_N$.

(ii) The estimate (2.16), i.e.

$$\sum_{i=1}^{N} a_i(x; \xi_0, \xi_1, \dots, \xi_N)\xi_i \geq \frac{\nu(x)}{\lambda(|\xi_0|)} \sum_{i=1}^{N} |\xi_i|^p \tag{2.29}$$

holds for a.e. $x \in \Omega$ and for all real $\xi_0, \xi_1, \dots, \xi_N$, where $\lambda = \lambda(t)$ is a nondecreasing function defined on $[0, \infty)$ and such that $\lambda(t) \geq 1$.

Now, we are able to formulate our boundary value problem:

Problem 2.1 Find a function $u \in W$ such that the identity

$$\int_\Omega \left(\sum_{i=1}^{N} a_i(x; u, \nabla u)\frac{\partial \varphi}{\partial x_i} + c_0|u|^{p-2}u\varphi + a_0(x; u, \nabla u)\varphi \right) dx = 0 \tag{2.30}$$

holds for every $\varphi \in W$.

Remark 2.6 The function $u \in W$ will be our V-solution in the sense of Section 2.1. The functions φ appearing in (2.30) will be called *test functions*. Note also that (2.30) corresponds to (2.12) where the "right-hand side" f^* is contained in $a_0(x, u, \nabla u)$.

The first of the main results can be formulated as follows.

Theorem 2.1 *Let Ω be a bounded domain. Under the above assumptions* (*i.e.* (2.17) *and* A–F) *Problem* 2.1 *has at least one solution.*

To be able to give a proof, we need several lemmas. The first and the second lemma deal with *apriori estimates* for a solution of Problem 2.1.

Lemma 2.1 *Let u be a solution of Problem* 2.1 *and suppose that assumptions* (2.17), B, F *are satisfied. Then*

$$\|u\|_\infty \leq L \tag{2.31}$$

where

$$L = \left(\frac{1}{c_0 - c_3} \|f\|_\infty \right)^{\frac{1}{p}}, \tag{2.32}$$

and the function f appears in assumption F.

Proof. Due to assumption B, we can take for the test function φ the function $u|u|^\gamma$. Then (2.30) becomes

$$\int_\Omega |u|^\gamma \left((\gamma+1) \sum_{i=1}^N a_i(x; u, \nabla u) \frac{\partial u}{\partial x_i} + c_0|u|^p + a_0(x; u, \nabla u) u \right) dx = 0$$

and it follows from assumption F that

$$\begin{aligned}\int_\Omega |u|^\gamma \Big(\frac{\nu(x)(\gamma+1)}{\lambda(\|u\|_\infty)} |\nabla u|^p + c_0|u|^p - c_3|u|^p \\ - \lambda(\|u\|_\infty)\nu(x)|\nabla u|^p - f(x) \Big) dx \le 0 ,\end{aligned} \tag{2.33}$$

i.e.

$$\begin{aligned}(c_0 - c_3) \int_\Omega |u|^{p+\gamma} dx \\ \le \int_\Omega f(x)|u(x)|^\gamma dx - \int_\Omega |u|^\gamma \left(\frac{\gamma+1}{\lambda(\|u\|_\infty)} - \lambda(\|u\|_\infty) \right) \nu(x) |\nabla u|^p dx .\end{aligned}$$

Consequently, for γ sufficiently large, we have by the Hölder inequality

$$\begin{aligned}(c_0 - c_3) \int_\Omega |u|^{p+\gamma} dx &\le \int_\Omega |f(x)||u(x)|^\gamma dx \\ &\le \left(\int_\Omega |f(x)|^{\frac{\gamma}{p}+1} dx \right)^{\frac{p}{\gamma+p}} \left(\int_\Omega |u(x)|^{p+\gamma} dx \right)^{\frac{\gamma}{\gamma+p}} ,\end{aligned}$$

i.e.

$$(c_0 - c_3)\|u\|_{p+\gamma}^p \le \|f\|_{\frac{\gamma}{p}+1} \tag{2.34}$$

and (2.31) follows with L from (2.32) by letting γ tend to infinity. Indeed, we have

$$\left(\int_\Omega |f(x)|^{\frac{\gamma}{p}+1} dx \right)^{\frac{p}{\gamma+p}} \le \|f\|_\infty (\text{meas}\, \Omega)^{\frac{p}{\gamma+p}}$$

and hence

$$\limsup_{\gamma\to\infty} \|f\|_{\frac{\gamma}{p}+1} \le \|f\|_\infty .$$

It follows from (2.34) that

$$\limsup_{\gamma\to\infty} \|u\|_{p+\gamma} \le \left(\frac{1}{c_0 - c_3} \|f\|_\infty \right)^{\frac{1}{p}} .$$

Assume that (2.31) were not true. Then there is $\eta > 0$ and a set $\mathcal{A} \subset \Omega$, meas $\mathcal{A} > 0$ such that

$$u(x) \ge L + \eta \quad \text{for} \quad x \in \mathcal{A} .$$

So, we get

$$\liminf_{\gamma\to\infty}\left(\int_\Omega |u(x)|^{p+\gamma}\,dx\right)^{\frac{1}{p+\gamma}} \geq \liminf_{\gamma\to\infty}\left(\int_{\mathcal{A}} |u(x)|^{p+\gamma}\,dx\right)^{\frac{1}{p+\gamma}}$$
$$\geq \liminf_{\gamma\to\infty}(L+\eta)(\operatorname{meas}\mathcal{A})^{\frac{1}{p+\gamma}} = L+\eta\,,$$

a contradiction. This proves (2.31). □

Lemma 2.2 *Suppose that assumptions* (2.17), B, C, D, F *are satisfied. Let l be such that*

$$l \geq \max\{\|f\|_1 + \|f\|_\infty, \|g_0\|_{L^1(\Omega)+L^{p'}(\Omega)}\}\,. \tag{2.35}$$

Then there exists a constant $\mathcal{M} > 0$ (depending on $c_0, c_3, \sigma, l, \lambda, p, f, g_0$) such that for every solution u of Problem 2.1,

$$\|u\|_{1,p,\nu} \leq \mathcal{M}\,. \tag{2.36}$$

Proof. If we take for the test function φ in (2.30) our solution u, we obtain

$$\int_\Omega \left(\sum_{i=1}^N a_i(x;u,\nabla u)\frac{\partial u}{\partial x_i} + c_0|u|^p + a_0(x;u,\nabla u)u\right)dx = 0\,.$$

Using assumption F (ii) and Lemma 2.1, we obtain

$$0 \geq \int_\Omega \left(\frac{\nu(x)}{\lambda(L)}|\nabla u(x)|^p + c_0|u(x)|^p + a_0(x;u,\nabla u)u(x)\right)dx\,,$$

i.e. due to assumption D (i),

$$\min\left\{c_0, \frac{1}{\lambda(L)}\right\}\|u\|_{1,p,\nu}^p \leq \int_\Omega |u(x)||a_0(x;u,\nabla u)|\,dx$$
$$\leq \lambda(L)\int_\Omega |u(x)|\left[g_0(x) + |u(x)|^{p-1+\sigma} + \left(\nu^{\frac{1}{p}}(x)|\nabla u(x)|\right)^{p-1+\sigma}\right.$$
$$\left. + \nu(x)|\nabla u(x)|^p\right]dx$$
$$= \lambda(L)\int_\Omega \left[|u|g_0 + |u|^{p+\sigma} + |u|^{1-\frac{\sigma}{2}}|u|^{\frac{\sigma}{2}}(\nu^{\frac{1}{p}}|\nabla u|)^{p-1+\sigma} + |u|\nu|\nabla u|^p\right]dx\,.$$

Using for the third term on the right-hand side the inequality $ab \leq a^{1+\varepsilon} + b^{1+\frac{1}{\varepsilon}}$ with $\varepsilon = \dfrac{p-1+\sigma}{1-\sigma}$, $a = |u|^{1-\frac{\sigma}{2}}$, $b = \left(|u|^{\frac{\sigma}{2}\frac{p}{p-1+\sigma}}\nu|\nabla u|^p\right)^{\frac{p-1+\sigma}{p}}$, we obtain

$$\min\left\{c_0, \frac{1}{\lambda(L)}\right\}\|u\|_{1,p,\nu}^p$$
$$\leq \lambda(L)\int_\Omega \left(|u|g_0 + |u|^{p+\sigma} + |u|^{(1-\frac{\sigma}{2})\frac{p}{1-\sigma}} + |u|^{\frac{\sigma}{2}\frac{p}{p-1+\sigma}}\nu|\nabla u|^p + |u|\nu|\nabla u|^p\right)dx\,.$$

It follows from (2.24) that $\left(1-\frac{\sigma}{2}\right)\frac{p}{1-\sigma} > p+\sigma$ and $\frac{\sigma}{2}\frac{p}{p-1+\sigma} < 1$. Hence, we can use the estimates $|u|^{(1-\frac{\sigma}{2})\frac{p}{1-\sigma}} \le L^{(1-\frac{\sigma}{2})\frac{p}{1-\sigma}-(p+\sigma)}|u|^{p+\sigma}$ and $|u|\nu|\nabla u|^p \le L^{1-\frac{\sigma}{2}\frac{p}{p-1+\sigma}}|u|^{\frac{\sigma}{2}\frac{p}{p-1+\sigma}}\nu|\nabla u|^p$ in order to get

$$\|u\|^p_{1,p,\nu} \le K\int_\Omega \left(|u|g_0 + |u|^{p+\sigma} + |u|^\tau \nu|\nabla u|^p\right) dx \tag{2.37}$$

with $\tau = \frac{\sigma}{2}\frac{p}{p-1+\sigma}$ and with $K > 0$ depending on $c_0, c_3, \sigma, L, \lambda, p$.

The first term on the right-hand side of the last inequality can be estimated as follows. Applying the Hölder inequality, we get

$$\int_\Omega |u|g_0\, dx \le (\|u\|_\infty + \|u\|_p)\|g_0\|_{L^1(\Omega)+L^{p'}(\Omega)} \le c_1 + c_2\|u\|_{1,p,\nu}\,. \tag{2.38}$$

In order to estimate the second and the third term we need to derive some auxiliary inequalities. Let us proceed as in the proof of Lemma 2.1 but replacing $\|u\|_\infty$ in (2.33) by $K_1 = \left(\frac{1}{c_0-c_3}l\right)^{\frac{1}{p}}$. Then we obtain

$$\begin{aligned}\int_\Omega |u|^\gamma\left[\left(\frac{\gamma+1}{\lambda(K_1)} - \lambda(K_1)\right)\nu|\nabla u|^p + (c_0-c_3)|u|^p\right] dx &\le \int_\Omega f(x)|u(x)|^\gamma\, dx\\ &\le (K_1)^\gamma l\,.\end{aligned}$$

Since $\gamma > 0$ was arbitrary, we can take $\gamma = \lambda^2(K_1) + \lambda(K_1) - 1$. We have $\gamma \ge 1$ (notice that $\lambda(t) \ge 1$!) and

$$\int_\Omega \left[|u|^\gamma \nu|\nabla u|^p + (c_0-c_3)|u|^{\gamma+p}\right] dx \le (K_1)^\gamma l\,. \tag{2.39}$$

It follows from the Hölder inequality and (2.39) that

$$\begin{aligned}\int_\Omega |u|^{p+\sigma}\, dx &= \int_\Omega |u|^{(p+\gamma)\frac{\sigma}{\gamma}}|u|^{(\gamma-\sigma)\frac{p}{\gamma}}\, dx\\ &\le \left(\int_\Omega |u|^{p+\gamma}\, dx\right)^{\frac{\sigma}{\gamma}}\left(\int_\Omega |u|^p\, dx\right)^{1-\frac{\sigma}{\gamma}} \qquad (2.40)\\ &\le \left(\frac{l}{c_0-c_3}\right)^{\frac{\sigma}{\gamma}} K_1^\sigma \|u\|^{p\left(1-\frac{\sigma}{\gamma}\right)}_{1,p,\nu}\,,\end{aligned}$$

and

$$\begin{aligned}\int_\Omega |u|^\tau \nu |\nabla u|^p\,dx &= \int_\Omega |u|^\tau \nu^{\frac{\tau}{\gamma}} |\nabla u|^{\frac{p\tau}{\gamma}} \nu^{\frac{\gamma-\tau}{\gamma}} |\nabla u|^{\frac{p(\gamma-\tau)}{\gamma}}\,dx \\ &\le \left(\int_\Omega |u|^\gamma \nu |\nabla u|^p\,dx\right)^{\frac{\tau}{\gamma}} \left(\int_\Omega \nu |\nabla u|^p\,dx\right)^{1-\frac{\tau}{\gamma}} \\ &\le K_1^\tau l^{\frac{\tau}{\gamma}} \|u\|_{1,p,\nu}^{p\left(1-\frac{\tau}{\gamma}\right)}.\end{aligned} \tag{2.41}$$

Now, the estimate (2.36) follows from (2.37) due to (2.38), (2.40) and (2.41), since $p > 1, \tau < 1, \gamma \ge 1$. □

The next two lemmas are connected with the application of the Leray–Lions theorem.

Lemma 2.3 *Suppose that the assumptions* C, D, E, F *are satisfied. Let* $u \in V$ *and let* $\{u_n\}$ *be a sequence in* V *such that* $\|u_n\|_{1,p,\nu} \le K$ *and* $\lambda(|u_n(x)|) \le K$ *for a.e.* $x \in \Omega$ *and every* $n = 1, 2, \dots$ *with* $K \ge 1$ *a fixed constant. Moreover, suppose that*

$$\lim_{n\to\infty} \|u_n - u\|_p = 0 \tag{2.42}$$

and

$$\lim_{n\to\infty} \int_\Omega \sum_{i=1}^N (a_i(x; u_n, \nabla u_n) - a_i(x; u_n, \nabla u)) \frac{\partial(u_n - u)}{\partial x_i}\,dx = 0\,. \tag{2.43}$$

Then

$$\lim_{n\to\infty} \|u_n - u\|_{1,p,\nu} = 0\,. \tag{2.44}$$

Proof. It is sufficient to prove that there exists a *subsequence* of $\{u_n\}$ which converges to u in the norm of $W^{1,p}(\nu, \Omega)$. Indeed, the sequence $\{\|u_n - u\|_{1,p,\nu}\}$ is bounded since

$$\|u_n - u\|_{1,p,\nu} \le \|u_n\|_{1,p,\nu} + \|u\|_{1,p,\nu} \le K + \|u\|_{1,p,\nu} = K_2\,.$$

Denote by h the $\limsup_{n\to\infty} \|u_n - u\|_{1,p,\nu}$. Consequently, there exists a subsequence u_{z_n} of $\{u_n\}$ such that

$$\lim_{n\to\infty} \|u_{z_n} - u\|_{1,p,\nu} = h\,.$$

Since $\{u_{z_n}\}$ satisfies all assumptions of our lemma and we have supposed that the lemma is true for a subsequence, we can choose a subsequence $\{u_{\lambda_n}\}$ of $\{u_{z_n}\}$ such that

$$\lim_{n\to\infty} \|u_{\lambda_n} - u\|_{1,p,\nu} = 0\,.$$

Consequently, $h = 0$ and

$$0 \leq \liminf_{n\to\infty} \|u_n - u\|_{1,p,\nu} \leq \limsup_{n\to\infty} \|u_n - u\|_{1,p,\nu} = 0$$

and finally we get (2.44).

Due to (2.42) and (2.43) and E, there is a subsequence of $\{u_n\}$ (which we will denote again by $\{u_n\}$) such that

$$\lim_{n\to\infty} u_n(x) = u(x) \quad \text{a.e. in} \quad \Omega \tag{2.45}$$

and

$$\lim_{n\to\infty} \sum_{i=1}^{N} [a_i(x; u_n(x), \nabla u_n(x)) - a_i(x; u_n(x), \nabla u(x))] \frac{\partial (u_n - u)}{\partial x_i} = 0 \tag{2.46}$$

a.e. in Ω.

Let us fix the value of x. The ellipticity condition (2.29) implies that

$$\sum_{i=1}^{N} a_i(x; u_n(x), \nabla u_n(x)) \frac{\partial u_n}{\partial x_i} \geq \frac{\nu(x)}{\lambda(|u_n(x)|)} |\nabla u_n(x)|^p ,$$

$$\sum_{i=1}^{N} a_i(x; u_n(x), \nabla u(x)) \frac{\partial u}{\partial x_i} \geq \frac{\nu(x)}{\lambda(|u_n(x)|)} |\nabla u(x)|^p .$$

Furthermore, due to (2.26) we have

$$\begin{aligned}
&\sum_{i=1}^{N} \left| a_i(x; u_n, \nabla u_n) \frac{\partial u}{\partial x_i} \right| \\
&\qquad \leq \sum_{i=1}^{N} \nu^{\frac{1}{p}}(x) \lambda(|u_n(x)|)(g(x) + |u_n(x)|^{p-1} + \nu^{1-\frac{1}{p}}(x) |\nabla u_n|^{p-1}) \left| \frac{\partial u}{\partial x_i} \right| \\
&\qquad = \lambda(|u_n(x)|)(g(x) + |u_n(x)|^{p-1} + \nu^{1-\frac{1}{p}}(x) |\nabla u_n|^{p-1}) \sum_{i=1}^{N} \nu^{\frac{1}{p}}(x) \left| \frac{\partial u}{\partial x_i} \right| \\
&\qquad \leq N\lambda(|u_n(x)|)\{g(x) + |u_n(x)|^{p-1} + \nu^{1-\frac{1}{p}}(x) |\nabla u_n|^{p-1}\} \nu^{\frac{1}{p}}(x) |\nabla u|
\end{aligned}$$

and similarly

$$\begin{aligned}
&\sum_{i=1}^{N} \left| a_i(x; u_n, \nabla u) \frac{\partial u_n}{\partial x_i} \right| \\
&\qquad \leq N\lambda(|u_n(x)|)\{g(x) + |u_n(x)|^{p-1} + \nu^{1-\frac{1}{p}}(x) |\nabla u|^{p-1}\} \nu^{\frac{1}{p}}(x) |\nabla u_n| .
\end{aligned}$$

Consequently, we have derived the estimate

$$\begin{aligned}
\lambda(|u_n(x)|) \sum_{i=1}^{N} [a_i(x; u_n, \nabla u_n) - a_i(x; u_n, \nabla u)] \frac{\partial(u_n - u)}{\partial x_i} \\
\geq \nu(x)|\nabla u_n|^p - N\lambda^2(|u_n(x)|) \\
\cdot \left[g(x) + |u_n|^{p-1} + \nu^{1-\frac{1}{p}}(x)|\nabla u_n|^{p-1}\right] \\
\cdot \nu^{\frac{1}{p}}(x)|\nabla u| - N\lambda^2(|u_n(x)|) \left[g(x) + |u_n|^{p-1} + \nu^{1-\frac{1}{p}}(x)|\nabla u|^{p-1}\right] \\
\cdot \nu^{\frac{1}{p}}(x)|\nabla u_n| + \nu(x)|\nabla u|^p .
\end{aligned} \tag{2.47}$$

This relation implies that the sequence $\nabla u_n(x)$ is bounded. This can be shown by contradiction: Assume that we have a subsequence (again denoted by $\{u_n\}$) such that $|\nabla u_n(x)|$ tends to infinity. Using the boundedness of $\lambda(|u_n(x)|)$, we can rewrite (2.47) into the form

$$\begin{aligned}
K_1 \sum_{i=1}^{N} [a_i(x; u_n, \nabla u_n) - a_i(x; u_n, \nabla u)] \frac{\partial(u_n - u)}{\partial x_i} \\
\geq \nu(x)|\nabla u_n|^p \Bigg[1 - K_2 \frac{|\nabla u(x)|}{|\nabla u_n(x)|} \\
- \frac{K_2 \nu^{\frac{1}{p}-1}(x)}{|\nabla u_n(x)|^{p-1}} \left(g(x) + |u_n(x)|^{p-1} + \nu^{\frac{1}{p'}}(x)|\nabla u(x)|^{p-1}\right)\Bigg] \\
- K_2 \left(g(x) + |u_n(x)|^{p-1}\right) \nu^{\frac{1}{p}}(x)|\nabla u(x)| + \nu(x)|\nabla u(x)|^p
\end{aligned}$$

with some suitable positive constants K_1, K_2. Here the left-hand side tends to zero due to (2.46) while the right-hand side tends to infinity (notice that we can suppose $u_n(x) \to u(x)$ due to (2.45)).

Consequently, from the bounded sequence $\{\nabla u_n(x)\}$ we can select a *convergent* subsequence (again denoted by $\{u_n\}$):

$$\lim_{n\to\infty} \nabla u_n(x) = \kappa = (\kappa_1, \kappa_2, \ldots, \kappa_N) .$$

Then we have from (2.46) (and from (2.45)) that

$$\sum_{i=1}^{N} [a_i(x; u, \kappa) - a_i(x; u, \nabla u)] \left(\kappa_i - \frac{\partial u}{\partial x_i}\right) = 0 ,$$

which implies, due to assumption E, that

$$\kappa = \nabla u(x) ,$$

i.e.

$$\lim_{n\to\infty} \nabla u_n(x) = \nabla u(x)\,. \tag{2.48}$$

Now, let us investigate the sequence $\{\nu|\nabla u_n|^p\}$. It follows from our assumptions and from (2.17) that for every measurable subset E of Ω, the following estimate holds:

$$\begin{aligned}
\int_E \nu(x)|\nabla u_n|^p\,dx &\le K\int_E \sum_{i=1}^N a_i(x;u_n,\nabla u_n)\frac{\partial u_n}{\partial x_i}\,dx\\
&\le K\int_E \left|\sum_{i=1}^N \left(a_i(x;u_n,\nabla u_n)\frac{\partial u_n}{\partial x_i} - a_i(x;u,\nabla u)\frac{\partial u}{\partial x_i}\right)\right|\,dx\\
&\quad + K\int_E \left|\sum_{i=1}^N a_i(x;u,\nabla u)\frac{\partial u}{\partial x_i}\right|\,dx\,.
\end{aligned} \tag{2.49}$$

Further,

$$\begin{aligned}
&\left|\sum_{i=1}^N \left(a_i(x;u_n,\nabla u_n)\frac{\partial u_n}{\partial x_i} - a_i(x;u,\nabla u)\frac{\partial u}{\partial x_i}\right)\right|\\
&\qquad\le \sum_{i=1}^N (a_i(x;u_n,\nabla u_n) - a_i(x;u_n,\nabla u))\frac{\partial(u_n-u)}{\partial x_i}\\
&\qquad\quad + \left|\sum_{i=1}^N a_i(x;u_n,\nabla u)\frac{\partial(u_n-u)}{\partial x_i}\right|\\
&\qquad\quad + \left|\sum_{i=1}^N [a_i(x;u_n,\nabla u_n) - a_i(x;u,\nabla u)]\frac{\partial u}{\partial x_i}\right|\,.
\end{aligned} \tag{2.50}$$

Due to assumption D, we have

$$\begin{aligned}
&\left|\int_E \sum_{i=1}^N a_i(x;u_n,\nabla u_n)\frac{\partial u}{\partial x_i}\,dx\right|\\
&\le \int_E \sum_{i=1}^N \lambda(|u_n(x)|)[g(x) + |u_n|^{p-1} + \nu^{\frac{1}{p'}}(x)|\nabla u_n|^{p-1}]\nu^{\frac{1}{p}}(x)\left|\frac{\partial u}{\partial x_i}\right|\,dx\\
&\le K_3\left(\int_E \nu(x)|\nabla u(x)|^p\,dx\right)^{\frac{1}{p}}\left(\int_E [g(x) + |u_n|^{p-1} + \nu^{\frac{1}{p'}}|\nabla u_n|^{p-1}]^{\frac{p}{p-1}}\,dx\right)^{\frac{p-1}{p}}\\
&\le K_3(\|g\|_{p'} + 2\|u_n\|_{1,p,\nu}^{p-1})\left(\int_E \nu(x)|\nabla u(x)|^p\,dx\right)^{\frac{1}{p}}\\
&\le K_4\left(\int_E \nu(x)|\nabla u(x)|^p\,dx\right)^{\frac{1}{p}}\,.
\end{aligned}$$

Consequently, the functions $\sum_{i=1}^{N} a_i(x; u_n, \nabla u_n)\frac{\partial u}{\partial x_i}$ are equiintegrable, and by the Vitali theorem we have

$$\lim_{n\to\infty} \int_{\Omega} \left| \sum_{i=1}^{N} [a_i(x; u_n, \nabla u_n) - a_i(x; u, \nabla u)] \frac{\partial u}{\partial x_i} \right| dx = 0 . \tag{2.51}$$

Similarly, we have

$$\lim_{n\to\infty} \int_{\Omega} \left| \sum_{i=1}^{N} a_i(x; u_n, \nabla u) \frac{\partial (u_n - u)}{\partial x_i} \right| dx = 0 . \tag{2.52}$$

Indeed,

$$\begin{aligned}
&\int_E \sum_{i=1}^{N} |a_i(x; u_n, \nabla u)| \left| \frac{\partial (u_n - u)}{\partial x_i} \right| \\
&\quad \le \int_E \lambda(|u_n|) \left(g + |u_n|^{p-1} + \nu^{\frac{1}{p'}} |\nabla u|^{p-1} \right) \sum_{i=1}^{N} \nu^{\frac{1}{p}} \left| \frac{\partial (u_n - u)}{\partial x_i} \right| dx \\
&\quad \le K_4 \left[\left(\int_E |g|^{p'} dx \right)^{\frac{1}{p'}} + \left(\int_E |u_n|^p dx \right)^{\frac{1}{p'}} + \left(\int_E \nu |\nabla u|^p dx \right)^{\frac{p-1}{p}} \right] \\
&\qquad \cdot \left(\int_E \nu |\nabla (u_n - u)|^p dx \right)^{\frac{1}{p}} \\
&\quad \le K_5 \left(\|g\|_{p'} + \|u_n\|_p^{p-1} + \|u\|_{1,p,\nu}^{p-1} \right) \left(\|u\|_{1,p,\nu} + \|u_n\|_{1,p,\nu} \right) .
\end{aligned}$$

This proves the equiintegrability of $\sum_{i=1}^{N} a_i(x; u_n, \nabla u)\frac{\partial (u_n - u)}{\partial x_i}$, and since the function itself converges to zero a.e., we have (2.52).

Since

$$\lim_{n\to\infty} \int_E \sum_{i=1}^{N} [a_i(x; u_n, \nabla u_n) - a_i(x; u_n, \nabla u)] \frac{\partial (u_n - u)}{\partial x_i} dx = 0$$

due to (2.43) and E, it follows by (2.49)–(2.52) that we have

$$\lim_{n\to\infty} \int_E \nu(x) |\nabla u_n(x)|^p dx \le K \int_E \left| \sum_{i=1}^{N} a_i(x; u, \nabla u) \frac{\partial u}{\partial x_i} \right| dx ,$$

and consequently, the function $\nu|\nabla u|^p$ is integrable over Ω and

$$\lim_{n\to\infty}\int_\Omega \nu|\nabla u_n - \nabla u|^p\,dx = 0\,.$$

This together with (2.42) proves the relation (2.44). □

Now, we are able to solve Problem 2.1 under some additional assumptions:

Lemma 2.4 *Suppose, in addition, that the function $\lambda = \lambda(t)$ from condition* (2.16) *is a constant λ_0 and that*

$$|a_0(x;\xi)| \le \lambda_0 \tag{2.53}$$

for a.e. $x \in \Omega$ and all $\xi \in \mathbb{R}^{N+1}$. Then Problem 2.1 *has at least one solution $u \in W$.*

Proof. Let us define an operator Φ: $V \times V \to V^*$ by the formula

$$\langle \Phi(u,v), w\rangle = \int_\Omega \left(\sum_{i=1}^N a_i(x;u,\nabla u)\frac{\partial w}{\partial x_i} + c_0|v|^{p-2}vw + a_0(x;v,\nabla v)w\right)dx \tag{2.54}$$

for every $w \in V$ and an operator $T: V \to V^*$ by

$$Tu = \Phi(u,u)\,, \qquad u \in V\,. \tag{2.55}$$

It is easy to verify that, due to all our assumptions, the operator T satisfies all conditions of the Leray–Lions theorem (see Theorem 1.4), and consequently, there exists at least one solution $u \in V$ of the equation

$$Tu = 0\,. \tag{2.56}$$

Now it remains only to show that this solution is bounded. So, let u be a solution of (2.56) and denote

$$w = u - \min\{u, K\}\,, \qquad K \ge K_0$$

(see assumption A). If we denote

$$\Omega_K = \{x \in \Omega;\ u(x) > K\}$$

then $u = w + K$ in Ω_K and $w = 0$ in $\Omega \setminus \Omega_K$. Equation (2.56) means that

$$\langle \Phi(u,u), \varphi\rangle = 0 \quad \text{for every } \varphi \in V\,,$$

and since w belongs to V due to assumption A, we have

$$\begin{aligned} 0 &= \langle \Phi(u,u), w\rangle \\ &= \int_\Omega \left(\sum_{i=1}^N a_i(x;u,\nabla u)\frac{\partial w}{\partial x_i} + c_0|u|^{p-2}uw + a_0(x;u,\nabla u)w\right)dx = \end{aligned}$$

$$= \int_{\Omega_K} \left(\sum_{i=1}^N a_i(x; u, \nabla u) \frac{\partial (u-K)}{\partial x_i} + c_0 |u|^{p-2} u(u-K) \right.$$

$$\left. + a_0(x; u, \nabla u)(u-K) \right) dx$$

$$= \int_{\Omega_K} \left(\sum_{i=1}^N a_i(x; w+K, \nabla w) \frac{\partial w}{\partial x_i} + c_0 |w+K|^{p-2} w^2 \right.$$

$$\left. + c_0 |w+K|^{p-2} wK + a_0(x; w+K, \nabla w) w \right) dx .$$

Using the fact that w is positive in Ω_K ($w = u - K$ and $u(x) > K$ in Ω_K), the estimate (2.16) and the assumption (2.53), we obtain that

$$\int_{\Omega_K} \left(\frac{\nu(x)}{\lambda_0} |\nabla w|^p + c_0 |w|^p \right) dx$$

$$\leq \int_{\Omega_K} \left[c_0 |w|^p - c_0 |w+K|^{p-2} w(w+K) - a_0(x; w+K, \nabla w) w \right] dx$$

$$\leq \int_{\Omega_K} |a_0(x; w+K, \nabla w)| |w| \, dx \leq \lambda_0 \int_{\Omega_K} |w| \, dx ,$$

and the Hölder inequality with any exponent $r > 1$ yields

$$\min \left\{ \frac{1}{\lambda_0}, c_0 \right\} \int_{\Omega_K} \left(|\nu(x)| |\nabla w|^p + |w|^p \right) dx \leq \lambda_0 \|w\|_r (\operatorname{meas} \Omega_K)^{\frac{r-1}{r}} ,$$

i.e.

$$\|w\|_{1,p,\nu}^p \leq \frac{\lambda_0}{\min \left\{ \frac{1}{\lambda_0}, c_0 \right\}} \|w\|_r (\operatorname{meas} \Omega_K)^{\frac{r-1}{r}} .$$

If we take for r the parameter p_s^* from Remark 2.4, we obtain in view of (2.23) the estimates

$$\begin{aligned} \|w\|_{1,p,\nu}^{p-1} &\leq \frac{\lambda_0 c_2}{\min \left\{ \frac{1}{\lambda_0}, c_0 \right\}} (\operatorname{meas} \Omega_K)^{\frac{p_s^*-1}{p_s^*}} , \\ \|w\|_{p_s^*}^{p-1} &\leq \frac{\lambda_0 c_2^p}{\min \left\{ \frac{1}{\lambda_0}, c_0 \right\}} (\operatorname{meas} \Omega_K)^{\frac{p_s^*-1}{p_s^*}} . \end{aligned} \tag{2.57}$$

For $h > K$ we have

$$\|w\|_{p_s^*}^{p-1} = \left(\int_{\Omega_K} |u - K|^{p_s^*}\,dx\right)^{\frac{p-1}{p_s^*}} \geq \left(\int_{\Omega_h} |u - K|^{p_s^*}\,dx\right)^{\frac{p-1}{p_s^*}}$$
$$\geq \left(\int_{\Omega_h} |h - K|^{p_s^*}\,dx\right)^{\frac{p-1}{p_s^*}} \geq (h-K)^{p-1}(\operatorname{meas}\Omega_h)^{\frac{p-1}{p_s^*}} \tag{2.58}$$

and it follows from (2.58), (2.57) that

$$(h-K)^{p-1}(\operatorname{meas}\Omega_h)^{\frac{p-1}{p_s^*}} \leq \frac{\lambda_0 c_2^p}{\min\left\{\frac{1}{\lambda_0}, c_0\right\}}(\operatorname{meas}\Omega_K)^{\frac{p_s^*-1}{p_s^*}}$$

whenever $h > K \geq K_0$ with $p_s^* > p$, i.e.

$$\operatorname{meas}\Omega_h \leq \left[\frac{\lambda_0 c_2^p}{\min\left\{\frac{1}{\lambda_0}, c_0\right\}}\right]^{\frac{p_s^*}{p-1}} \frac{1}{(h-K)^{p_s^*}}(\operatorname{meas}\Omega_K)^{\frac{p_s^*-1}{p-1}}. \tag{2.59}$$

Now, we use Lemma 1.4 which asserts that a nonnegative nonincreasing function ζ, satisfying for $0 < K_0 < K < h$ the estimate

$$\zeta(h) \leq \frac{\tau}{(h-K)^{p_s^*}}\,[\zeta(K)]^r \tag{2.60}$$

with suitable constants $\tau = c$, $p_s^* = \sigma$ and $r = \delta > 1$, has the property that $\zeta(K_0 + d) = 0$ with some suitable $d > 0$. If we take $\zeta(h) = \operatorname{meas}\Omega_h$, we see that (2.59) is nothing else then (2.60) since $r = \dfrac{p_s^* - 1}{p-1} > 1$ due to the fact that $p_s^* > p$.

Consequently, we have

$$\operatorname{meas}\Omega_{K_0+d} = 0$$

where d depends on u, λ_0, c_0, p_s^*, c_2, K_0 and p.

Since the same procedure applies also to meas $\tilde{\Omega}_K$, where $\tilde{\Omega}_K = \{x \in \Omega;\ -u > K\}$, we finally obtain

$$|u(x)| \leq K_0 + d \quad \text{a.e. in } \Omega\,.$$

So we have shown that the solution $u \in V$ of (2.56) is bounded, i.e. belongs to W, and since (2.56) is nothing else than (2.30), the lemma is proved. □

2.3 Second order equations (proof of Theorem 2.1)

Now, we are prepared to prove Theorem 2.1. For this purpose, let us introduce new "coefficients" $b_i(x;\xi) = b_i(x;\xi_0,\xi_1,\dots,\xi_N)$ and $a_{0,n}(x;\xi)$ by the formulas

$$b_i(x;\xi_0,\xi_1,\dots,\xi_N) = \begin{cases} a_i(x;-L,\xi_1,\dots,\xi_N) & \text{for} \quad \xi_0 < -L\,, \\ a_i(x;\xi_0,\xi_1,\dots,\xi_N) & \text{for} \quad |\xi_0| \le L\,, \\ a_i(x;L,\xi_1,\dots,\xi_N) & \text{for} \quad \xi_0 > L\,, \end{cases} \tag{2.61}$$

$i = 1, 2, \dots, N$, where $L > 0$ is the constant from (2.32) (Lemma 2.1) and

$$a_{0,n}(x;\xi_0,\xi_1,\dots,\xi_N) = \begin{cases} a_0(x;\xi_0,\xi_1,\dots,\xi_N) & \text{if} \quad |a_0(x;\xi)| \le n\,, \\ n\dfrac{a_0(x;\xi_0,\dots,\xi_N)}{|a_0(x;\xi_0,\dots,\xi_N)|} & \text{if} \quad |a_0(x;\xi)| > n\,, \end{cases} \tag{2.62}$$

with $n = 1, 2, \dots$.

It is easy to verify that for every fixed $n \in \mathbb{N}$, the functions b_i and $a_{0,n}$ satisfy the assumptions of Lemma 2.4 (notice that we can choose $\lambda_0 = \max\{\lambda(L), n\}$). Consequently, according to Lemma 2.4, for every $n \in \mathbb{N}$ there exists a function $u_n \in W$ such that the relation

$$\int_\Omega \left(\sum_{i=1}^N b_i(x;u_n,\nabla u_n)\frac{\partial \varphi}{\partial x_i} + c_0|u_n|^{p-2}u_n\varphi + a_{0,n}(x;u_n,\nabla u_n)\varphi \right) dx = 0 \tag{2.63}$$

holds for every $\varphi \in W$. The apriori estimate (2.31) from Lemma 2.1 yields

$$\|u_n\|_\infty \le L\,, \tag{2.64}$$

and consequently, due to (2.61), we can write (2.63) in an equivalent form

$$\int_\Omega \left(\sum_{i=1}^N a_i(x;u_n,\nabla u_n)\frac{\partial \varphi}{\partial x_i} + c_0|u_n|^{p-2}u_n\varphi + a_{0,n}(x;u_n,\nabla u_n)\varphi \right) dx = 0\,. \tag{2.65}$$

The apriori estimate from Lemma 2.2 yields

$$\|u_n\|_{1,p,\nu} \le \mathcal{M}\,. \tag{2.66}$$

It follows from (2.64) and (2.66) that there exists a subsequence of the sequence $\{u_n\}$ (we will denote it again by $\{u_n\}$) such that

$$\begin{aligned} &\{u_n\} \text{ converges weakly to } u \text{ in } V\,, \\ &\{u_n\} \text{ converges weakly* to } u \text{ in } L^\infty(\Omega)\,, \end{aligned} \tag{2.67}$$

where $\|u\|_\infty \le L$, and consequently,

$$u \in W\,.$$

We shall prove that this element u is a solution of Problem 2.1, i.e., that the relation

$$\int_\Omega \left(\sum_{i=1}^n a_i(x; u, \nabla u) \frac{\partial \varphi}{\partial x_i} + c_0 |u|^{p-2} u \varphi + a_0(x; u, \nabla u) \varphi \right) dx = 0 \tag{2.68}$$

holds for every $\varphi \in W$.

Relation (2.68) can be obtained by passing to the limit for $n \to \infty$ in (2.65). However, to be able to do it, we need several *auxiliary assertions*:

(i) *The sequence* $\{u_n\}$ *converges strongly to* u *in* V, *i.e.*

$$\lim_{n \to \infty} \|u_n - u\|_{1,p,\nu} = 0\,. \tag{2.69}$$

(ii) *For every* $\varphi \in W$, *we have*

$$\lim_{n \to \infty} \int_\Omega |a_{0,n}(x; u_n, \nabla u_n) - a_0(x; u, \nabla u)| |\varphi| \, dx = 0\,. \tag{2.70}$$

(iii) *For every* $\varphi \in W$, *we have*

$$\lim_{n \to \infty} \int_\Omega c_0 \left| |u_n|^{p-2} u_n - |u|^{p-2} u \right| |\varphi| \, dx = 0\,. \tag{2.71}$$

(iv) *For every* $\varphi \in W$, *we have*

$$\lim_{n \to \infty} \int_\Omega \sum_{i=1}^N |a_i(x; u_n, \nabla u_n) - a_i(x; u, \nabla u)| \left| \frac{\partial \varphi}{\partial x_i} \right| dx = 0\,. \tag{2.72}$$

Proof of (i). To be able to apply Lemma 2.3, we need only to verify that the assumption (2.43) is satisfied, since we have (2.66) and (2.64). If we choose $\varphi = u_n - u$ in (2.65), we get

$$\begin{aligned} &\int_\Omega \sum_{i=1}^N (a_i(x; u_n, \nabla u_n) - a_i(x; u_n, \nabla u)) \frac{\partial (u_n - u)}{\partial x_i} \\ &\quad = - \int_\Omega c_0 |u_n|^{p-2} u_n (u_n - u) \, dx - \int_\Omega a_{0,n}(x; u_n \nabla u_n)(u_n - u) \, dx \\ &\quad - \int_\Omega \sum_{i=1}^N a_i(x; u, \nabla u) \frac{\partial (u_n - u)}{\partial x_i} \, dx \\ &\quad + \int_\Omega \sum_{i=1}^N (a_i(x; u, \nabla u) - a_i(x; u_n, \nabla u)) \frac{\partial (u_n - u)}{\partial x_i} \, dx\,. \end{aligned} \tag{2.73}$$

Since V is compactly imbedded into $L^p(\Omega)$ (see Remark 2.4), the boundedness of $\{\|u_n\|_{1,p,\nu}\}$ implies the *strong* convergence of u_n to u in $L^p(\Omega)$ and consequently, also a.e. in Ω. Then we have, due to (2.64), that

$$\lim_{n\to\infty}\int_\Omega c_0|u_n|^{p-2}u_n(u_n-u)\,dx=0\,. \tag{2.74}$$

The weak convergence of $\{u_n\}$ to u in W implies that

$$\lim_{n\to\infty}\int_\Omega\sum_{i=1}^N a_i(x;u,\nabla u)\frac{\partial(u_n-u)}{\partial x_i}=0\,. \tag{2.75}$$

Notice that $\dfrac{a_i(x;u,\nabla u)}{\nu^{\frac{1}{p}}(x)}$ belongs to $L^{p'}(\Omega)$ due to the growth assumptions on a_i. The continuity of the Nemytskij operator given by the function

$$\frac{a_i(x;\,.\,,\nabla u)}{\nu^{\frac{1}{p}}(x)}$$

as an operator from $L^p(\Omega)$ into $L^{p'}(\Omega)$ (see Section 1.3) implies that for a fixed $u\in W$

$$\lim_{n\to\infty}\int_\Omega\sum_{i=1}^N(a_i(x;u,\nabla u)-a_i(x;u_n,\nabla u))\frac{\partial(u_n-u)}{\partial x_i}\,dx=0\,. \tag{2.76}$$

It remains to check the second term on the right-hand side of (2.73). From the growth conditions (assumption D(i)) we obtain

$$\begin{aligned}&\left|\int_\Omega a_{0,n}(x;u_n,\nabla u_n)(u_n-u)\,dx\right|\\&\quad\le\lambda(L)\int_\Omega\Big(g_0(x)+L^{p-1+\sigma}+\Big(\nu^{\frac{1}{p}}(x)|\nabla u_n(x)|\Big)^{p-1+\sigma}\\&\qquad+\nu(x)|\nabla u_n(x)|^p\Big)|u_n-u|\,dx\end{aligned}$$

and since

$$\Big(\nu^{\frac{1}{p}}(x)|\nabla u_n(x)|\Big)^{p-1+\sigma}\le1+\nu(x)|\nabla u_n(x)|^p$$

(here we use the inequality $|\Psi(x)|^a\le1+|\Psi(x)|^b$ provided $0<a<b$), we get

$$\begin{aligned}&\left|\int_\Omega a_{0,n}(x;u_n,\nabla u_n)(u_n-u)\,dx\right|\\&\quad\le\lambda(L)\int_\Omega\Big(L^{p-1+\sigma}+1+g_0(x)\Big)|u_n-u|\,dx\\&\qquad+2\lambda(L)\int_\Omega\nu(x)|\nabla u_n(x)|^p|u_n-u|\,dx\,.\end{aligned} \tag{2.77}$$

Since we immediately have

$$\lim_{n\to\infty}\int_\Omega (L^{p-1+\sigma}+1+g_0(x))|u_n-u|\,dx=0\,, \tag{2.78}$$

we have to show that also

$$\lim_{n\to\infty}\int_\Omega \nu(x)|\nabla u_n|^p|u_n-u|\,dx=0\,. \tag{2.79}$$

For this purpose, let us take $\varphi=|u_n-u|^\gamma(u_n-u)$ as a test function in (2.65), with some positive exponent γ which will be fixed later. We then have

$$\begin{aligned}\int_\Omega\Bigg[&\sum_{i=1}^N a_i(x;u_n,\nabla u_n)(\gamma+1)|u_n-u|^\gamma\frac{\partial(u_n-u)}{\partial x_i}\\ &+c_0|u_n|^{p-2}u_n|u_n-u|^\gamma(u_n-u)\\ &+a_{0,n}(x;u_n,\nabla u_n)|u_n-u|^\gamma(u_n-u)\Bigg]\,dx=0\,.\end{aligned} \tag{2.80}$$

It follows from (2.17) and from (2.64) that

$$\begin{aligned}\sum_{i=1}^N &a_i(x;u_n,\nabla u_n)(\gamma+1)|u_n-u|^\gamma\left(\frac{\partial u_n}{\partial x_i}-\frac{\partial u}{\partial x_i}\right)\\ &\geq\frac{1}{\lambda(L)}\nu(x)|u_n-u|^\gamma(\gamma+1)|\nabla u_n|^p\\ &\quad-\sum_{i=1}^N a_i(x;u_n,\nabla u_n)(\gamma+1)|u_n-u|^\gamma\frac{\partial u}{\partial x_i}\,.\end{aligned} \tag{2.81}$$

Similarly as when deriving (2.77) we obtain that

$$\begin{aligned}a_{0,n}(x;u_n,\nabla u_n)(u_n-u)&\geq-|a_{0,n}(x;u_n,\nabla u_n)||u_n-u|\\ &\geq-2\lambda(L)L\left[g_0(x)+L^{p-1+\sigma}+1+2\nu(x)|\nabla u_n|^p\right]\end{aligned} \tag{2.82}$$

(we use the fact that $|u_n-u|\leq\|u_n\|_\infty+\|u\|_\infty\leq 2L$), and also

$$c_0|u_n|^{p-2}u_n|u_n-u|^\gamma(u_n-u)\geq-c_0L^{p-1}|u_n-u|^{\gamma+1}\,. \tag{2.83}$$

From (2.80)–(2.83) we obtain

$$\int_\Omega |u_n - u|^\gamma \nu(x)|\nabla u_n|^p \left(\frac{\gamma+1}{\lambda(L)} - 4L\lambda(L)\right) dx$$
$$\leq \int_\Omega \sum_{i=1}^N a_i(x; u_n, \nabla u_n)(\gamma+1)|u_n - u|^\gamma \frac{\partial u}{\partial x_i}\, dx$$
$$+ \int_\Omega c_0 L^{p-1}|u_n - u|^{\gamma+1}\, dx + 2L\lambda(L)\int_\Omega \left(g_0(x) + L^{p-1+\sigma} + 1\right)|u_n - u|^\gamma\, dx.$$

If we choose γ in such a way that

$$\frac{\gamma+1}{\lambda(L)} - 4L\lambda(L) > 1\,,$$

we obtain the inequality

$$\begin{aligned}
&\int_\Omega |u_n - u|^\gamma \nu(x)|\nabla u_n(x)|^p\, dx \\
&\qquad\leq \int_\Omega \sum_{i=1}^N a_i(x; u_n, \nabla u_n)(\gamma+1)|u_n - u|^\gamma \frac{\partial u}{\partial x_i}\, dx \\
&\qquad\quad + \int_\Omega c_0 L^{p-1}|u_n - u|^{\gamma+1}\, dx \\
&\qquad\quad + \int_\Omega 2L\lambda(L)\left(g_0(x) + L^{p-1+\sigma} + 1\right)|u_n - u|^\gamma\, dx\,.
\end{aligned} \tag{2.84}$$

The second and the third term on the right-hand side tend to zero for $n \to \infty$ due to the Lebesgue theorem. Let us estimate the first term on the right-hand side:

$$\left|\int_\Omega \sum_{i=1}^N a_i(x; u_n, \nabla u_n)(\gamma+1)|u_n - u|^\gamma \frac{\partial u}{\partial x_i}\, dx\right|$$
$$\leq (\gamma+1)\left(\int_\Omega |u_n - u|^{\gamma p}\nu(x)|\nabla u|^p\, dx\right)^{\frac{1}{p}} \left(\sum_{i=1}^N \int_\Omega \frac{|a_i(x; u_n, \nabla u_n)|^{p'}}{\nu^{\frac{1}{p-1}}(x)}\, dx\right)^{\frac{1}{p'}}$$
$$\leq (\gamma+1)N^{\frac{1}{p'}}\lambda(L)\left[\|g\|_{p'} + \left(\int_\Omega |u_n(x)|^p\, dx\right)^{\frac{1}{p'}} + \int_\Omega \nu(x)|\nabla u_n|^p\, dx\right]$$
$$\cdot\left(\int_\Omega |u_n - u|^{\gamma p}\nu(x)|\nabla u|^p\, dx\right)^{\frac{1}{p}}$$
$$\leq (\gamma+1)N^{\frac{1}{p'}}\lambda(L)(\|g\|_{p'} + 2\mathcal{M}^{p-1})\left(\int_\Omega |u_n - u|^{\gamma p}\nu(x)|\nabla u(x)|^p\, dx\right)^{\frac{1}{p}}$$

(we have used the growth conditions (2.26) and the apriori estimate (2.66)).

By the Lebesgue theorem,

$$\lim_{n\to\infty} \int_\Omega |u_n - u|^{\gamma p} \nu(x)\, |\nabla u(x)|^p \, dx = 0$$

and consequently, also the first term on the right-hand side od (2.84) tends to zero, which implies that

$$\lim_{n\to\infty} \int_\Omega |u_n - u|^{\gamma} \nu(x) |\nabla u_n(x)|^p \, dx = 0 .$$

However, this result implies condition (2.79) since by the Hölder inequality we get

$$\begin{aligned} &\int_\Omega |u_n - u| \nu(x) |\nabla u_n|^p \, dx \\ &\quad \le \left(\int_\Omega |u_n - u|^\gamma \nu(x) |\nabla u_n|^p \, dx \right)^{\frac{1}{\gamma}} \left(\int_\Omega \nu(x) |\nabla u_n|^p \, dx \right)^{1-\frac{1}{\gamma}} \\ &\quad \le \mathcal{M}^{p\left(1-\frac{1}{\gamma}\right)} \left(\int_\Omega |u_n - u|^\gamma \nu(x) |\nabla u_n(x)|^p \, dx \right)^{\frac{1}{\gamma}} . \end{aligned}$$

Thus, (2.69) holds due to Lemma 2.3 and (i) is proved. □

Proof of (ii). Since $\varphi \in W$ is fixed (and, in particular, $\varphi \in L^\infty(\Omega)$) it suffices to show that

$$\lim_{n\to\infty} \int_\Omega |a_{0,n}(x; u_n, \nabla u_n) - a_0(x; u, \nabla u)| \, dx = 0 . \tag{2.85}$$

But for $u_n \in W$, i.e. satisfying (2.64), we have due to the definition of $a_{0,n}$ — see (2.62) — and due to assumption D(i) — see (2.25) — that

$$\begin{aligned} \int_\Omega |a_{0,n}(x; u_n, \nabla u_n)| \, dx &\le \int_\Omega |a_0(x; u_n, \nabla u_n)| \, dx \\ &\le \lambda(L) \int_\Omega \Big[g_0(x) + |u_n|^{p-1+\sigma} \\ &\qquad + (\nu^{\frac{1}{p}}(x) |\nabla u_n|)^{p-1+\sigma} + \nu(x) |\nabla u_n|^p \Big] \, dx \\ &\le \lambda(L) \Big\{ \|g_0\|_1 + L^{p-1+\sigma} \operatorname{meas} \Omega \\ &\qquad + \int_\Omega (1 + \nu(x) |\nabla u_n|^p + \nu(x) |\nabla u_n|^p) \, dx \Big\} \\ &\le \lambda(L) [\|g_0\|_1 + (L^{p-1+\sigma} + 1) \operatorname{meas} \Omega + 2\mathcal{M}^p] \end{aligned} \tag{2.86}$$

(we have also used the apriori estimate (2.66)). Consequently, the sequence

$$\left\{ \int_\Omega |a_{0,n}(x; u_n, \nabla u_n) - a_0(x; u, \nabla u)| \, dx \right\}$$

is bounded. Therefore, there exists

$$\limsup_{n\to\infty} \int_\Omega |a_{0,n}(x; u_n, \nabla u_n) - a_0(x; u, \nabla u)|\, dx = h$$

and we can choose a *subsequence* $\{u_{z_n}\}$ of $\{u_n\}$ such that

$$\lim_{n\to\infty} \int_\Omega |a_{0,z_n}(x; u_{z_n}, \nabla u_{z_n}) - a_0(x; u, \nabla u)|\, dx = h$$

and simultaneously

$$\lim_{n\to\infty} u_{z_n}(x) = u(x)\,, \qquad \lim_{n\to\infty} \nabla u_{z_n}(x) = \nabla u(x) \quad \text{a.e. in } \Omega\,.$$

Let us fix the value of x such that the last relations hold and that at the same time $a_0(x; \xi)$ is continuous in $\xi \in \mathbb{R}^{N+1}$. For every $\varepsilon > 0$ there exists a $\delta = \delta(\varepsilon) > 0$ such that

$$|a_0(x; \xi_0, \xi_1, \dots, \xi_N) - a_0(x; u(x), \nabla u(x))| < \varepsilon \tag{2.87}$$

whenever $|\xi_0 - u(x)| < \delta$, $\left|\xi_i - \dfrac{\partial u}{\partial x_i}(x)\right| < \delta$ for any i. On the other hand, we have (for z_n sufficiently large)

$$|u_{z_n}(x) - u(x)| < \delta\,, \qquad \left|\frac{\partial u_{z_n}}{\partial x_i}(x) - \frac{\partial u}{\partial x_i}(x)\right| < \delta\,,$$

$$a_0(x; u(x), \nabla u(x)) + \varepsilon < z_n\,, \qquad -z_n < a_0(x; u(x), \nabla u(x)) - \varepsilon\,,$$

i.e.

$$|a_0(x; u(x), \nabla u(x))| < z_n - \varepsilon < z_n\,.$$

But due to the definition of $a_{0,n}$ we then have

$$a_{0,z_n}(x; u_{z_n}, \nabla u_{z_n}(x)) = a_0(x; u_{z_n}(x), \nabla u_{z_n}(x))$$

and so we conclude from (2.87) that

$$|a_{0,z_n}(x; u_{z_n}(x), \nabla u_{z_n}(x)) - a_0(x; u(x), \nabla u(x))| < \varepsilon$$

if n is sufficiently large. Hence the sequence $\{a_{0,z_n}(x; u_{z_n}(x), \nabla u_{z_n}(x))\}$ converges a.e. in Ω to $a_0(x; u(x), \nabla u(x))$, and since the estimate (2.86) holds for every subset E of Ω,

$$\int_E |a_{0,z_n}(x; u_{z_n}(x), \nabla u_{z_n}(x))|\, dx$$
$$\leq \lambda(L)\left[\int_E |g_0|\, dx + (1 + L^{p-1+\sigma}) \operatorname{meas} E + 2\int_E \nu(x)|\nabla u|^p\, dx\right],$$

we can use Vitali's theorem and claim

$$h = 0.$$

Since

$$\begin{aligned} 0 &\leq \liminf_{n\to\infty} \int_\Omega |a_{0,n}(x; u_n(x), \nabla u_n(x)) - a_0(x; u(x), \nabla u(x))|\, dx \\ &\leq \limsup_{n\to\infty} \int_\Omega |a_{0,n}(x; u_n(x), \nabla u_n(x)) - a_0(x; u(x), \nabla u(x))|\, dx = 0, \end{aligned}$$

due to (2.85) we have proved (2.70). □

Proof of (iii). It follows immediately from the Lebesgue theorem since

$$c_0 ||u_n|^{p-2}u_n - |u|^{p-2}u| \leq 2c_0 L^{p-1}.$$

□

Proof of (iv). It follows from (i) that $u_n(x) \to u(x)$ and $\nabla u_n(x) \to \nabla u(x)$ a.e. in Ω and consequently, we also have that

$$\lim_{n\to\infty} \sum_{i=1}^N (a_i(x; u_n(x), \nabla u_n(x)) - a_i(x; u(x), \nabla u(x))) \left| \frac{\partial \varphi}{\partial x_i} \right| = 0 \tag{2.88}$$

a.e. in Ω.

Let E be a measurable subset of Ω. Then

$$\begin{aligned} &\int_E \sum_{i=1}^N |a_i(x; u_n(x), \nabla u_n(x))| \left| \frac{\partial \varphi}{\partial x_i} \right| dx \\ &\leq \sum_{i=1}^N \left(\int_E \frac{|a_i(x; u_n(x), \nabla u_n(x)|^{p'}}{\nu^{\frac{p'}{p}}(x)} \right)^{\frac{1}{p'}} \left(\int_E \left| \frac{\partial \varphi}{\partial x_i} \right|^p \nu(x)\, dx \right)^{\frac{1}{p}} \\ &\leq N\lambda(L) \left[\left(\int_E |g(x)|^{p'}\, dx \right)^{\frac{1}{p'}} + \left(\int_E |u_n(x)|^p\, dx \right)^{\frac{1}{p'}} \right. \\ &\quad \left. + \left(\int_E \nu(x) |\nabla u_n(x)|^p \right)^{\frac{1}{p'}} \right] \|\varphi\|_{1,p,\nu} \end{aligned}$$

due to the growth assumption D(ii). Since $u_n \to u$ in V, $u_n \to u$ in $L^p(\Omega)$ and $\nu^{\frac{1}{p}} |\nabla u_n| \to \nu^{\frac{1}{p}} |\nabla u|$ in $L^p(\Omega)$, we conclude that $|u_n|^p$ and $\nu(x)|\nabla u_n|^p$ are equiintegrable. Consequently, also the sequence

$$\left\{ \sum_{i=1}^N |a_i(x; u_n, \nabla u_n)| \left| \frac{\partial \varphi}{\partial x_i} \right| \right\}$$

is equiintegrable, and (2.72) follows, due to (2.88), by Vitali's theorem.

Consequently, Theorem 2.1 is proved. □

Example 2.1 Let $\Omega = \{x \in \mathbb{R}^N;\ |x| < 1\}$. Put

$$v(x) = [\mathrm{dist}(x, \partial\Omega)]^{\varrho} = \left| x - \frac{x}{|x|} \right|^{\varrho}$$

with $\varrho < \frac{p}{N}$. Consider the following BVP:

$$-\operatorname{div}\left(\frac{\left| x - \frac{x}{|x|} \right|^{\varrho}}{1 + |u|^p} |\nabla u|^{p-2} \nabla u \right) + e^u - |u|^p + \left| x - \frac{x}{|x|} \right|^{\varrho} |\nabla u|^p = g(x) \quad \text{in} \quad \Omega\,,$$

$$u = 0 \quad \text{on} \quad \partial\Omega\,,$$

where $g(x) \in L^{\infty}(\Omega)$.

This corresponds to the operator from (2.13) with

$$a_i(x; u, \nabla u) = \frac{\left| x - \frac{x}{|x|} \right|^{\varrho}}{1 + |u|^p} |\nabla u|^{p-2} \frac{\partial u}{\partial x_i}\,, \qquad i = 1, \dots, N\,,$$

$$a_0(x; u, \nabla u) = e^u - |u|^p - u|u|^{p-2} + \left| x - \frac{x}{|x|} \right|^{\varrho} |\nabla u|^p - g(x)\,, \qquad c_0 = 1\,,$$

and (2.16) holds if we put $\lambda(|u|) = e^{|u|^p}$. It is possible to verify also all other assumptions mentioned in Section 2.2 and hence our BVP has at least one $W_0^{1,p}(v, \Omega)$-*solution* in the sense of (2.30), i.e. there exists at least one $u \in W = W_0^{1,p}(v, \Omega) \cap L^{\infty}(\Omega)$ such that

$$\int_{\Omega} \frac{\left| x - \frac{x}{|x|} \right|^{\varrho}}{1 + |u|^p} |\nabla u|^{p-2} \nabla u \nabla \varphi \, dx + \int_{\Omega} \left(e^u - |u|^p + \left| x - \frac{x}{|x|} \right|^{\varrho} |\nabla u|^p \right) \varphi \, dx = \int_{\Omega} g(x) \varphi(x) \, dx$$

holds for any $\varphi \in W$. ◇

Example 2.2 Let Ω be a bounded open set in $\mathbb{R}^N$ and let $x_0 \in \Omega \cup \partial\Omega$ be a fixed point. Put

$$v(x) = [\mathrm{dist}(x, x_0)]^{\varrho} = |x - x_0|^{\varrho}$$

with

$$-N < \varrho < \min(N(p-1), p)\,.$$

Let $g \in L^\infty(\Omega)$. Consider the following BVP:

$$-\operatorname{div}\left(\frac{|x-x_0|^\varrho}{1+e^u}|\nabla u|^{p-2}\nabla u\right) + |x-x_0|^\varrho|\nabla u|^p + \cosh u - |u|^p = g(x) \quad \text{in} \quad \Omega\,,$$

$$u = 0 \quad \text{in} \quad \partial\Omega\,.$$

It is possible to verify all the assumptions from Section 2.2 and hence this BVP has at least one $W_0^{1,p}(\nu,\Omega)$-*solution* according to Theorem 2.1. ◇

Example 2.3 Let $\Omega \subset \mathbb{R}^N$ be an open bounded set and let $g \in L^\infty(\Omega)$. Put $\nu(x) \equiv 1$ in Ω and consider BVP

$$-\operatorname{div}((e^u+1)|\nabla u|^{p-2}\nabla u) + |\nabla u|^p = g(x) \quad \text{in} \quad \Omega\,,$$

$$u = 0 \quad \text{on} \quad \partial\Omega\,.$$

According to Theorem 2.1 this BVP has at least one $W_0^{1,p}(\Omega)$-*solution.* ◇

2.4 Second order equations (unbounded domains)

In this section we assume that Ω is an unbounded domain in $\mathbb{R}^N$ and consider $V = W_0^{1,p}(\nu,\Omega)$. Put $d = \inf\{|x|\,;\ x \in \Omega\}$ and for any positive integer n define

$$\Omega_n = \{x \in \Omega\,;\ |x| < d+n\}\,.$$

Assumption G Let $\nu \in L^1_{\mathrm{loc}}(\Omega)$. Moreover, for any positive integer n let there exist a real number $s_n > \max\left\{1, \frac{N}{p}, \frac{1}{p-1}\right\}$ such that $\nu^{-s_n} \in L^1(\Omega_n)$.

The main result of this section is the following assertion:

Theorem 2.2 *Let Ω be unbounded, $V = W_0^{1,p}(\nu,\Omega)$ and let assumptions* (2.17), D, E, F *and* G *be satisfied. Then Problem* 2.1 *has at least one solution.*

Proof. Put $V_n = W_0^{1,p}(\nu,\Omega_n)$ and $W_n = V_n \cap L^\infty(\Omega_n)$. Then it follows from assumption G that assumption C is fulfilled for any Ω_n (with $s = s_n$) and assumptions A and B are fulfilled automatically with $V = V_n$. Then it follows from Theorem 2.1 that for any positive integer n there exists at least one $u_n \in W_n$ such that

$$\int_{\Omega_n}\left(\sum_{i=1}^N a_i(x;u_n,\nabla u_n)\frac{\partial\varphi}{\partial x_i} + c_0|u_n|^{p-2}u_n\varphi + a_0(x;u_n,\nabla u_n)\varphi\right)dx = 0 \tag{2.89}$$

holds for any $\varphi \in W_n$. We will extend the function u_n outside of Ω_n defining $u_n(x) = 0$ for $x \in \Omega \setminus \Omega_n$. With respect to (2.64) and (2.66) we have $\|u_n\|_\infty \le L$, $\|u_n\|_{1,p,\nu} \le \mathcal{M}$ for any n (note that these estimates are independent of Ω_n according to Lemmas 2.1 and 2.2), hence it is possible to find a subsequence (denoted again by $\{u_n\}$) which converges weakly in $W_0^{1,p}(\nu, \Omega)$ and weakly* in $L^\infty(\Omega)$ to an element $u \in W_0^{1,p}(\nu, \Omega) \cap L^\infty(\Omega)$ for which $\|u\|_\infty \le L$ and $\|u\|_{1,p,\nu} \le \mathcal{M}$. We will show that u satisfies (2.30) for any $\varphi \in W_0^{1,p}(\nu, \Omega) \cap L^\infty(\Omega)$, which means that u is *a solution* of Problem 2.1. With respect to the definition of $W_0^{1,p}(\nu, \Omega)$ it is sufficient to prove that (2.30) holds for any fixed $\varphi \in C_0^\infty(\Omega)$.

Let us take an arbitrary but fixed $\varphi \in C_0^\infty(\Omega)$ and denote by C the support of φ. Because $\varphi = 0$ in $\Omega \setminus C$ the relation (2.30) is equivalent to

$$\int_C \left(\sum_{i=1}^N a_i(x; u, \nabla u) \frac{\partial \varphi}{\partial x_i} + c_0 |u|^{p-2} u \varphi + a_0(x; u, \nabla u) \varphi \right) dx = 0 \,. \tag{2.90}$$

The relation (2.90) follows from

$$\lim_{n\to\infty} \int_C (|u_n - u|^p + \nu |\nabla u_n - \nabla u|^p)\, dx = 0 \tag{2.91}$$

by the same procedure as in the proof of Theorem 2.1. Hence the remaining part of the proof consists in establishing (2.91).

Let us introduce open bounded sets A and B such that $C \subset B \subset \overline{B} \subset A \subset \Omega$ and A has the cone property. By assumption G, for any positive integer n there exists a real number

$$s_n > \max \left\{ 1, \frac{N}{p}, \frac{1}{p-1} \right\}$$

such that $\nu^{-s_n} \in L^1(\Omega_n)$. Choose n_0 in such a way that $A \subset \Omega_n$ for any $n > n_0$. Then also $\nu^{-s_n} \in L^1(A)$ for any $n > n_0$. Hence assumption (2.17) is verified for $\Omega = A$. Then it follows from Remark 2.4 that the imbedding of $W^{1,p}(\nu, A)$ in $L^p(A)$ is compact and hence

$$\lim_{n\to\infty} \int_A |u_n - u|^p \, dx = 0 \,. \tag{2.92}$$

We shall prove that

$$\lim_{n\to\infty} \int_B \nu |\nabla u_n - \nabla u|^p \, dx = 0 \tag{2.93}$$

and the relation (2.91) will be the consequence of (2.92), (2.93) and $C \subset B \subset A$.

In order to prove (2.93) we apply Lemma 2.3. With respect to (2.92) we have

$$\lim_{n\to\infty} \|u_n - u\|_{L^p(B)} = 0 \,;$$

also

$$\|u_n\|_{W^{1,p}(\nu,B)} \leq \mathcal{M}$$

and

$$\|u_n\|_{L^\infty(B)} \leq L .$$

It remains to verify the assumption

$$\lim_{n\to\infty} \int_B \sum_{i=1}^N (a_i(x; u_n, \nabla u_n) - a_i(x; u_n, \nabla u)) \frac{\partial(u_n - u)}{\partial x_i} \, dx = 0 . \tag{2.94}$$

It follows from the partition of unity that there exists a function $\psi \in C_0^\infty(\mathbb{R}^N)$ such that $0 \leq \psi(x) \leq 1$ for any $x \in \mathbb{R}^N$, $\operatorname{supp}\psi \subset A$ and $\psi(x) = 1$ for any $x \in B$. Note that for $n > n_0$ we have $A \subset \Omega_n$. Take in (2.89) a test function $\varphi(x) = \psi(x)(u_n(x) - u(x))|u_n(x) - u(x)|^\gamma$ with $\gamma > 0$ which will be fixed later. We get

$$\begin{aligned} \int_A \Bigg\{ & \sum_{i=1}^N a_i(x; u_n, \nabla u_n) \left[\frac{\partial \psi}{\partial x_i} (u_n - u)|u_n - u|^\gamma \right. \\ & \left. + \psi(\gamma + 1)|u_n - u|^\gamma \frac{\partial(u_n - u)}{\partial x_i} \right] + c_0 |u_n|^{p-2} u_n \psi (u_n - u)|u_n - u|^\gamma \\ & + a_0(x; u_n, \nabla u_n) \psi (u_n - u)|u_n - u|^\gamma \Bigg\} \, dx = 0 . \end{aligned}$$

By the same procedure as in the proof of (i) in Section 2.3 (see pages 57–61) we derive from here:

$$\lim_{n\to\infty} \int_A \psi(x)\nu(x)|\nabla u_n|^p |u_n - u| \, dx = 0 . \tag{2.95}$$

Taking in (2.89) $\varphi(x) = \psi(x)(u_n(x) - u(x))$ as a test function, we obtain

$$\begin{aligned} & \int_A \psi \sum_{i=1}^N (a_i(x; u_n, \nabla u_n) - a_i(x; u_n, \nabla u)) \frac{\partial(u_n - u)}{\partial x_i} \, dx \\ & = - \int_A \sum_{i=1}^N a_i(x; u_n, \nabla u_n) \frac{\partial \psi}{\partial x_i} (u_n - u) \, dx \\ & \quad - \int_A c_0 |u_n|^{p-2} u_n \psi (u_n - u) \, dx - \int_A a_0(x; u_n, \nabla u_n) \psi (u_n - u) \, dx \\ & \quad - \int_A \psi \sum_{i=1}^N a_i(x; u, \nabla u) \frac{\partial(u_n - u)}{\partial x_i} \, dx \\ & \quad + \int_A \psi \sum_{i=1}^N (a_i(x; u, \nabla u) - a_i(x; u_n, \nabla u)) \frac{\partial(u_n - u)}{\partial x_i} \, dx , \end{aligned}$$

which together with

$$\int_B \sum_{i=1}^N (a_i(x; u_n, \nabla u_n) - a_i(x; u_n, \nabla u)) \frac{\partial(u_n - u)}{\partial x_i}\, dx$$
$$\leq \int_A \psi \sum_{i=1}^N (a_i(x; u_n, \nabla u_n) - a_i(x; u_n, \nabla u)) \frac{\partial(u_n - u)}{\partial x_i}\, dx$$

(we have used assumption E) yields the estimate

$$\int_B \sum_{i=1}^N (a_i(x; u_n, \nabla u_n) - a_i(x; u_n, \nabla u)) \frac{\partial(u_n - u)}{\partial x_i}\, dx$$
$$\leq -\int_A \sum_{i=1}^N a_i(x; u_n, \nabla u_n) \frac{\partial \psi}{\partial x_i}(u_n - u)\, dx - \int_A c_0 |u_n|^{p-2} u_n \psi (u_n - u)\, dx$$
$$- \int_A a_0(x; u_n, \nabla u_n)\psi(u_n - u)\, dx - \int_A \psi \sum_{i=1}^N a_i(x; u, \nabla u) \frac{\partial(u_n - u)}{\partial x_i}\, dx$$
$$+ \int_A \psi \sum_{i=1}^N (a_i(x; u, \nabla u) - a_i(x; u_n, \nabla u)) \frac{\partial(u_n - u)}{\partial x_i}\, dx\,. \tag{2.96}$$

Then (2.95) together with the Lebesgue theorem, the Vitali theorem and continuity of the Nemytskij operator yields that all terms on the right-hand side of (2.96) tend to zero. The proof of these facts is the same as in Section 2.3. Hence (2.94) holds and Lemma 2.3 implies that (2.93) is true. This completes the proof of Theorem 2.2. □

Example 2.4 Let $\Omega = \{x \in \mathbb{R}^N\,;\ |x| > 1\}$. Put

$$v(x) = [\operatorname{dist}(x, \partial\Omega)]^{\varrho} = \left| x - \frac{x}{|x|} \right|^{\varrho}$$

with $\varrho < \min\left\{1, \frac{p}{N}, p - 1\right\}$. Suppose that

$$g \in [L^1(\Omega) + L^{p'}(\Omega)] \cap L^{\infty}(\Omega)$$

and

$$\lambda(|u|) = c e^{|u|^p}$$

with c large enough. Then it is possible to verify all the assumptions for BVP

$$-\operatorname{div}\left(\frac{\left|x-\frac{x}{|x|}\right|^{\varrho}}{1+|u|^{\varrho}}|\nabla u|^{p-2}\nabla u\right)+\frac{e^{u}}{|x|^{N+1}}-|u|^{p}+\left|x-\frac{x}{|x|}\right|^{\varrho}|\nabla u|^{p}=g(x) \quad \text{in} \quad \Omega,$$

$$u = 0 \quad \text{on} \quad \partial\Omega$$

and hence it has at least one $W_0^{1,p}(\nu, \Omega)$-*solution* according to Theorem 2.2. ◇

2.5 Higher order equations (growth conditions)

Now, let us consider a general differential operator A of order $2k$ ($k \geq 1$) of the form (2.1). We will look for a weak solution $u \in W^{k,p}(\Omega, w)$ — a V-solution in the sense of Definition 2.1 — with a space V satisfying (2.6).

The choice of an appropriate weighted Sobolev space is closely connected with the properties of the “coefficients” $a_\alpha(x; \xi)$ which appear in the definition of the differential operator A in (2.1).

Let $m = m(k, N)$ be the number of all (different) multiindices α, where $\alpha = (\alpha_1, \alpha_2, \ldots, \alpha_N)$ is of the length $|\alpha| = \alpha_1 + \alpha_2 + \cdots + \alpha_N$ not exceeding k. We will assume throughout the next sections that $a_\alpha(x; \xi)$, defined for $x \in \Omega$ and $\xi \in \mathbb{R}^m$, $\xi = \{\xi_\alpha\,;\ |\alpha| \leq k\}$, satisfies the Carathéodory condition:

$$a_\alpha \in \mathrm{CAR}\,. \tag{2.97}$$

Before formulating growth conditions for $a_\alpha(x; \xi)$, let us formulate an important assumption which will be used throughout the next sections.

Assumption H For any multiindex β, $|\beta| < k$, there is a parameter $q(\beta) > 1$ and a weight function $\omega_\beta = \omega_\beta(x)$ such that for every $u \in W^{k,p}(\Omega, w)$ (or $u \in V$ or $u \in W_0^{k,p}(\Omega, w)$) with $w = \{w_\alpha = w_\alpha(x), x \in \Omega; |\alpha| \leq k\}$,

$$D^\beta u \in L^{q(\beta)}(\Omega, \omega_\beta) \tag{2.98}$$

and, moreover,

$$\|D^\beta u\|_{q(\beta),\omega_\beta} \leq \tilde{c}_\beta \|u\|_{k,p,w} \tag{2.99}$$

with $\tilde{c}_\beta$ independent of u, $\tilde{c}_\beta > 0$.

Remark 2.7 (i) Assumption H is obviously fulfilled if we choose $q(\beta) = p$ and $\omega_\beta = w_\beta$, but the more important case occurs if we have $q(\beta) \neq p$ (usually $q(\beta) > p$, but not necessarily) and a new weight function ω_β different from w_β.

(ii) Assumption H expresses in fact the claim that a certain *imbedding* for weighted Sobolev space takes place, in the sense mentioned in Section 1.5. For a fixed multi-index β, $|\beta| < k$, inequality (2.99) is in fact a generalized (higher order) Hardy type inequality, see Example 1.7, and if we consider, e.g., (2.98) for all β with $|\beta| \le n$ where $n < k$ and with $q(\beta) = q$ for all β, $|\beta| \le n$, then (2.99) describes the continuous imbedding

$$V \hookrightarrow W^{n,q}(\Omega, \omega)$$

with the family $\omega = \{\omega_\beta\ ;\ |\beta| \le n\}$, where V is $W^{k,p}(\Omega, w)$ or $W_0^{k,p}(\Omega, w)$ or some subspace according to (2.6).

Now, let us consider several types of *growth conditions* for $a_\alpha(x; \xi)$.

Definition 2.2 For $p > 1$ denote by $p' = \dfrac{p}{p-1}$ the exponent conjugate to p. Let $g_\alpha \in L^{p'}(\Omega)$ and let $c_\alpha \ge 0$ be a constant. We say that the coefficients a_α of the differential operator A from (2.1) satisfy the *growth conditions of type* (A) if the estimate

$$|a_\alpha(x;\xi)| \le w_\alpha^{\frac{1}{p}}(x)\left(g_\alpha(x) + c_\alpha \sum_{|\beta|\le k} w_\beta^{\frac{1}{p'}}(x)|\xi_\beta|^{p-1}\right) \tag{2.100}$$

holds for a.e. $x \in \Omega$ and every $\xi \in \mathbb{R}^m$; $|\alpha| \le k$.

Definition 2.3 Let $q(\beta)$ and ω_β, $|\beta| \le k-1$, be the parameters appearing in assumption H, and let g_α and c_α be as in the foregoing Definition 2.2. We say that the coefficients a_α satisfy the *growth conditions of type* (B) if the estimate

$$|a_\alpha(x;\xi)| \le w_\alpha^{\frac{1}{p}}(x)\left(g_\alpha(x) + c_\alpha \sum_{|\beta|=k} w_\beta^{\frac{1}{p'}}(x)|\xi_\beta|^{p-1} + c_\alpha \sum_{|\beta|\le k-1} \omega_\beta^{\frac{1}{p'}}(x)|\xi_\beta|^{\frac{q(\beta)}{p'}}\right) \tag{2.101}$$

holds for a.e. $x \in \Omega$ and every $\xi \in \mathbb{R}^m$; $|\alpha| \le k$.

Definition 2.4 Let g_α, c_α, $q(\beta)$ and ω_β be as in the foregoing Definition 2.3. Moreover, let $\tilde{g}_\alpha \in L^{q(\alpha)'}$ for $|\alpha| \le k-1$, $q(\alpha)' = q(\alpha)/(q(\alpha)-1)$. We say that the coefficients a_α satisfy the *growth conditions of type* (C) if the following estimates hold for a.e. $x \in \Omega$ and every $\xi \in \mathbb{R}^m$:

(i) For $|\alpha| = k$,

$$\begin{aligned} |a_\alpha(x;\xi)| \le w_\alpha^{\frac{1}{p}}(x)\Bigg(g_\alpha(x) &+ c_\alpha \sum_{|\beta|=k} w_\beta^{\frac{1}{p'}}(x)|\xi_\beta|^{p-1} \\ &+ c_\alpha \sum_{|\beta|\le k-1} \omega_\beta^{\frac{1}{p'}}(x)|\xi_\beta|^{\frac{q(\beta)}{p'}}\Bigg)\ ; \end{aligned} \tag{2.102}$$

(ii) for $|\alpha| \le k-1$,

$$|a_\alpha(x;\xi)| \le \omega_\alpha^{\frac{1}{q(\alpha)}}(x)\left(\tilde{g}_\alpha(x) + c_\alpha \sum_{|\beta|=k} w_\beta^{\frac{1}{q(\alpha)'}}(x)|\xi_\beta|^{\frac{p}{q(\alpha)'}} + c_\alpha \sum_{|\beta|\le k-1} \omega_\beta^{\frac{1}{q(\alpha)'}}(x)|\xi_\beta|^{\frac{q(\beta)}{q(\alpha)'}}\right). \tag{2.103}$$

Let us denote

$$\kappa_1 = k - \frac{N}{p} \tag{2.104}$$

and suppose that $\kappa_1 > 0$, i.e., $kp > N$. In this case we can — similarly as in the case of classical Sobolev spaces, see Theorem 1.2 (c) — consider also imbeddings of $W^{k,p}(\Omega, w)$ into weighted spaces of continuous functions, more precisely, into

$$C(\Omega,\omega) = \{u = u(x) \quad \text{continuous on} \quad \Omega; \quad \|u\|_\omega^* = \operatorname*{supess}_{x\in\Omega} |u(x)\omega(x)| < \infty\} \tag{2.105}$$

with a suitable weight function ω (see also Section 1.5, part (iv)). Therefore, we will make the following

Assumption J Let $kp > N$ and let κ_1 be defined by (2.104). For any multiindex β, $|\beta| < \kappa_1$, there is a weight function $\omega_\beta = \omega_\beta(x)$ such that for every $u \in W^{k,p}(\Omega, w)$ (or $u \in W_0^{k,p}(\Omega, w)$ or $u \in V$),

$$D^\beta u \in C(\Omega, \omega_\beta) \tag{2.106}$$

and, moreover,

$$\operatorname*{supess}_{x\in\Omega} |D^\beta u(x)\omega_\beta(x)| \le \tilde{c}_\beta \|u\|_{k,p,w} \tag{2.107}$$

with $\tilde{c}_\beta$ independent of u, $\tilde{c}_\beta > 0$.

Inequality (2.107) states again the validity of a certain *imbedding* of weighted Sobolev spaces, this time into a space of continuous functions. Taking into account assumption J, we can modify our foregoing growth conditions of type (A)–(C). For this purpose, let us write $\xi \in \mathbb{R}^m$ in the form $\xi = (\kappa, \tilde{\xi})$, where $\kappa = \{\xi_\beta;\ |\beta| < \kappa_1\}$, and denote

$$h(x,\kappa) = \sum_{|\beta|<\kappa_1} |\omega_\beta(x)\xi_\beta|. \tag{2.108}$$

If we denote by $h(x, u(x))$ the expression

$$\sum_{|\beta|<\kappa_1} |\omega_\beta(x) D^\beta u(x)|$$

then, in view of (2.107),

$$|h(x, u(x))| \leq c\|u\|_{k,p,w} \tag{2.109}$$

for $u \in W^{k,p}(\Omega, w)$ (or $u \in V$).

Definition 2.5 Let us use the same notation as in the preceding Definitions 2.2–2.4. Suppose, moreover, that $\hat{g}_\alpha \in L^1(\Omega)$ for $|\alpha| < \kappa_1$. We say that the coefficients a_α satisfy the *growth conditions of type* (D), if there is a positive, continuous, nondecreasing function $G(t)$, $t \geq 0$, such that the following estimates hold for a.e. $x \in \Omega$ and every $\xi \in \mathbb{R}^m$:

(i) For $|\alpha| = k$,

$$\begin{aligned}|a_\alpha(x;\xi)| \leq G(h(x,\kappa))w_\alpha^{\frac{1}{p}}(x)\Bigg(g_\alpha(x) &+ c_\alpha \sum_{|\beta|=k} w_\beta^{\frac{1}{p'}}(x)|\xi_\beta|^{p-1}\\ &+ c_\alpha \sum_{\kappa_1\leq|\beta|\leq k-1} \omega_\beta^{\frac{1}{p'}}(x)|\xi_\beta|^{\frac{q(\beta)}{p'}}\Bigg);\end{aligned} \tag{2.110}$$

(ii) for $\kappa_1 \leq |\alpha| \leq k-1$,

$$\begin{aligned}|a_\alpha(x;\xi)| \leq G(h(x,\kappa))\omega_\alpha^{\frac{1}{q(\alpha)}}(x)\Bigg(\tilde{g}_\alpha(x) &+ c_\alpha \sum_{|\beta|=k} w_\beta^{\frac{1}{q(\alpha)'}}(x)|\xi_\beta|^{\frac{p}{q(\alpha)'}}\\ &+ c_\alpha \sum_{\kappa_1\leq|\beta|\leq k-1} \omega_\beta^{\frac{1}{q(\alpha)'}}(x)|\xi_\beta|^{\frac{q(\beta)}{q(\alpha)'}}\Bigg);\end{aligned} \tag{2.111}$$

(iii) for $|\alpha| < \kappa_1$,

$$\begin{aligned}|a_\alpha(x;\xi)| \leq G(h(x,\kappa))\omega_\alpha(x)\Bigg(\hat{g}_\alpha(x) &+ c_\alpha \sum_{|\beta|=k} w_\beta(x)|\xi_\beta|^{p}\\ &+ c_\alpha \sum_{\kappa_1\leq|\beta|\leq k-1} \omega_\beta(x)|\xi_\beta|^{q(\beta)}\Bigg).\end{aligned} \tag{2.112}$$

Remark 2.8 We can concentrate on the growth conditions of type (D) since the previous ones can be considered as special cases when for $\kappa_1 \leq 0$ the set of multiindices ξ_β with $|\beta| < \kappa_1$ is empty by definition. Then we set $G(t) \equiv 1$, and since the case (iii) in Definition 2.5 is irrelevant, we obtain from (2.110) and (2.111) immediately (2.102) and (2.103), i.e. growth conditions of type (C). Further, if we do not differ between $|\alpha| = k$ and $|\alpha| < k$ in $a_\alpha(x;\xi)$ (and use, instead of (2.103), the same estimate (2.102) also for $|\alpha| \le k-1$), we immediately obtain (2.101), i.e. growth conditions of type (B). Finally, we can use the natural imbedding

$$W^{k,p}(\Omega, w) \hookrightarrow W^{k-1,p}(\Omega, w)$$

(i.e. the estimates

$$\|D^\beta u\|_{p,w_\beta} \le \|u\|_{k,p,w}$$

for $|\beta| \le k-1$, $u \in W^{k,p}(\Omega, w)$) which follows directly from the definition of the norm in $W^{k,p}(\Omega, w)$. Consequently, we can choose $q(\beta) = p$ and $\omega_\beta = w_\beta$ also in (2.98). If we do this in (2.101), we immediately obtain (2.100), i.e. growth conditions of type (A).

In connection with the facts just mentioned, the question arises *why* we consider four different types of growth conditions if a condition of type (D) covers the foregoing ones. Therefore, let us emphasize an important point in our approach: While the natural way should be to find for a boundary value problem, i.e. for a differential operator A and the boundary conditions, the appropriate space V (i.e., the weights w_α and the growth parameter p) in which the solution should be looked for, here we go in the *opposite direction*: We start with the space V (more precisely, with the weighted space $W^{k,p}(\Omega, w)$) and try to find appropriate differential operators A which allow to derive existence theorems for the corresponding boundary value problem. And here we have a choice which depends on our knowledge of the space (Assumptions H and J) and of our willingness to use more complicated and sophisticated functional analytic tools. Thus, growth conditions of type (A) are the simplest ones: in this case we can obtain an existence theorem almost immediately, using the Leray–Lions theorem, and we need no knowledge about the space V except its definition. On the contrary, conditions of type (D) are connected with the use of all the tools which assumptions H and J offer (see Section 2.6).

2.6 Higher order equations (operator representation)

Let V be the space from (2.6), expressing — roughly speaking — the boundary conditions associated with our partial differential equation

$$Au = f \quad \text{in} \quad \Omega\,, \tag{2.113}$$

and let $a(u, \varphi)$ be the form from (2.9), i.e.

$$a(u,\varphi) = \sum_{|\alpha|\le k} \int_\Omega a_\alpha(x; u(x), \nabla u(x), \dots, \nabla^k u(x)) D^\alpha \varphi(x)\, dx\,, \tag{2.114}$$

where $\nabla^j u(x) = \{D^\beta u(x)\,;\ |\beta| = j\}$, $j \in \mathbb{N}$.

First, let us slightly modify the definition of the solution of a boundary value problem.

Definition 2.6 Let $u_0 \in W^{k,p}(\Omega, w)$ be a fixed function. *A* function $u = u_0 + v$, $v \in V$, satisfying the integral identity

$$a(u_0 + v, \varphi) = \int_\Omega f(x)\varphi(x)\,dx \tag{2.115}$$

for every $\varphi \in V$, will be called a *weak solution of the nonhomogeneous boundary value problem* for the equation (2.113).

Remark 2.9 As was already mentioned at the end of Introduction, in order to avoid tedious calculations with expressions involving the terms $u_0 + v$, $D^\alpha(u_0 + v)$, etc., we will suppose in the sequel that $u_0 \equiv 0$, i.e. instead of (2.115) we will be looking for a *weak solution of the homogeneous boundary value problem, i.e. for a function* $u \in V$ satisfying the integral identity

$$a(u, \varphi) = \int_\Omega f(x)\varphi(x)dx \tag{2.116}$$

for every $\varphi \in V$. The reader should bear in mind that the simple shift $u \mapsto u_0 + u$ transfers existence results for the homogeneous problem to existence results for the nonhomogeneous one.

So, we have again arrived at the concept of a V-solution from Definition 2.1, with the small change that for simplicity we consider a concrete form of the duality expression $\langle f, \varphi \rangle$.

Lemma 2.5 *Let functions $a_\alpha(x; \xi)$ satisfy one of the growth conditions of type (A), (B), (C) or (D) (see Definitions 2.2–2.5). Then the form $a(u, \varphi)$ from (2.114) is well-defined for any $u, \varphi \in W^{k,p}(\Omega, w)$. Moreover, there exists a positive continuous function $H(t)$ defined in $[0, \infty)$ such that*

$$|a(u, \varphi)| \le H(\|u\|_{k,p,w})\|\varphi\|_{k,p,w}\,. \tag{2.117}$$

Proof. It is sufficient to consider the growth conditions of type (D)(see Remark 2.4). Then it follows from (2.110) (for $|\alpha| = k$) that

$$|a_\alpha(x; u(x), \dots, \nabla^k u(x))| \le G\,(h(x, u(x)))\, w_\alpha^{\frac{1}{p}}(x)$$
$$\cdot \left(|g_\alpha(x)| + c_\alpha \sum_{|\beta|=k} w_\beta^{\frac{1}{p'}}(x)|D^\beta u(x)|^{p-1} + c_\alpha \sum_{\kappa_1 \le |\beta| \le k-1} \omega_\beta^{\frac{1}{p'}}(x)|D^\beta u(x)|^{\frac{q(\beta)}{p'}} \right)$$

and consequently

$$
\begin{aligned}
&\int_\Omega \left| a_\alpha(x; u(x), \dots, \nabla^k u(x)) D^\alpha \varphi(x) \right| dx \\
&\quad \le \int_\Omega G\,(h(x, u(x)))\, w_\alpha^{\frac{1}{p}}(x) |D^\alpha \varphi(x)| |g_\alpha(x)|\, dx \qquad (2.118)\\
&\qquad + c_\alpha \sum_{|\beta|=k} \int_\Omega G\,(h(x, u(x)))\, w_\alpha^{\frac{1}{p}}(x) |D^\alpha \varphi(x)| w_\beta^{\frac{1}{p'}}(x) |D^\beta u(x)|^{\frac{p}{p'}}\, dx \\
&\qquad + c_\alpha \sum_{\kappa_1 \le |\beta| \le k-1} \int_\Omega G\,(h(x, u(x)))\, w_\alpha^{\frac{1}{p}}(x) |D^\alpha \varphi(x)| \omega_\beta^{\frac{1}{p'}}(x) |D^\beta u(x)|^{\frac{q(\beta)}{p'}}\, dx\,.
\end{aligned}
$$

It follows from the monotonicity of the function G and from (2.109) that

$$
G(h\,(x, u(x))) \le G(c\|u\|_{k,p,w})\,. \qquad (2.119)
$$

We will estimate the integrals in (2.118) using (2.119) and the Hölder inequality with the exponents p and p':

$$
\begin{aligned}
&\int_\Omega |a_\alpha \left(x; u(x), \dots, \nabla^k u(x)\right) D^\alpha \varphi(x)|\, dx \\
&\quad \le G(c\|u\|_{k,p,w}) \|D^\alpha \varphi\|_{p,w_\alpha} \Bigg(\|g_\alpha\|_{p'} + c_\alpha \sum_{|\beta|=k} \|D^\beta u\|_{p,w_\beta}^{p-1} \qquad (2.120)\\
&\qquad\qquad + c_\alpha \sum_{\kappa_1 \le |\beta| \le k-1} \|D^\beta u\|_{q(\beta),\omega_\beta}^{\frac{q(\beta)}{p'}} \Bigg)\,.
\end{aligned}
$$

Since we have $\|D^\beta u\|_{p,w_\beta} \le \|u\|_{k,p,w}$ by definition of the norm in $W^{k,p}(\Omega, w)$ and $\|D^\beta u\|_{q(\beta),\omega_\beta} \le \tilde{c}_\beta \|u\|_{k,p,w}$ due to assumption H, we obtain from (2.120)

$$
\begin{aligned}
&\int_\Omega |a_\alpha(x; u(x), \dots, \nabla^k u(x)) D^\alpha \varphi(x)|\, dx \qquad (2.121)\\
&\quad \le \|D^\alpha \varphi\|_{p,w_\alpha} H_\alpha(\|u\|_{k,p,w}) \le \|\varphi\|_{k,p,w} H_\alpha(\|u\|_{k,p,w})
\end{aligned}
$$

for $|\alpha| = k$, where $H_\alpha(t)$ is a positive continuous function.

A similar estimate can be derived for $\kappa_1 \leq |\alpha| \leq k-1$ from the condition (2.111): we get

$$\begin{aligned}
&\int_\Omega |a_\alpha(x; u(x), \ldots, \nabla^k u(x)) D^\alpha \varphi(x)|\, dx \\
&\leq \int_\Omega G\,(h(x, u(x)))\, \omega_\alpha^{\frac{1}{q(\alpha)}}(x)|D^\alpha \varphi(x)||\tilde{g}_\alpha(x)|\, dx \qquad (2.122)\\
&\quad + c_\alpha \sum_{|\beta|=k} \int_\Omega G\,(h(x, u(x)))\, \omega_\alpha^{\frac{1}{q(\alpha)}}(x)|D^\alpha \varphi(x)| w_\beta^{\frac{1}{q(\alpha)'}}(x)|D^\beta u(x)|^{\frac{p}{q(\alpha)'}}\, dx \\
&\quad + c_\alpha \sum_{\kappa_1 \leq |\beta| \leq k-1} \int_\Omega G\,(h(x, u(x)))\, \omega_\alpha^{\frac{1}{q(\alpha)}}(x)|D^\alpha \varphi(x)| \omega_\beta^{\frac{1}{q(\alpha)'}}(x)|D^\beta u(x)|^{\frac{q(\beta)}{q(\alpha)'}}\, dx .
\end{aligned}$$

Applying the Hölder inequality with exponents $q(\alpha), q(\alpha)'$ in (2.112) we obtain

$$\begin{aligned}
&\int_\Omega |a_\alpha(x; u(x), \ldots, \nabla^k u(x)) D^\alpha \varphi(x)|\, dx \\
&\quad \leq G(c\|u\|_{k,p,w}) \|D^\alpha \varphi\|_{q(\alpha),\omega_\alpha} \Bigg(\|\tilde{g}_\alpha\|_{q(\alpha)'} + c_\alpha \sum_{|\beta|=k} \|D^\beta u\|_{p,w_\beta}^{\frac{p}{q(\alpha)'}} \qquad (2.123)\\
&\qquad + c_\alpha \sum_{\kappa_1 \leq |\beta| \leq k-1} \|D^\beta u\|_{q(\beta),\omega_\beta}^{\frac{q(\beta)}{q(\alpha)'}} \Bigg) .
\end{aligned}$$

Now, using the fact that $\|D^\alpha \varphi\|_{q(\alpha),\omega_\alpha} \leq \tilde{c}_\alpha \|\varphi\|_{k,p,w}$ (see assumption H) in (2.123), we obtain an estimate of the form (2.121), but for $|\alpha| \in [\kappa_1, k-1]$.

Finally, for $|\alpha| < \kappa_1$, we use the condition (2.112) which leads to the estimate

$$\begin{aligned}
&\int_\Omega |a_\alpha(x; u(x), \ldots, \nabla^k u(x)) D^\alpha \varphi(x)|\, dx \\
&\leq \int_\Omega G\,(h(x, u(x)))\, \omega_\alpha(x)|D^\alpha \varphi(x)||\hat{g}_\alpha(x)|\, dx \qquad (2.124)\\
&\quad + c_\alpha \sum_{|\beta|=k} \int_\Omega G\,(h(x, u(x)))\, \omega_\alpha(x)|D^\alpha \varphi(x)| w_\beta(x)|D^\beta u(x)|^p\, dx \\
&\quad + c_\alpha \sum_{\kappa_1 \leq \beta \leq k-1} \int_\Omega G\,(h(x, u(x)))\, \omega_\alpha(x)|D^\alpha \varphi(x)| \omega_\beta(x)|D^\beta u(x)|^{q(\beta)}\, dx .
\end{aligned}$$

Since $D^\alpha \varphi \in C(\Omega, \omega_\alpha)$ for $|\alpha| < \kappa_1$ (see (2.106)), we obtain from (2.124), (2.106)

$$\begin{aligned}\int_\Omega |a_\alpha(x; u(x), \dots, \nabla^k u(x)) D^\alpha \varphi(x)|\, dx \\ \le G(c\|u\|_{k,p,w}) \operatorname*{supess}_{x\in\Omega} |\omega_\alpha(x) D^\alpha \varphi(x)| \\ \cdot \left(\|\hat{g}_\alpha\|_1 + c_\alpha \sum_{|\beta|=k} \|D^\beta u\|^p_{p,w_\beta} + c_\alpha \sum_{\kappa_1 \le |\beta| \le k-1} \|D^\beta u\|^{q(\beta)}_{q(\beta),\omega_\beta} \right).\end{aligned} \tag{2.125}$$

Now, using the fact that $\operatorname*{supess}_{x\in\Omega} |\omega_\alpha(x) D^\alpha \varphi(x)| \le \tilde{c}_\alpha \|\varphi\|_{k,p,w}$ for $|\alpha| < \kappa_1$ (see (2.107)), we obtain from (2.125) an estimate of the form (2.121), but for $|\alpha| < \kappa_1$.

Since

$$|a(u,\varphi)| \le \sum_{|\alpha|\le k} \int_\Omega |a_\alpha(x; u(x), \dots, \nabla^k u(x)) D^\alpha \varphi(x)|\, dx\,,$$

we finally obtain the estimate (2.117) taking

$$H(t) = \sum_{|\alpha|\le k} H_\alpha(t)\,.$$

□

Since $a(u, \varphi)$ is *linear* with respect to φ, the expression $a(u, \cdot)$ represents the value of a continuous linear functional $F \in V^*$:

$$a(u,\varphi) := \langle F, \varphi\rangle\,. \tag{2.126}$$

It follows from Lemma 2.5 that

$$\|F\|_{V^*} \le H(\|u\|_V)\,. \tag{2.127}$$

The dependence of F on u (cf. (2.126)) will be expressed by writing

$$F = Tu\,.$$

Hence we have defined the operator $T : V \to V^*$ by the relation

$$a(u,\varphi) = \langle Tu, \varphi\rangle\,, \quad u, \varphi \in V\,. \tag{2.128}$$

It follows from (2.127) that

$$\|Tu\|_{V^*} \le H(\|u\|_V)\,,$$

which means that the *operator* T is *bounded* (i.e. T maps bounded sets in V to bounded sets in V^*).

Using the properties of the Nemytskij operator acting between weighted spaces we are also able to prove the *continuity* of T.

Lemma 2.6 *Let functions $a_\alpha = a_\alpha(x;\xi)$ defined on $\Omega \times \mathbb{R}^m$ satisfy the Carathéodory condition and one of the growth conditions of the type (A), (B), (C) or (D). Then the operator $T : V \to V^*$ defined by (2.128) is bounded and continuous.*

Proof. We can identify the weighted Sobolev space $W^{k,p}(\Omega, w)$ with a closed subspace of the product space

$$P_0 = \prod_{|\beta|\le k} L^p(\Omega, w_\beta)$$

(notice that by definition $D^\beta u \in L^p(\Omega, w_\beta)$ for $u \in W^{k,p}(\Omega, w)$, $|\beta| \le k$). But due to assumptions H and J we have for $u \in V$

$$D^\beta u \in L^{q(\beta)}(\Omega, \omega_\beta) \quad \text{for} \quad \kappa_1 \le |\beta| \le k-1$$

and

$$D^\beta u \in C(\Omega, \omega_\beta) \quad \text{for} \quad |\beta| < \kappa_1 .$$

Hence we can identify the space V with a closed subspace of the product space

$$P = \prod_{|\beta|\le k} X_\beta , \tag{2.129}$$

where

$$\begin{aligned} X_\beta &= L^p(\Omega, w_\beta) && \text{for} \quad |\beta| = k ,\\ X_\beta &= L^{q(\beta)}(\Omega, \omega_\beta) && \text{for} \quad \kappa_1 \le |\beta| \le k-1 ,\\ X_\beta &= C(\Omega, \omega_\beta) && \text{for} \quad |\beta| < \kappa_1 . \end{aligned} \tag{2.130}$$

Note that $\kappa_1 = k - \dfrac{N}{p}$ and if $\kappa_1 \le 0$ then we use only the first two spaces in (2.130). Let us also recall that the norm in $C(\Omega, \omega_\beta)$ is defined by (2.105) and in fact it is the L^∞-norm of $\omega_\beta u$.

Now we use Theorem 1.1, which can be rewritten for our purposes as follows:

The Nemytskij operator $\mathcal{N}$, defined on the product space

$$P_1 = \prod_{i=1}^m L^{p_i}(\Omega, w_i)$$

(of weighted Lebesgue spaces) by the formula

$$\mathcal{N}(u_1, u_2, \dots, u_m)(x) = h(x, u_1(x), u_2(x), \dots, u_m(x)) ,$$

where $h = h(x, s_1, s_2, \dots, s_m)$ is defined on $\Omega \times \mathbb{R}^m$ and satisfies the Carathéodory condition, maps P_1 continuously into the dual space $(L^{p_0}(\Omega, w_0))^* = L^{p_0'}(\Omega, w_0^{1-p_0'})$ if and only if the function h satisfies the growth condition

$$|h(x, s_1, s_2, \dots, s_m)| \le w_0^{\frac{1}{p_0}}(x) \left(b(x) + c \sum_{i=1}^m w_i^{\frac{1}{p_0'}}(x) |s_i|^{\frac{p_i}{p_0'}} \right) \tag{2.131}$$

with $b \in L^{p_0'}(\Omega)$ *and* $c \geq 0$.

Also the following property of $\mathcal{N}$ can be easily verified.

Let $\mathcal{N}$ *be the Nemytskij operator defined on (a little modified) product space*

$$P_2 = C(\Omega, w_1) \times L^{p_2}(\Omega, w_2) \times \cdots \times L^{p_m}(\Omega, w_m)\,, \tag{2.132}$$

where now the function h satisfies the growth conditions

$$\begin{aligned} &|h(x, s_1, s_2, \ldots, s_m)| \\ &\leq g_0(|w_1(x)s_1|)w_0^{\frac{1}{p_0}}(x)\left(b(x) + c\sum_{i=2}^{k} w_i^{\frac{1}{p_0'}}(x)|s_i|^{\frac{p_i}{p_0'}}\right) \end{aligned} \tag{2.133}$$

with $g_0 = g_0(t)$ *positive, continuous and nondecreasing on* $[0, \infty)$, $b \in L^{p_0'}(\Omega)$ *and* $c \geq 0$. *Then* $\mathcal{N}$ *maps* P_2 *continuously into the dual space* $(L^{p_0}(\Omega, w_0))^*$.

Since the growth conditions of type (A), (B) and (C) are of the form (2.131) and the growth conditions of type (D) are of the form (2.133) (see Definitions 2.2–2.5), we can use the above mentioned results for the Nemytskij operators generated by the "coefficients" $a_\alpha(x; \xi)$ of our differential operator A. For simplicity, we will deal only with the growth conditions of type (D) which are the most general ones.

The product space P from (2.129) is of the type (2.132), and consequently it follows that

(i) for $|\alpha| = k$, the Nemytskij operator associated with $a_\alpha(x; \xi)$ (satisfying (2.110)) maps P *continuously* into $(L^p(\Omega, w_\alpha))^* = X_\alpha^*$ (see (2.130));

(ii) for $\kappa_1 \leq |\alpha| \leq k - 1$, the Nemytskij operator associated with $a_\alpha(x; \xi)$ (satisfying (2.111)) maps P *continuously* into $(L^{q(\alpha)}(\Omega, \omega_\alpha))^* = X_\alpha^*$ (see (2.130) again);

(iii) for $|\alpha| < \kappa_1$, we can see from (2.112) that the Nemytskij operator associated with $a_\alpha(x; \xi)$ maps P *continuously* into $L^1(\Omega, \omega_\alpha^{-1})$, and we have

$$L^1(\Omega, \omega_\alpha^{-1}) \hookrightarrow (L^\infty(\Omega, \omega_\alpha))^* = X_\alpha^*\,,$$

where $L^\infty(\Omega, \omega_\alpha) = \{u = u(x)\,;\ \|\omega_\alpha u\|_\infty = \operatorname*{supess}_{x\in\Omega} |\omega_\alpha(x)u(x)| < \infty\}$.

Since $P^* = \prod\limits_{|\beta|\leq k} X_\beta^*$, we can identify P^* with a closed subspace of V^* (note that $V \subset P$). We can also write

$$\begin{aligned} \langle Tu, \varphi\rangle &= \sum_{|\alpha|\leq k} \int_\Omega a_\alpha(x; u(x), \ldots, \nabla^k u(x)) D^\alpha\varphi(x)\,dx \\ &= \sum_{|\alpha|\leq k} \langle \mathcal{N}_\alpha(u, \ldots, \nabla^k u), D^\alpha\varphi\rangle_\alpha\,, \end{aligned}$$

where $\mathcal{N}_\alpha$ is the Nemytskij operator associated with a_α and $\langle\cdot,\cdot\rangle_\alpha$ denotes the duality between X_α^* and X_α. Hence we have obtained that T (expressed in terms of the set $\{\mathcal{N}_\alpha\,;\ |\alpha|\le k\}$) maps V *continuously* into V^*. □

Remark 2.10 According to the definition of the operator T (see (2.128)) a function $u\in V$ is a weak solution of our homogeneous boundary value problem for the differential equation (2.113) if the identity

$$\langle Tu,\varphi\rangle=\int_\Omega f(x)\varphi(x)\,dx \tag{2.134}$$

holds for every $\varphi\in V$. Moreover, we can suppose that $f\in V^*$, replace the integral on the right-hand side of (2.134) by $\langle f,\varphi\rangle$ and look for $u\in V$ satisfying

$$\langle Tu,\varphi\rangle=\langle f,\varphi\rangle \tag{2.135}$$

for any $\varphi\in V$. Since (2.135) can be rewritten as *the operator equation*

$$Tu=f \tag{2.136}$$

for our operator $T\colon V\to V^*$, we thus obtain a *more general setting* of our problem:

"Find $u\in V$ such that (2.135) holds for any $\varphi\in V$."

Remark 2.11 To be more specific, we will suppose that the right-hand side f in (2.113) is of the form

$$f=\sum_{|\alpha|\le k}(-1)^{|\alpha|}D^\alpha f_\alpha\,, \tag{2.137}$$

where the family $\{f_\alpha\,;\ |\alpha|\le k\}$ satisfies the conditions

$$\begin{aligned}
&f_\alpha\in\big(L^p(\Omega,w_\alpha)\big)^*=L^{p'}(\Omega,w_\alpha^{1-p'}) && \text{for}\quad |\alpha|=k\,;\\
&f_\alpha\in\big(L^{q(\alpha)}(\Omega,\omega_\alpha)\big)^*=L^{q(\alpha)'}(\Omega,\omega_\alpha^{1-q(\alpha)'}) && \text{for}\quad \kappa_1\le|\alpha|\le k-1\,;\\
&f_\alpha\in L^1(\Omega,\omega_\alpha^{-1})\subset\big(L^\infty(\Omega,\omega_\alpha)\big)^* && \text{for}\quad |\alpha|<\kappa_1\,.
\end{aligned} \tag{2.138}$$

In the case of the Dirichlet problem where we take for V the space $W_0^{k,p}(\Omega,w)$, the operator equation (2.136) means that the integral identity

$$\sum_{|\alpha|\le k}\int_\Omega a_\alpha(x;u(x),\dots,\nabla^k u(x))D^\alpha\varphi(x)\,dx=\sum_{|\alpha|\le k}\int_\Omega f_\alpha(x)D^\alpha\varphi(x)\,dx$$

holds for every $\varphi\in V$.

2.7 Higher order equations (degree of the mapping T)

Our existence results are based on the solvability of the operator equation (2.136). The main tool for proving the existence of a solution of (2.136) will be the *degree theory for generalized monotone mappings* developed by F. E. Browder [10] and I. V. Skrypnik [61]. However, in order to apply this theory, we need to prove that the operator $T: V \to V^*$ satisfies the *condition* $\alpha(V)$ (sometimes also called *condition* $(S)_+$).

Recall (see Definition 1.1) that the operator $T: V \to V^*$ is said to satisfy *condition* $\alpha(V)$ if the assumptions

$$u_n \rightharpoonup u_0 \quad \text{(weakly) in} \quad V \quad \text{for} \quad n \to \infty, \tag{2.139}$$

$$\limsup_{n\to\infty} \langle Tu_n, u_n - u_0 \rangle \leq 0 \tag{2.140}$$

imply the strong convergence $u_n \to u_0$ in V.

We will need several assumptions.

Assumption K The imbedding of V into $W^{k-1,p}(\Omega, w)$ is compact, i.e.

$$V \hookrightarrow\hookrightarrow W^{k-1,p}(\Omega, w) \tag{2.141}$$

holds.

Note that in (2.141), the symbol w means $\{w_\alpha\,;\ |\alpha| \leq k\}$ for the space on the left-hand side in (2.141) and $\{w_\alpha\,;\ |\alpha| \leq k-1\}$ in the space on the right-hand side in (2.141). The imbedding (2.141) holds in the case of classical Sobolev spaces (i.e. if $w_\alpha(x) \equiv 1$, $|\alpha| \leq k$) if Ω is bounded and its boundary $\partial\Omega$ is lipschitzian.

Let us denote

$$H^{k-1,p}(\Omega, \omega) = \{u\,;\ D^\beta u \in X_\beta\,,\ |\beta| \leq k-1\}\,,$$

where X_β are defined in (2.130).

Assumption L The imbedding of V into $H^{k-1,p}(\Omega, \omega)$ is compact, i.e.

$$V \hookrightarrow\hookrightarrow H^{k-1,p}(\Omega, \omega) \tag{2.142}$$

holds.

Remark 2.12 The imbedding $V \hookrightarrow W^{k-1,p}(\Omega, w)$ is a consequence of the imbedding $V \hookrightarrow W^{k,p}(\Omega, w)$ required in (2.6); it should be emphasized that in assumption K, we require its *compactness*. The same holds for (2.142) since the imbedding $V \hookrightarrow H^{k-1,p}(\Omega, \omega)$ is required in assumptions H and J.

In further hypotheses, we will write the basic vector $\xi = \{\xi_\beta\,;\ |\beta| \le k\} \in \mathbb{R}^m$ in the form

$$\xi = (\kappa, \eta, \zeta)\,,$$

where

$$\begin{aligned} \zeta &= \{\zeta_\beta\,;\ |\beta| = k\} \in \mathbb{R}^{m_1} && \text{with} \quad \zeta_\beta = \xi_\beta\,, \\ \eta &= \{\eta_\beta\,;\ \kappa_1 \le |\beta| \le k-1\} \in \mathbb{R}^{m_2} && \text{with} \quad \eta_\beta = \xi_\beta\,, \\ \kappa &= \{\kappa_\beta\,;\ |\beta| < \kappa_1\} \in \mathbb{R}^{m_3} && \text{with} \quad \kappa_\beta = \xi_\beta\,. \end{aligned}$$

If $\kappa_1 = k - \dfrac{N}{p} \le 0$, we will write simply

$$\xi = (\eta, \zeta)$$

and the same notation will be used if it is not necessary to distinguish ξ_β with $|\beta| < \kappa_1$ and ξ_β with $|\beta| \in [\kappa_1, k-1]$.

Let us recall that in (2.108) we have introduced the function $h(x, \kappa)$ and denoted

$$h(x, u(x)) = \sum_{|\beta| < \kappa_1} |\omega_\beta(x) D^\beta u(x)|\,.$$

Assumption M Let G_1 be a continuous, positive, nonincreasing function on $[0, \infty)$, G_2 a continuous, positive, nondecreasing function on $[0, \infty)$. We will suppose that for every $\xi = (\kappa, \eta, \zeta) \in \mathbb{R}^m$ and for a.e. $x \in \Omega$, the following *ellipticity condition* holds:

$$\begin{aligned} \sum_{|\alpha|=k} a_\alpha(x; \kappa, \eta, \zeta)\zeta_\alpha \ge G_1(h(x, \kappa)) &\sum_{|\beta|=k} w_\beta(x) |\zeta_\beta|^p \\ &- G_2(h(x, \kappa)) \sum_{\kappa_1 \le |\beta| \le k-1} \omega_\beta(x) |\eta_\beta|^{q(\beta)}\,. \end{aligned} \tag{2.143}$$

If $\kappa_1 < 0$, we put $G_1(t) = G_2(t) \equiv \text{const} > 0$.

Assumption N We will suppose that the differential operator A from (2.1) is monotone in its principal part, i.e. for every $(\kappa, \eta, \zeta) \in \mathbb{R}^m$, $(\kappa, \eta, \hat{\zeta}) \in \mathbb{R}^m$ with $\zeta \ne \hat{\zeta}$ and for a.e. $x \in \Omega$, the following *monotonicity condition* holds:

$$\sum_{|\alpha|=k} (a_\alpha(x; \kappa, \eta, \zeta) - a_\alpha(x; \kappa, \eta, \hat{\zeta}))(\zeta_\alpha - \hat{\zeta}_\alpha) > 0\,. \tag{2.144}$$

Assumption P The set Ω has *a finite Lebesgue measure.*

Lemma 2.7 *Let assumptions* K, L, M, N, P *be satisfied and let the coefficients* $a_\alpha = a_\alpha(x; \xi)$ *satisfy the growth conditions of type* (*D*). *Then the operator* T *defined by* (2.128) *satisfies condition* $\alpha(V)$.

Proof. Assume (2.139) and (2.140). The compactness of the imbedding (2.141) implies that

$$D^\beta u_n \to D^\beta u_0 \quad \text{(strongly)} \quad \text{in} \quad L^p(\Omega, w_\beta) \quad \text{for} \quad |\beta| \le k-1 . \tag{2.145}$$

Consequently, it remains to prove that

$$D^\beta u_n \to D^\beta u_0 \quad \text{(strongly)} \quad \text{in} \quad L^p(\Omega, w_\beta) \quad \text{for} \quad |\beta| = k . \tag{2.146}$$

The assertion (2.146) will be proved in the following *Steps* 1–6.

Step 1 Let us consider the numbers

$$\mathcal{I}_n = \sum_{|\alpha|\le k-1} \int_\Omega a_\alpha(x; u_n, \dots, \nabla^k u_n) D^\alpha(u_n - u_0)\, dx . \tag{2.147}$$

Recall that in Lemma 2.6 we have shown that the Nemytskij operator $\mathcal{N}_\alpha u = \mathcal{N}_\alpha(u, \dots, \nabla^k u)$ associated with the Carathéodory function $a_\alpha(x; \xi)$ maps V continuously into X_α^*, where X_α is described in (2.130). Hence we have

$$\begin{aligned} |\mathcal{I}_n| \le & \sum_{\kappa_1 \le |\alpha| \le k-1} \|D^\alpha(u_n - u_0)\|_{q(\alpha),\omega_\alpha} H_\alpha(\|u_n\|_{k,p,w}) \\ & + \sum_{|\alpha|<\kappa_1} \|D^\alpha(u_n - u_0)\|_{C(\Omega,\omega_\alpha)} H_\alpha(\|u_n\|_{k,p,w}) \end{aligned}$$

(cf. formulas (2.123) and (2.125)). The compactness of the imbedding (2.142) implies that

$$\|D^\alpha(u_n - u_0)\|_{q(\alpha),\omega_\alpha} \to 0 \quad \text{and} \quad \|D^\alpha(u_n - u_0)\|_{C(\Omega,\omega_\alpha)} \to 0$$

for $n \to \infty$ and for corresponding α. Since $\|u_n\|_{k,p,w}$ as well as $H_\alpha(\|u_n\|_{k,p,w})$ are bounded due to the weak convergence of u_n, we conclude that

$$\lim_{n\to\infty} \mathcal{I}_n = 0 . \tag{2.148}$$

Step 2 Let us consider the numbers

$$\mathcal{J}_n = \sum_{|\alpha|=k} \int_\Omega a_\alpha(x; u_n, \dots, \nabla^{k-1} u_n, \nabla^k u_0) D^\alpha(u_n - u_0)\, dx .$$

We can write $\mathcal{J}_n$ in the form

$$\mathcal{J}_n = \mathcal{J}_{n,1} + \mathcal{J}_{n,2} ,$$

where

$$\begin{aligned} \mathcal{J}_{n,1} = \sum_{|\alpha|=k} \int_\Omega & (a_\alpha(x; u_n, \dots, \nabla^{k-1} u_n, \nabla^k u_0) \\ & - a_\alpha(x; u_0, \dots, \nabla^{k-1} u_0, \nabla^k u_0)) D^\alpha(u_n - u_0)\, dx , \end{aligned}$$

$$\mathcal{J}_{n,2} = \sum_{|\alpha|=k} \int_\Omega a_\alpha(x; u_0, \dots, \nabla^{k-1} u_0, \nabla^k u_0) D^\alpha(u_n - u_0)\, dx .$$

It follows from (2.110) that the Nemytskij operator $\mathcal{N}_\alpha$ maps V into $X_\alpha^* = (L^p(\Omega, w_\alpha))^*$ for $|\alpha| = k$, and consequently

$$\mathcal{J}_{n,2} = \sum_{|\alpha|=k} \langle \mathcal{N}_\alpha(u_0, \dots, \nabla^k u_0), D^\alpha(u_n - u_0)\rangle_\alpha .$$

Since u_0 is fixed, the weak convergence of u_n to u_0 implies that

$$\lim_{n\to\infty} \mathcal{J}_{n,2} = 0 . \tag{2.149}$$

Similarly, we can write

$$\mathcal{J}_{n,1} = \sum_{|\alpha|=k} \langle \mathcal{N}_\alpha(u_n, \dots, \nabla^{k-1}u_n, \nabla^k u_0) - \mathcal{N}_\alpha(u_0, \dots, \nabla^{k-1}u_0, \nabla^k u_0), D^\alpha(u_n - u_0)\rangle_\alpha .$$

Since here the "last variable" $\nabla^k u_0$ in the Nemytskij operator $\mathcal{N}_\alpha$ is fixed, we in fact consider $\mathcal{N}_\alpha$ not on the whole space $\prod_{|\beta|\le k} X_\beta$ but only on $\prod_{|\beta|<k} X_\beta$. Hence due to the continuity of $\mathcal{N}_\alpha$ and the strong convergence of $D^\beta u_n$ to $D^\beta u_0$ in X_β for $|\beta| < k$, we get

$$\mathcal{N}_\alpha(u_n, \dots, \nabla^{k-1}u_n, \nabla^k u_0) \to \mathcal{N}_\alpha(u_0, \dots, \nabla^{k-1}u_0, \nabla^k u_0)$$

(strongly) in $(L^p(\Omega, w_\alpha))^*$. Consequently, also

$$\lim_{n\to\infty} \mathcal{J}_{n,1} = 0 . \tag{2.150}$$

Finally, it follows from (2.149) and (2.150) that

$$\lim_{n\to\infty} \mathcal{J}_n = 0 . \tag{2.151}$$

Step 3 Let E be a measurable subset of Ω and denote

$$K_n(E) = \sum_{|\alpha|=k} \int_E (a_\alpha(x; u_n, \dots, \nabla^{k-1}u_n, \nabla^k u_n) - a_\alpha(x; u_n, \dots, \nabla^{k-1}u_n, \nabla^k u_0)) D^\alpha(u_n - u_0)\, dx . \tag{2.152}$$

Then

$$\langle Tu_n, u_n - u_0\rangle = \sum_{|\alpha|\le k} \int_\Omega a_\alpha(x; u_n, \dots, \nabla^k u_n) D^\alpha(u_n - u_0)\, dx = \mathcal{I}_n + \mathcal{J}_n + K_n(\Omega)$$

and due to (2.140), (2.148) and (2.151), we have

$$\limsup_{n\to\infty} K_n(\Omega) \le 0 . \tag{2.153}$$

The monotonicity condition (2.144) where we put $(\kappa, \eta) = \{D^\beta u_n\,;\ |\beta| \le k-1\}$, $\zeta = \nabla^k u_n$ and $\hat{\zeta} = \nabla^k u_0$, implies that

$$0 \le K_n(E) \le K_n(\Omega) \quad \text{for} \quad E \subset \Omega\,. \tag{2.154}$$

Hence

$$\lim_{n\to\infty} K_n(E) = 0 \tag{2.155}$$

for every measurable set $E \subset \Omega$ due to (2.153) and (2.154). Further,

$$\begin{aligned} K_n(E) > \sum_{|\alpha|=k} \int_E a_\alpha(x; u_n, \dots, \nabla^k u_n) D^\alpha u_n\, dx \\ - \sum_{|\alpha|=k} \int_E |a_\alpha(x; u_n, \dots, \nabla^{k-1} u_n, \nabla^k u_0)|(|D^\alpha u_n| + |D^\alpha u_0|)\, dx \\ - \sum_{|\alpha|=k} \int_E |a_\alpha(x; u_n, \dots, \nabla^k u_n)||D^\alpha u_0|\, dx\,. \end{aligned}$$

Estimating the first term on the right-hand side by the ellipticity condition (2.143) and the second and third terms by the growth condition (2.110), we obtain

$$\begin{aligned} K_n(E) \ge \int_E G_1(h(x,u_n)) \sum_{|\beta|=k} w_\beta |D^\beta u_n|^p\, dx \\ - \int_E G_2(h(x,u_n)) \sum_{\kappa_1 \le |\beta| \le k-1} \omega_\beta |D^\beta u_n|^{q(\beta)}\, dx \\ - \sum_{|\alpha|=k} \int_E G(h(x,u_n)) w_\alpha^{\frac{1}{p}} \Bigg(|g_\alpha| + c_\alpha \sum_{|\beta|=k} w_\beta^{\frac{1}{p'}} |D^\beta u_0|^{p-1} \\ + c_\alpha \sum_{\kappa_1 \le |\beta| \le k-1} \omega_\beta^{\frac{1}{p'}} |D^\beta u_n|^{\frac{q(\beta)}{p'}} \Bigg) (|D^\alpha u_n| + |D^\alpha u_0|)\, dx \\ - \sum_{|\alpha|=k} \int_E G(h(x,u_n)) w_\alpha^{\frac{1}{p}} \Bigg(|g_\alpha| + c_\alpha \sum_{|\beta|=k} w_\beta^{\frac{1}{p'}} |D^\beta u_n|^{p-1} \\ + c_\alpha \sum_{\kappa_1 \le |\beta| \le k-1} \omega_\beta^{\frac{1}{p'}} |D^\beta u_n|^{\frac{q(\beta)}{p'}} \Bigg) |D^\alpha u_0|\, dx\,. \end{aligned} \tag{2.156}$$

Since $h(x,u_n) \le c\|u_n\|_{k,p,w} \le c_0$ (due to (2.109)) and due to the weak convergence of the sequence u_n (in $W^{k,p}(\Omega, w)$), it follows from the monotonicity properties of the functions G_1, G_2 and G that

$$G_1(h(x,u_n)) \ge c_1\,, \quad G_2(h(x,u_n)) \le c_2\,, \quad G(h(x,u_n)) \le c_3\,.$$

Using these estimates and putting the negative terms to the left-hand side, we finally obtain from (2.156) that

$$\begin{aligned}
&\int_E \sum_{|\beta|=k} w_\beta |D^\beta u_n|^p \, dx \\
&\quad \le c_4 K_n(E) + c_5 \sum_{\kappa_1 \le |\beta| \le k-1} \int_E \omega_\beta |D^\beta u_n|^{q(\beta)} \, dx \\
&\quad + c_6 \sum_{|\alpha|=k} \sum_{|\beta|=k} \int_E w_\alpha^{\frac{1}{p}} w_\beta^{\frac{1}{p'}} \Big[|D^\beta u_0|^{p-1} (|D^\alpha u_n| + |D^\alpha u_0|) \\
&\quad + |D^\beta u_n|^{p-1} |D^\alpha u_0| \Big] \, dx \qquad (2.157) \\
&\quad + c_7 \sum_{|\alpha|=k} \sum_{\kappa_1 \le |\beta| \le k-1} \int_E w_\alpha^{\frac{1}{p}} \omega_\beta^{\frac{1}{p'}} |D^\beta u_n|^{\frac{q(\beta)}{p'}} (|D^\alpha u_n| + |D^\alpha u_0|) \, dx \\
&\quad + c_8 \sum_{|\alpha|=k} \int_E w_\alpha^{\frac{1}{p}} |g_\alpha| (|D^\alpha u_n| + |D^\alpha u_0|) \, dx
\end{aligned}$$

with some positive constants c_i, $i = 5, \ldots, 8$.

Step 4 We will prove that the left-hand side in (2.157) approaches zero uniformly with respect to n when meas $E \to 0$. To this end we use Young's inequality in the form

$$ab \le \frac{\varepsilon^p}{p} a^p + \frac{\varepsilon^{-p'}}{p'} b^{p'} \qquad (2.158)$$

($a, b \ge 0$, $\varepsilon > 0$, $p > 1$, $p' = \dfrac{p}{p-1}$) in order to estimate some terms on the right-hand side of (2.157). Taking $a = w_\alpha^{\frac{1}{p}} |D^\alpha u_n|$ or $a = w_\alpha^{\frac{1}{p}} |D^\alpha u_0|$ and $b = w_\beta^{\frac{1}{p'}} |D^\beta u_0|^{p-1}$ or $b = w_\beta^{\frac{1}{p'}} |D^\beta u_n|^{p-1}$ or $b = \omega_\beta^{\frac{1}{p'}} |D^\beta u_n|^{\frac{q(\beta)}{p'}}$ or $b = |g_\alpha|$, we obtain the following estimates:

$$\begin{aligned}
&\int_E w_\alpha^{\frac{1}{p}} w_\beta^{\frac{1}{p'}} |D^\beta u_0|^{p-1} |D^\alpha u_n| \, dx \\
&\qquad \le \frac{\varepsilon^p}{p} \int_E w_\alpha |D^\alpha u_n|^p \, dx + \frac{\varepsilon^{-p'}}{p'} \int_E w_\beta |D^\beta u_0|^p \, dx \, , \\
&\int_E w_\alpha^{\frac{1}{p}} w_\beta^{\frac{1}{p'}} |D^\beta u_0|^{p-1} |D^\alpha u_0| \, dx \\
&\qquad \le \frac{\varepsilon^p}{p} \int_E w_\alpha |D^\alpha u_0|^p \, dx + \frac{\varepsilon^{-p'}}{p'} \int_E w_\beta |D^\beta u_0|^p \, dx \, ,
\end{aligned}$$

$$\int_E w_\alpha^{\frac{1}{p}} w_\beta^{\frac{1}{p'}} |D^\beta u_n|^{p-1} |D^\alpha u_0|\,dx \le \frac{\varepsilon^{-p}}{p}\int_E w_\alpha |D^\alpha u_0|^p\,dx + \frac{\varepsilon^{p'}}{p'}\int_E w_\beta |D^\beta u_n|^p\,dx$$

for $|\alpha| = |\beta| = k$ (note that in the last case we have used ε^{-1} instead of ε, cf. (2.158));

$$\int_E w_\alpha^{\frac{1}{p}} \omega_\beta^{\frac{1}{p'}} |D^\beta u_n|^{\frac{q(\beta)}{p'}} |D^\alpha u_n|\,dx \le \frac{\varepsilon^{p}}{p}\int_E w_\alpha |D^\alpha u_n|^p\,dx + \frac{\varepsilon^{-p'}}{p'}\int_E \omega_\beta |D^\beta u_n|^{q(\beta)}\,dx\,,$$

$$\int_E w_\alpha^{\frac{1}{p}} \omega_\beta^{\frac{1}{p'}} |D^\beta u_n|^{\frac{q(\beta)}{p'}} |D^\alpha u_0|\,dx \le \frac{\varepsilon^{p}}{p}\int_E w_\alpha |D^\alpha u_0|^p\,dx + \frac{\varepsilon^{-p'}}{p'}\int_E \omega_\beta |D^\beta u_n|^{q(\beta)}\,dx$$

for $|\alpha| = k$, $\kappa_1 \le |\beta| \le k-1$;

$$\int_E w_\alpha^{\frac{1}{p}} |g_\alpha| (|D^\alpha u_n| + |D^\alpha u_0|)\,dx \le \frac{\varepsilon^{p}}{p}\int_E w_\alpha |D^\alpha u_n|^p\,dx + \frac{\varepsilon^{p}}{p}\int_E w_\alpha |D^\alpha u_0|^p\,dx + \frac{2\varepsilon^{-p'}}{p'}\int_E |g_\alpha|^{p'}\,dx$$

for $|\alpha| = k$. Using all these estimates in (2.157), we obtain

$$\begin{aligned}\int_E \sum_{|\beta|=k} w_\beta |D^\beta u_n|^p\,dx &\le c_4 K_n(E) + (c_5 + c_9\varepsilon^{-p'}) \sum_{\kappa_1\le|\beta|\le k-1} \int_E \omega_\beta |D^\beta u_n|^{q(\beta)}\,dx \\ &+ (c_{10}\varepsilon^p + c_{11}\varepsilon^{p'}) \sum_{|\beta|=k} \int_E w_\beta |D^\beta u_n|^p\,dx \\ &+ c_{12}(\varepsilon) \sum_{|\beta|=k} \int_E w_\beta |D^\beta u_0|^p\,dx + c_{13}(\varepsilon) \sum_{|\beta|=k} \int_E |g_\beta|^{p'}\,dx\end{aligned} \tag{2.159}$$

with some positive constants $c_4, c_5, c_9, c_{10}, c_{11}, c_{12} = c_{12}(\varepsilon)$ and $c_{13} = c_{13}(\varepsilon)$ (note that we are allowed to denote all multiindices by β in the last expression). Taking now $\varepsilon > 0$ small enough (e.g. such that $c_{10}\varepsilon^p + c_{11}\varepsilon^{p'} < \frac{1}{2}$), we obtain from (2.159) the estimate

$$\int_E \sum_{|\beta|=k} w_\beta(x)|D^\beta u_n(x)|^p \, dx \tag{2.160}$$

$$\leq 2c_4 K_n(E) + c_{14} \sum_{|\beta|=k} \left(\int_E w_\beta(x)|D^\beta u_0(x)|^p \, dx + \int_E |g_\beta(x)|^{p'} \, dx \right)$$

$$+ c_{15} \sum_{\kappa_1 \leq |\beta| \leq k-1} \int_E \omega_\beta(x)|D^\beta u_n(x)|^{q(\beta)} \, dx \, .$$

Notice that all three terms on the right-hand side of (2.160) approach zero uniformly with respect to n if meas $E \to 0$. Indeed, for the first term this follows from (2.154), (2.155), for the second term from the absolute continuity of the integral, for the third term from the equiintegrability of the sequence $\omega_\beta(x)|D^\beta u_n(x)|^{q(\beta)}$ (which follows from the strong convergence $D^\beta u_n \to D^\beta u_0$ in $L^{q(\beta)}(\Omega, \omega_\beta)$ in view of the imbedding (2.142)). Consequently, we obtain from (2.160) that

$$\lim_{\text{meas } E \to 0} \int_E \sum_{|\beta|=k} w_\beta(x)|D^\beta u_n(x)|^p \, dx = 0 \tag{2.161}$$

uniformly with respect to $n \in \mathbb{N}$, i.e. the sequence $(w_\beta|D^\beta u_n|^p)$ is *equiintegrable.*

Step 5 In this step we will show that for $|\beta| = k$,

$$w_\beta^{\frac{1}{p}}(x) D^\beta u_n(x) \to w_\beta^{\frac{1}{p}}(x) D^\beta u_0(x) \tag{2.162}$$

in *measure.* For this purpose let us denote

$$F_{\delta,n} = \{x \in \Omega \, ; \, \sum_{|\beta|=k} w_\beta^{\frac{1}{p}}(x)|D^\beta u_n(x) - D^\beta u_0(x)| \geq \delta\} \tag{2.163}$$

with $\delta > 0$. Because of the strong convergence (2.145), we have $D^\beta u_n(x) \to D^\beta u_0(x)$ in measure for all $|\beta| \leq k-1$. Since meas $\Omega < \infty$, it follows from Luzin's theorem (see e.g. E. Hewitt, K. Stromberg [33]) that for $\varepsilon > 0$ there exists an open set $E_{1,\varepsilon} \subset \Omega$ such that meas $E_{1,\varepsilon} < \frac{\varepsilon}{4}$ and

$$K_\varepsilon = \sup_{x \in \Omega \setminus E_{1,\varepsilon}} \left(\sum_{|\beta|=k} |D^\beta u_0(x)| + \sum_{|\beta| \leq k-1} |D^\beta u_n(x)| \right) < \infty \tag{2.164}$$

(independently of $n \in \mathbb{N}$). Further, from Luzin's theorem, assumptions M, N and the growth conditions of type (D), it follows that there exists an open set $E_{2,\varepsilon} \subset \Omega$ such that meas $E_{2,\varepsilon} < \frac{\varepsilon}{4}$ and

$$K_{\varepsilon,\delta} = \inf \sum_{|\alpha|=k} [a_\alpha(x; \eta, \zeta) - a_\alpha(x; \eta, \hat{\zeta})](\zeta_\alpha - \hat{\zeta}_\alpha) > 0 \, ,$$

where the infimum is taken over the set

$$\{x \in \Omega \setminus E_{2,\varepsilon};\ |\eta| \le K_\varepsilon\,,\ |\hat{\zeta}| \le K_\varepsilon\,,\ \sum_{|\beta|=k} w_\beta^{\frac{1}{p}}(x)|\zeta_\beta - \hat{\zeta}_\beta| \ge \delta\}$$

with K_ε from (2.164). Now, it follows from (2.154) that

$$\begin{aligned} K_n(\Omega) &\ge K_n(F_{\delta,n} \setminus (E_{1,\varepsilon} \cup E_{2,\varepsilon})) \\ &= \int\limits_{F_{\delta,n}\setminus(E_{1,\varepsilon}\cup E_{2,\varepsilon})} \sum_{|\alpha|=k} (a_\alpha(x; u_n, \dots, \nabla^{k-1}u_n, \nabla^k u_n) \\ &\qquad - a_\alpha(x; u_n, \dots, \nabla^{k-1}u_n, \nabla^k u_0)) D^\alpha(u_n - u_0)\, dx \\ &\ge K_{\varepsilon,\delta}\left(\text{meas } F_{\delta,n} - \frac{\varepsilon}{2}\right) \end{aligned}$$

and consequently,

$$\text{meas } F_{\delta,n} \le \frac{K_n(\Omega)}{K_{\varepsilon,\delta}} + \frac{\varepsilon}{2}\,.$$

But due to (2.155), we can find an $n_0 \in \mathbb{N}$ such that $K_n(\Omega) < \frac{\varepsilon}{2} K_{\varepsilon,\delta}$ for $n \ge n_0$. Hence we obtain

$$\text{meas } F_{\delta,n} < \varepsilon\,. \tag{2.165}$$

Since $\varepsilon > 0$ and $\delta > 0$ were arbitrary, the convergence in measure (2.162) follows from (2.165).

Step 6 Since meas $\Omega < \infty$, the equiintegrability (2.161) and the convergence in measure (2.162) of the sequence $D^\beta u_n$ imply the strong convergence of $D^\beta u_n$ to $D^\beta u_0$ in $L^p(\Omega, w_\beta)$ for $|\beta| = k$. So we obtain (2.146) and the assertion is proved. □

Corollary 2.1 *The properties of the mapping $T\colon V \to V^*$ just derived allow us to introduce its degree*

$$\text{Deg}[T;\, B_R, 0] \tag{2.166}$$

with respect to the ball $B_R = \{u \in V\,;\ \|u\|_{k,p,w} \le R\}$ and the point $0 \in V^*$.

For details concerning the degree (2.166) see Section 1.7. In the foregoing steps we have shown that $\text{Deg}[T;\, B_R, 0]$ is well-defined and has the properties stated in the section mentioned.

2.8 Higher order equations (existence results)

In this section we present some existence results concerning weak solvability of the homogeneous BVP for the equation (2.113). Recall that the right-hand side f is assumed to be of the form (2.137). Recall also that the results of this section can be easily extended to the case of the nonhomogeneous BVP (cf. Remark 2.9).

Let us point out that we have dealt only with the most general growth conditions of type (D) (see Definition 2.5). However, existence results hold also (with obvious modifications) for the weaker growth conditions — compare with Remark 2.8. For the simplest growth conditions of type (A), existence of a (weak) solution was proved (for the Dirichlet problem) in A. Kufner, B. Opic [41] (see also A. Kufner, A. M. Sändig [42], Section 17).

Note also that Ω may be possibly an unbounded domain in $\mathbb{R}^N$. However, we have to assume that meas $\Omega < \infty$.

Theorem 2.3 *Let $T: V \to V^*$ be defined by* (2.128)*, let the coefficients $a_\alpha(x;\xi)$ satisfy the growth conditions of type (D) and let the right-hand side f be of the form* (2.137)*. Let us suppose that assumptions* K, L, M, N, P *are satisfied and, moreover, that there exists a number $R > 0$ such that*

$$\sum_{|\alpha|\le k} \int_\Omega (a_\alpha(x; u(x), \dots, \nabla^k u(x)) - f_\alpha(x)) D^\alpha u(x)\, dx \ge 0 \tag{2.167}$$

holds for every $u \in V$, $\|u\|_{k,p,w} = R$.

Then the boundary value problem for the differential equation (2.113) *has at least one weak solution $u \in V$ such that $\|u\|_{k,p,w} \le R$.*

Proof. Let us introduce an operator $T_f: V \to V^*$ by

$$T_f u = Tu - f\,.$$

Then condition (2.167) can be written as

$$\langle T_f u, u\rangle \ge 0\,. \tag{2.168}$$

If there is $u \in V$ with $\|u\|_{k,p,w} = R$ such that $T_f u = 0$, then $Tu = f$ and our assertion is proved.

If there is no such $u \in V$ then condition (2.168) and Theorem 1.6 imply that the degree of T_f satisfies

$$\operatorname{Deg}[T_f; B_R(0), 0] = 1\,.$$

However, then it follows from the basic property of the degree that there exists at least one $u \in B_R(0)$ such that $T_f u = 0$, i.e.

$$Tu = f\,.$$

□

Remark 2.13 A useful tool in the Leray–Lions theory of monotone operators is the concept of *coercivity*. An operator $T\colon V \to V^*$ is called coercive on V if

$$\lim_{\|u\|_V\to\infty} \frac{\langle Tu, u\rangle}{\|u\|_V} = \infty\,. \tag{2.169}$$

Obviously, this condition can help us to overcome the difficulties arising if we like to verify condition (2.167). Therefore, let us introduce an additional assumption.

Assumption Q Suppose that there are positive constants c_{16}, c_{17}, c_{18} such that the following *coercivity condition* holds for all $\xi \in \mathbb{R}^m$ and a.e. $x \in \Omega$:

$$\sum_{|\alpha|\le k} a_\alpha(x;\xi)\xi_\alpha \ge c_{16} \sum_{|\alpha|=k} w_\alpha(x)|\xi_\alpha|^p + c_{17} w_\Theta(x)|\xi_\Theta|^p - c_{18}\,, \tag{2.170}$$

where Θ is the zero multiindex $(0,0,\dots,0)$.

Lemma 2.8 *Let assumptions* K *and* Q *be fulfilled. Then the operator* $T\colon V \to V^*$ *defined by* (2.128) *satisfies* (2.169).

Proof. It follows from assumption K that the following imbeddings hold:

$$V \hookrightarrow\hookrightarrow W^{k-1,p}(\Omega, w) \hookrightarrow L^p(\Omega, w_\Theta)\,.$$

Applying Theorem 1.12, we obtain that for any $\varepsilon > 0$ there exists $c(\varepsilon) > 0$ such that for every $u \in V$

$$\|u\|_{k-1,p,w} \le \varepsilon \|u\|_{k,p,w} + c(\varepsilon)\|u\|_{p,w_\Theta}\,,$$

i.e.

$$\begin{aligned} &\sum_{|\alpha|\le k-1} \int_\Omega |D^\alpha u(x)|^p w_\alpha(x)\,dx \\ &\quad\le 2^{p-1}\varepsilon^p \left(\sum_{|\alpha|=k} \int_\Omega |D^\alpha u(x)|^p w_\alpha(x)\,dx + \sum_{|\alpha|\le k-1} \int_\Omega |D^\alpha u(x)|^p w_\alpha(x)\,dx \right) \\ &\qquad + 2^{p-1} c(\varepsilon)^p \int_\Omega |u(x)|^p w_\Theta(x)\,dx\,. \end{aligned}$$

Consequently, we obtain that for $\varepsilon > 0$ small $(0 < \varepsilon^p < 2^{1-p})$ there is a constant $c_{19} > 0$ such that

$$\begin{aligned} &\sum_{|\alpha|\le k-1} \int_\Omega |D^\alpha u(x)|^p w_\alpha(x)\,dx \\ &\quad\le c_{19} \left(\sum_{|\alpha|=k} \int_\Omega |D^\alpha u(x)|^p w_\alpha(x)\,dx + \int_\Omega |u(x)|^p w_\Theta(x)\,dx \right) \end{aligned}$$

and hence

$$\|u\|_{k,p,w}^p = \sum_{|\alpha|\le k} \|D^\alpha u\|_{p,w_\alpha}^p \le c_{20} \left(\sum_{|\alpha|=k} \|D^\alpha u\|_{p,w_\alpha}^p + \|u\|_{p,w_\Theta}^p \right)$$

with $c_{20} > 0$ independent of $u \in V$. This inequality together with (2.170) yields

$$\begin{aligned}
\langle Tu, u\rangle &= \sum_{|\alpha|\le k} \int_\Omega a_\alpha(x; u(x), \dots, \nabla^k u(x)) D^\alpha u(x)\, dx \\
&\ge c_{16} \sum_{|\alpha|=k} \int_\Omega w_\alpha(x) |D^\alpha u(x)|^p\, dx + c_{17} \int_\Omega w_\Theta(x) |u(x)|^p\, dx - c_{18} \\
&\ge \min(c_{16}, c_{17}) \left(\sum_{|\alpha|=k} \|D^\alpha u\|_{p,w_\alpha}^p + \|u\|_{p,w_\Theta}^p \right) - c_{18} \\
&\ge \frac{\min(c_{16}, c_{17})}{c_{20}} \|u\|_{k,p,w}^p - c_{18}
\end{aligned}$$

and since $p > 1$, we immediately have (2.169). □

Remark 2.14 Instead of (2.170) we can also use the *modified coercivity condition*

$$\sum_{|\alpha|\le k} a_\alpha(x; \xi)\xi_\alpha \ge c_{16} \sum_{|\alpha|=k} w_\alpha(x)|\xi_\alpha|^p + c_{17}\omega(x)|\xi_\Theta|^q - c_{18} \tag{2.171}$$

with $q > 1$ and a suitable weight function $\omega = \omega(x)$ provided the imbeddings

$$V \hookrightarrow\hookrightarrow W^{k-1,p}(\Omega, w) \hookrightarrow L^q(\Omega, \omega) \tag{2.172}$$

take place. Indeed, similarly as in Lemma 2.8 we obtain that

$$\begin{aligned}
&\sum_{|\alpha|\le k-1} \int_\Omega |D^\alpha u(x)|^p w_\alpha(x)\, dx \\
&\qquad \le c_{19} \left[\sum_{|\alpha|=k} \int_\Omega |D^\alpha u(x)|^p w_\alpha(x)\, dx + \left(\int_\Omega |u(x)|^q \omega(x)\, dx \right)^{\frac{p}{q}} \right]
\end{aligned}$$

and consequently

$$\|u\|_{k,p,w}^p \le c_{20} \left(\sum_{|\alpha|=k} \|D^\alpha u\|_{p,w_\alpha}^p + \|u\|_{q,\omega}^p \right). \tag{2.173}$$

Due to the second imbedding in (2.172) we have

$$\|u\|_{q,\omega} \leq c_{21}\|u\|_{k-1,p,w} = c_{21}\left(\sum_{|\alpha|\leq k-1} \|D^\alpha u\|^p_{p,w_\alpha}\right)^{\frac{1}{p}},$$

which together with (2.173) implies that the expression

$$\left(\sum_{|\alpha|=k} \|D^\alpha u\|^p_{p,w_\alpha} + \|u\|^p_{q,\omega}\right)^{\frac{1}{p}} = |||u|||$$

defines a norm in V equivalent to $\|u\|_{k,p,w}$.

We have

$$\langle Tu, u\rangle \geq c_{16} \sum_{|\alpha|=k} \int_\Omega w_\alpha(x)|D^\alpha u(x)|^p\,dx + c_{17}\int_\Omega |u(x)|^q \omega(x)\,dx - c_{18} \tag{2.174}$$

$$\geq \min(c_{16}, c_{17})\left(\sum_{|\alpha|=k} \|D^\alpha u\|^p_{p,w_\alpha} + \|u\|^q_{q,\omega}\right) - c_{18}\,.$$

For $p < q$ it follows from Young's inequality (2.158) (now in the form $ab \leq \dfrac{a^r}{r} + \dfrac{b^{r'}}{r'}$ with $r = \dfrac{q}{p}$, $a = \left(\int_\Omega |u(x)|^q \omega(x)\,dx\right)^{\frac{p}{q}} = \|u\|^p_{q,\omega}$, $b = 1$) that

$$\|u\|^p_{q,\omega} \leq \frac{p}{q}\|u\|^q_{q,\omega} + c_{22}\,. \tag{2.175}$$

Thus we obtain from (2.174) and (2.175) that

$$\langle Tu, u\rangle \geq \min(c_{16}, c_{17})\left(\sum_{|\alpha|=k} \|D^\alpha u\|^p_{p,w_\alpha} + \frac{q}{p}\|u\|^p_{q,\omega} - c_{23}\right) \geq c_{24}|||u|||^p - c_{25}\,,$$

which again implies (2.169).

So we have the following result:

Theorem 2.4 *Let us assume the same as in Theorem* 2.3 *but replace condition* (2.167) *by the coercivity condition* (2.170) *or — provided the imbeddings* (2.172) *hold with* $q \geq p$ *— by the modified coercivity condition* (2.171). *Then the boundary value problem for the differential equation* (2.113) *has at least one weak solution* $u \in V$.

Proof. Since (2.170) or (2.171) imply (2.169), we have that

$$\frac{\langle T_f u, u\rangle}{\|u\|_{k,p,w}} = \frac{\langle Tu - f, u\rangle}{\|u\|_{k,p,w}} \geq \frac{\langle Tu, u\rangle}{\|u\|_{k,p,w}} - \frac{\|f\|_{V^*}\|u\|_{k,p,w}}{\|u\|_{k,p,w}}$$
$$= \frac{\langle Tu, u\rangle}{\|u\|_{k,p,w}} - \|f\|_{V^*} \to \infty \quad \text{if} \quad \|u\|_{k,p,w} \to \infty$$

($f \in V^*$ is fixed). Consequently, condition (2.167) is satisfied for $R > 0$ large enough and the assertion follows from Theorem 2.3. □

Remark 2.15 The reader has already realized that the conditions in the part devoted to the second order equations (Sections 2.2–2.4) are slightly different from the assumptions in the higher order case (see, e.g., Remark 2.2). In particular, the growth conditions in Section 2.5 are more general then those in assumption D. Therefore, let us point out that in the second order case, also *unbounded* domains are allowed while in the higher order case, we have to assume that Ω is bounded or at least of finite measure (see assumption P, p. 82) and, moreover, in the second order case we consider even weak solutions which belong to $L^\infty(\Omega)$. On the other hand, the price we must pay for the study of higher order equations is, among other, the fact that the truncation method (see assumption A) cannot be applied here.

2.9 Examples, remarks, comments

In Section 2.2, we have considered the special weighted space $W^{1,p}(\nu, \Omega)$ where the same weight function ν appeared *only at the derivatives*. Now, let us consider a "higher order analogue" of this space, namely, the space denoted by

$$X = W^{k,p}(\nu, \Omega) \tag{2.176}$$

and defined as the weighted Sobolev space $W^{k,p}(\Omega, w)$ with the special choice of $w = \{w_\alpha ;\ |\alpha| \leq k\}$:

$$\begin{aligned} w_\alpha(x) &= 1 && \text{for} \quad |\alpha| \leq k-1\,, \\ w_\alpha(x) &= \nu(x) && \text{for} \quad |\alpha| = k\,, \end{aligned}$$

with ν a *fixed* weight function satisfying, as usual, the conditions

$$\nu \in L^1_{\text{loc}}(\Omega)\,, \quad \nu^{1-p'} \in L^1_{\text{loc}}(\Omega)\,. \tag{2.177}$$

So, we have *no weights* (or, more precisely, the weight 1) at lower order terms. The space X is normed by

$$\|u\|_X = \left(\sum_{|\alpha|\leq k-1} \int_\Omega |D^\alpha u(x)|^p\,dx + \sum_{|\alpha|=k} \int_\Omega |D^\alpha u(x)|^p \nu(x)\,dx \right)^{\frac{1}{p}}. \tag{2.178}$$

This space was used in P. Drábek, A. Kufner, F. Nicolosi [21] to derive existence theorems. The imbedding

$$X \hookrightarrow W^{k,p_s}(\Omega) \tag{2.179}$$

which holds if we assume in addition that

$$\nu^{-s} \in L^1(\Omega) \tag{2.180}$$

with a suitable $s > 0$ and

$$p_s = \frac{ps}{s+1} < p \tag{2.181}$$

was important there, since assumption (2.180) allows to reduce, by virtue of (2.179), all considerations from the *weighted* Sobolev space to a *nonweighted* one and thus to avoid the more complicated apparatus of weighted Sobolev spaces. Moreover, the assumption

$$s > \frac{N}{p} \tag{2.182}$$

ensures the compactness of the imbedding of $W^{k,p_s}(\Omega)$ into $W^{k-1,p}(\Omega)$ and consequently also the *compact* imbedding

$$X \hookrightarrow\hookrightarrow W^{k-1,p}(\Omega)\,. \tag{2.183}$$

(For a detailed description see Example 1.3.)

Example 2.5 Let us consider a special *plane* domain $\Omega \subset \mathbb{R}^2$, namely the square $\Omega = \{x = (x_1, x_2)\,;\ -1 < x_i < 1\,,\ i = 1, 2\}$, and the operator

$$(A_0 u)(x) = (-1)^k \sum_{|\alpha|=k} D^\alpha(\nu(x)|D^\alpha u(x)|^{p-2} D^\alpha u(x))\,, \tag{2.184}$$

where

$$\nu(x) = \nu(x_1, x_2) = \begin{cases} 1 & \text{for } \ x_1 \le 0\,, \\ x_2^\lambda (1-x_1)^\gamma & \text{for } \ x_1 > 0\,,\ x_2 > 0\,, \\ |x_2|^\mu (1-x_1)^\gamma & \text{for } \ x_1 > 0\,,\ x_2 < 0 \end{cases} \tag{2.185}$$

with real numbers λ, μ, γ. This choice indicates that the operator A_0 can become *degenerated* (if the corresponding sign of λ, μ or γ is positive) or *singular* (in the case of negative sign) on a part of the *boundary* $\partial\Omega$ of Ω (namely, on the segment $\Gamma = \{x = (1, x_2)\,;\ -1 < x_2 < 1\})$ as well as in the *interior* of Ω (namely, on the segment $S = \{x = (x_1, 0)\,;\ 0 < x_1 < 1\}$).

It follows from conditions (2.177) that we have to suppose

$$\lambda, \mu \in (-1, p-1)$$

(with *no condition* on γ) while condition (2.180) together with (2.182) implies

$$\lambda, \mu, \gamma < \frac{p}{2}\,.$$

Consequently, we have to assume that

$$-1 < \lambda, \mu < \min\left\{\frac{p}{2}, p-1\right\} = \begin{cases} p-1 & \text{for} \quad 1 < p \leq 2\,, \\ \dfrac{p}{2} & \text{for} \quad p \geq 2\,, \end{cases} \qquad \gamma < \min\left\{\frac{p}{2}, p-1\right\}. \tag{2.186}$$

This indicates that for λ, μ and γ *positive* we have a degeneration which will be very small for p close to 1. On the other hand, also a singularity may occur, in a limited extent at S (for λ or μ negative, but greater than -1) but big enough at Γ (for *any* $\gamma < 0$).

For the operator A_0 from (2.184) we have

$$\begin{aligned} a_\alpha(x;\xi) &= \nu(x)|\xi_\alpha|^{p-2}\xi_\alpha \quad && \text{for} \quad |\alpha| = k\,, \\ a_\alpha(x;\xi) &= 0 && \text{for} \quad |\alpha| < k\,, \end{aligned} \tag{2.187}$$

and obviously the growth condition (2.100), the ellipticity condition (2.143) (with functions $G_1(t) = G_2(t) \equiv 1$) and the monotonicity condition (2.144) are satisfied and the space $X = W^{k,p}(\nu, \Omega)$ from (2.176) is the appropriate space. It remains only to choose the subspace $V \subset X$ in accordance with the corresponding boundary conditions, and Theorem 2.3 guarantees the existence of a weak solution $u \in V$. $\diamond$

Remark 2.16 Let us again consider the operator A_0 from (2.184), but now on a general domain $\Omega \subset \mathbb{R}^N$ ($N \geq 2$) and with a general weight function ν (satisfying, of course, conditions (2.177) and (2.180) with $s > \dfrac{N}{p}$ according to (2.182)). The operator A_0 is very simple and transparent and does not contain lower order terms, but our choice of the space X allows to add terms of the form

$$(-1)^{|\beta|} D^\beta(|D^\beta u(x)|^{p-1} \operatorname{sgn} D^\beta u(x)) \quad \text{with} \quad |\beta| < k\,, \tag{2.188}$$

i.e., with the *same growth* as terms in A_0 but without weights (in this case, we immediately see that the growth conditions of type (A) from Definition 2.2 are satisfied with $w_\beta(x) \equiv 1$).

Moreover, we can add terms of the form

$$(-1)^{|\beta|} D^\beta(|D^\beta u(x)|^{\frac{q(\beta)}{p'}} \operatorname{sgn} D^\beta u(x)) \quad \text{with} \quad |\beta| < k\,, \tag{2.189}$$

again without weights, but with a *growth bigger* than that of the terms in A_0, since we have

$$\begin{aligned} &1 \leq \frac{q(\beta)}{p'} \leq \frac{p_s N}{N - p_s(k - |\beta|)} \quad \text{if} \quad p_s(k - |\beta|) < N\,, \\ &\frac{q(\beta)}{p'} \geq 1 \quad \text{arbitrary otherwise,} \end{aligned} \tag{2.190}$$

where p_s is given by (2.181). Indeed, the growth is bigger than the growth of the coefficients a_α from (2.187) since for $s > \frac{N}{p}$ we have $\frac{q(\beta)}{p'} > p-1$. In this case, the growth conditions of type (B) from Definition 2.3 are satisfied, again with $w_\beta(x) \equiv 1$ if $|\beta| < k$.

Let us recall that the number $q(\beta)$ in (2.190) is determined by the imbedding theorem for classical (nonweighted) Sobolev spaces since we have, due to the imbedding (2.179),

$$X \hookrightarrow W^{|\beta|,q(\beta)}(\Omega) \quad \text{for} \quad |\beta| < k$$

(see Theorem 1.2 (i)). Moreover, according to this theorem, we have the imbedding

$$X \hookrightarrow C^{|\beta|}(\overline{\Omega}) \quad \text{for} \quad |\beta| < \kappa_2$$

provided

$$\kappa_2 = k - \frac{N}{p_s} > 0\,. \tag{2.191}$$

In this case, we can add to the operator A_0 also terms of the form

$$(-1)^{|\beta|} D^\beta (g_\beta(|D^\beta u(x)|) \operatorname{sgn} D^\beta u(x)) \quad \text{with} \quad |\beta| < \kappa_2\,, \tag{2.192}$$

where $g_\beta \colon \mathbb{R} \to \mathbb{R}^+$ are continuous and nondecreasing functions. Indeed, since here

$$a_\beta(x;\xi) = g_\beta(|\xi_\beta|) \operatorname{sgn} \xi_\beta\,, \quad |\beta| < \kappa_2\,,$$

we immediately see that the growth condition of type (D) from Definition 2.5 is satisfied with κ_1 replaced by κ_2, as well as the coercivity condition (2.170).

Example 2.6 Let us consider the operator A_0 from (2.184) on the plane domain Ω as in Example 2.5. Checking the number κ_2 from (2.191) we can see that

$$k-2 < \kappa_2 < k - \frac{2}{p} \quad \text{for} \quad 1 < p < 2\,,$$

$$k-1-\frac{2}{p} < \kappa_2 < k - \frac{2}{p} \quad \text{for} \quad p > 2\,.$$

If we consider a second order operator, i.e. $k = 1$, we see that for $p \in (1,2)$ the value κ_2 is negative and we cannot add terms of the form (2.192) while, for $p > 2$, we choose β with $|\beta| = 0$ and, e.g. for $g_\beta(t) = te^{t^2}$, obtain existence results for the equation

$$-\sum_{i=1}^{2} \frac{\partial}{\partial x_i}\left(\nu(x_1,x_2)\left|\frac{\partial u}{\partial x_i}\right|^{p-2}\frac{\partial u}{\partial x_i}\right) + ue^{u^2} = f \quad \text{in} \quad \Omega\,.$$

The main "trick" in the foregoing approach was the introduction of the "intermediate" space $W^{k,p_s}(\Omega)$, which together with the imbeddings

$$W^{k,p}(\nu,\Omega) \hookrightarrow W^{k,p_s}(\Omega) \hookrightarrow W^{k-1,p}(\Omega) \tag{2.193}$$

allowed to reduce all considerations to the nonweighted Sobolev spaces which are easier to handle. This approach was ensured by the rather restrictive *additional* assumption (2.180) which, among other, excluded strong degenerations.

Comparing the norms of $X = W^{k,p}(\nu,\Omega)$ (see (2.178)) and of $W^{k-1,p}(\Omega)$ (see (1.5)) it is easy to realize that the resulting imbedding

$$W^{k,p}(\nu,\Omega) \hookrightarrow W^{k-1,p}(\Omega) \tag{2.194}$$

can be obtained, if we have an inequality of the form

$$\int_\Omega |v(x)|^p\,dx \le c\int_\Omega |\nabla v(x)|^p \nu(x)\,dx \tag{2.195}$$

and use it for $v = D^\beta u$, $|\beta| = k-1$.

The inequality (2.195) is a special form of the *Hardy-type inequality* (1.32) and it is valid under certain conditions on the weight function ν appearing on its right-hand side. These conditions on ν appear now instead of conditions (2.180), (2.182) and sometimes they allow to consider broader classes of differential operators A than the approach described in the beginning. We will illustrate it by some examples; a detailed description can be found in A. Kufner, S. Leonardi [39].

Remark 2.17 (i) Inequality (2.195) is a special case of inequality (1.32) (we take there $p = q$, $w_i(x) = \nu(x)$ for $i = 1, 2, \dots, N$ and $\omega(x) \equiv 1$) and describes in fact the imbedding

$$W \hookrightarrow L^p(\Omega)\,,$$

where W is a subspace of the weighted Sobolev space $W^{1,p}(\nu,\Omega)$. Since the validity of (2.195) needs certain zero boundary conditions for v on the boundary $\partial\Omega$ or on its part, we will use the approach just described mainly for boundary value problems, in which at least at some part M of the boundary $\partial\Omega$, a *Dirichlet boundary condition* (expressed by the claim that $D^\beta u = 0$ on M for $|\beta| \le k-1$) appears.

(ii) While in the first approach, we started with the Sobolev space $W^{k,p_s}(\Omega)$ with $p_s < p$, here we start with the Sobolev space

$$W^{k-1,p}(\Omega)\,.$$

Therefore, our growth conditions change slightly. Not going into details, let us only mention that if we consider higher order terms of the form A_0, i.e. with $a_\alpha(x;\xi)$ given by (2.187) for $|\alpha| = k$, lower order terms of the form (2.189) can appear, but this time with parameters $q(\beta)$ satisfying

$$1 \le \frac{q(\beta)}{p'} \le \frac{Np}{N-(k-|\beta|-1)p} \quad \text{if} \quad p(k-|\beta|-1) < N\,. \tag{2.196}$$

A comparison with (2.190) shows that sometimes we can have a little lower growth than in the first approach.

Example 2.7 Let us consider the operator A_0 from (2.184), this time on the cube

$$\Omega = (0, 1)^N , \tag{2.197}$$

and write $x = (x', x_N)$ with $x' \in \mathbb{R}^{N-1}$. Let M be the bottom of this cube,

$$M = \{x = (x', x_N) \in \partial\Omega ;\ x_N = 0\} ,$$

and let us consider a BVP in Ω with *Dirichlet data* on M (the boundary conditions on the remaining part $\partial\Omega \setminus M$ are not important). Further, suppose that the weight function $\nu(x)$ in A_0 has the special form

$$\nu(x) = \nu(x', x_N) = \tilde{\nu}(x_N) , \tag{2.198}$$

i.e., depends only on x_N, $0 < x_N < 1$. For this special choice, conditions (2.177) have the form

$$\tilde{\nu} \in L^1_{\mathrm{loc}}(0, 1) , \quad \tilde{\nu}^{1-p'} \in L^1_{\mathrm{loc}}(0, 1) , \tag{2.199}$$

and we will suppose that they are satisfied.

Since we have Dirichlet boundary conditions on M (which means that the values of $D^\beta u$ on M are prescribed for all β with $|\beta| \le k - 1$), we construct the subspace $V \subset W^{k,p}(\nu, \Omega)$ as the *closure* of the set

$$\mathcal{V} = \{u \in C^\infty(\overline{\Omega}) ;\ \operatorname{supp} u \cap M = \emptyset\}$$

in the norm of $W^{k,p}(\nu, \Omega)$. Due to the density of the set $\mathcal{V}$ in the space V, we can consider inequality (2.195) for $u \in \mathcal{V}$ instead of $v \in V$. But for $v \in \mathcal{V}$, we have $\operatorname{supp} v \cap M = \emptyset$, i.e.

$$v|_M = v(x', 0) = 0 . \tag{2.200}$$

For such functions v we can use the onedimensional Hardy inequality (1.34), i.e. in our case, the inequality

$$\int_0^1 |v(x', x_N)|^p \, dx_N \le c_0 \int_0^1 \left| \frac{\partial v}{\partial x_N}(x', x_N) \right|^p \tilde{\nu}(x_N) \, dx_N . \tag{2.201}$$

This inequality holds, with a constant c_0 independent of x', if and only if the function

$$B(t) = \left(\int_t^1 d\tau \right)^{\frac{1}{p}} \left(\int_0^t \tilde{\nu}^{1-p'}(\tau) \, d\tau \right)^{\frac{1}{p'}} = (1 - t)^{\frac{1}{p}} \left(\int_0^t \tilde{\nu}^{1-p'}(\tau) \, d\tau \right)^{\frac{1}{p'}}$$

is bounded on $(0, 1)$ (see Example 1.4). If we denote

$$\mathcal{N}(t) = \int_0^t \tilde{\nu}^{1-p'}(\tau) \, d\tau \tag{2.202}$$

then (2.201) holds for v satisfying (2.200) if and only if there is a constant $c_1 > 0$ such that

$$\mathcal{N}(t)(1-t)^{p'-1} \le c_1 \quad \text{for} \quad t \in (0,1)\,. \tag{2.203}$$

Moreover, the imbedding expressed by (2.201) is *compact* provided

$$\mathcal{N}(t) = o((1-t)^{1-p'}) \quad \text{for} \quad t \to 1\,. \tag{2.204}$$

The integration of (2.201) with respect to $x' \in M$ leads to the inequality (2.195) which holds not only for $v \in \mathcal{V}$ but also for $v \in V$, and taking $v = D^\beta u$, $|\beta| = k-1$, we finally arrive at

$$\sum_{|\beta|=k-1} \int_\Omega |D^\beta u(x)|^p\,dx \le c \sum_{|\alpha|=k} \int_\Omega |D^\alpha u(x)|^p \nu(x)\,dx\,.$$

Since the left hand side is obviously a norm in $W^{k-1,p}(\Omega)$, we have derived the imbedding

$$V \hookrightarrow W^{k-1,p}(\Omega)$$

for our subspace $V \subset W^{k,p}(\nu,\Omega) = X$. This imbedding is continuous (and compact) if and only if (2.203) (and (2.204)) holds.

The estimates (2.203) and (2.204) are the conditions which replace conditions (2.180) and (2.182) if we use the second approach avoiding the "intermediate" space $W^{k,p_s}(\Omega)$. Let us compare these two types of conditions: In our setting, conditions (2.180) and (2.182) read

$$\int_0^1 \nu^{-s}(\tau)\,d\tau < \infty \qquad \begin{cases} \text{for} \quad s \ge \dfrac{1}{p-1} & \text{if} \quad p < \dfrac{N}{N-1}\,, \\ \text{for} \quad s > \dfrac{N}{p} & \text{if} \quad p \ge \dfrac{N}{N-1}\,, \end{cases} \tag{2.205}$$

while (2.203) and (2.204) require only the finiteness of the integral $\mathcal{N}(t)$ from (2.202) with a certain "bad" behaviour for $t \to 1$. Thus the second approach offers more possibilities in the choice of admissible weight functions ν, mainly in the case $p \ge \dfrac{N}{N-1}$ since then we have

$$s > \frac{N}{p} \ge 1 - p' \left(= \frac{1}{1-p}\right).$$

◇

Remark 2.18 (i) In the foregoing example, the essential tool was the one-dimensional Hardy inequality (2.201), more precisely, the conditions of its validity and of the compactness of the corresponding imbedding. These conditions have a form which is easy to check and are expressed in terms of the function $\mathcal{N}(t)$ from (2.202). Moreover, the existence of $\mathcal{N}(t)$ at the same time guarantees also the second condition in (2.199) and consequently, it remains only to check whether $\tilde{\nu} \in L^1_{\mathrm{loc}}(0,1)$.

(ii) If we consider a boundary value problem with Dirichlet data *on the top of our cube* Ω, i.e., for $x \in \partial\Omega$, $x = (x', x_N)$ with $x_N = 1$, and suppose again that ν has the form (2.198), then we arrive again at the Hardy inequality (2.201) but now for functions $v = v(x', x_N)$ satisfying $v(x', 1) = 0$. In this case, the conditions of its validity and of the compactness are expressed in terms of the function

$$\hat{\mathcal{N}}(t) = \int_t^1 \tilde{\nu}^{1-p'}(\tau)\,d\tau \; : \tag{2.206}$$

we need

$$t^{p'-1}\hat{\mathcal{N}}(t) \le c_1 \quad \text{for} \quad t \in (0,1)$$

instead of (2.203) and

$$\hat{\mathcal{N}}(t) = o(t^{1-p'}) \quad \text{for} \quad t \to 0 \tag{2.207}$$

instead of (2.204).

In order to make the conditions just mentioned more transparent, let us consider a particular function ν.

Example 2.8 If we consider in Example 2.7 and in Remark 2.18

$$\tilde{\nu}(x_N) = x_N^{\lambda} \tag{2.208}$$

(i.e., we consider a degeneration of the type $(\mathrm{dist}(x, M))^{\lambda}$, $\lambda > 0$) then we have for $\mathcal{N}(t)$ from (2.202)

$$\mathcal{N}(t) = ct^{(1-p')\lambda+1} = ct^{1-\frac{\lambda}{p-1}}$$

provided

$$\lambda < p - 1\,, \tag{2.209}$$

and condition (2.204) guaranteeing compactness is obviously fulfilled. Consequently, for weight functions of the type (2.208) with λ satisfying (2.209) we can use our approach and derive results about the existence of weak solutions with this type of degeneracy in the highest order terms. Conditions (2.199) are in this case obviously satisfied. On the other hand, condition (2.205) used in the first approach admits *less values* of λ than (2.209) if $p > \dfrac{N}{N-1}$ since it reads

$$\lambda < \frac{1}{s} < \frac{p}{N}\,.$$

If we consider the BVP from Remark 2.18 (ii), i.e. the Dirichlet data on the top of the cube Ω, the situation improves more since condition (2.207) will now be satisfied for

$$\lambda < p\,, \tag{2.210}$$

i.e., we have *more admissible values* for λ. ◇

Remark 2.19 All considerations concerning Example 2.8 remain valid if we assume $\lambda < 0$, which means — due to (2.187) — that *singularities* of the type $(\mathrm{dist}(x, M))^\lambda$, $\lambda < 0$, may occur among the coefficients of the differential operator A_0. Since conditions (2.209) and (2.210) contain no lower bound for λ, we can immediately conclude that our approach is suitable also for certain singular elliptic differential operators without restriction on the rate of singularity (of course only when the singularity appears on $\partial\Omega$ — compare with the condition $\nu \in L^1_{\mathrm{loc}}(\Omega)$).

Example 2.9 The weight function $\nu(x)$ considered in Example 2.8 — see (2.208) — describes a degeneration or singularity *on the boundary* $\partial\Omega$ of Ω. Therefore, let us now consider a weight whose "bad behaviour" is concentrated *in the interior of* Ω.

Let Ω be the square $(-1, 1) \times (-1, 1)$ and consider the second order differential equation

$$-\sum_{i=1}^{2} \frac{\partial}{\partial x_i}\left(\nu(x_1, x_2)\left|\frac{\partial u}{\partial x_i}\right|^{p-2}\frac{\partial u}{\partial x_i}\right) + G(u(x_1, x_2)) = f \tag{2.211}$$

on Ω with $p > 2$ and with the weight function

$$\nu(x_1, x_2) = \begin{cases} 1 & \text{for } x_1 \le 0\,, \\ x_2^\lambda & \text{for } x_1 > 0\,,\ x_2 > 0\,, \\ |x_2|^\mu & \text{for } x_1 > 0\,,\ x_2 < 0\,. \end{cases} \tag{2.212}$$

This is in fact the weight from (2.185) with $\gamma = 0$. As in Example 2.5, the conditions (2.177) lead to the restriction

$$\lambda, \mu \in (-1, p-1) \tag{2.213}$$

since the function ν from (2.212) vanishes or becomes singular at the segment $S = \{x = (x_1, 0);\ 0 < x_1 < 1\} \subset \Omega$.

(i) As we have seen in Example 2.5, we have still *another restriction*

$$\lambda, \mu < \frac{p}{2} \tag{2.214}$$

(note that we take $p > 2$, see (2.186)), and as was shown in Example 2.6, we can choose the function $G = G(t)$ rather general, for example

$$G(t) = te^{t^2}\,. \tag{2.215}$$

(ii) If we use our second approach, the role of the number κ_2 from (2.191) is played by

$$\kappa_3 = k - 1 - \frac{N}{p}$$

since we started our imbedding theorems from the space $W^{k-1,p}(\Omega)$. But in our case we have $k = 1$, i.e., κ_3 is negative and we have no imbedding into the space $C(\overline{\Omega})$. We can only allow

$$|G(t)| \leq c|t|^p$$

which is *weaker* than (2.215). On the other hand, the Hardy type inequality can be used (with Dirichlet data on the segments $S_1 = \{x = (x_1, 1)\,;\ 0 < x_1 < 1\}$ and $S_2 = \{x = (x_1, -1)\,;\ 0 < x_1 < 1\}$ on $\partial\Omega$) and the corresponding imbedding is compact *for all values of* λ and μ *satisfying* (2.213), which is an improvement of (2.214). ◇

Up to now, we have described two approaches in which we used the *special* weighted space $X = W^{k,p}(\nu, \Omega)$ with weights $w_\alpha(x) \equiv 1$ in lower order terms, i.e. for $|\alpha| \leq k-1$. The use of this space allowed only changes in the *order* of the growth (like $\sum |\xi_\beta|^{r(\beta)}$ with $r(\beta) \geq p$ for $|\beta| < k$) while the general approach described in Sections 2.5–2.8 and using the *general* weighted Sobolev space $W^{k,p}(\Omega, w)$ allows both to change the *order* and to introduce *weights* ω_β different from $w_\beta(x) \equiv 1$ (e.g. terms like

$$\sum \omega_\beta(x)|\xi_\beta|^{q(\beta)}$$

in (2.112)).

Example 2.10 Let us again consider the operator A_0 from (2.184) with the weight function $\nu = \nu(x)$ from (2.185). As was pointed out in Remark 2.16, if we use the space $X = W^{k,p}(\nu, \Omega)$ from (2.176) and apply the first approach, we can add lower order terms (say of order $2k - 2$) of the form

$$(-1)^{k-1} \sum_{|\beta|=k-1} D^\beta(|D^\beta u(x)|^{q-2} D^\beta u(x)) \tag{2.216}$$

with $1 < q < 2p-1$ for $1 < p < 2$. However, the approach described in Sections 2.5–2.8 allows to add, e.g., a lower order term of the form

$$(-1)^{k-1} \sum_{|\beta|=k-1} D^\beta(\omega(x)|D^\beta u(x)|^{q-2} D^\beta u(x))\,, \tag{2.217}$$

where the weight function ω has a similar structure as ν, i.e.

$$\omega(x) = \begin{cases} 1 & \text{for } x_1 \leq 0, \\ x_2^{\lambda_0}(1-x_1)^{\gamma_0} & \text{for } x_1 > 0,\ x_2 > 0, \\ |x_2|^{\mu_0}(1-x_1)^{\gamma_0} & \text{for } x_1 > 0,\ x_2 < 0 \end{cases}$$

with values $\lambda_0, \gamma_0, \mu_0$ connected with the values $p, q, \lambda, \gamma, \mu$. Roughly speaking, the inequality

$$\lambda_0 > \lambda\frac{q}{p} + \frac{q}{p} - q - 1$$

(with λ from (2.185)) must hold and similarly also for μ_0 a γ_0. ◇

Remark 2.20 It should be emphasized that the decision about the weight functions $w_\alpha(x)$ which appear in the definition of the space $W^{m,p}(\Omega, w)$ in terms of lower orders ($|\alpha| \leq k-1$) depends on the weight functions $w_\alpha(x)$ for $|\alpha| = k$ (i.e. those appearing at the derivatives of order k). An important role is played here by *Hardy-type inequalities*, i.e. inequalities of the form

$$\left(\int_\Omega |v(x)|^q \omega(x)\,dx\right)^{\frac{1}{q}} \leq c\left(\int_\Omega \sum_{i=1}^N \left|\frac{\partial v}{\partial x_i}(x)\right|^p w_i(x)\,dx\right)^{\frac{1}{p}}. \tag{2.218}$$

We proceed in the following way: We use inequality (2.218) (say, for $v \in C_0^\infty(\Omega)$, if we consider the Dirichlet problem) so that we have the imbedding

$$W_0^{1,p}(\Omega, w) \hookrightarrow L^q(\Omega, \omega). \tag{2.219}$$

Moreover, we assume that the imbedding (2.219) is *compact.*

Now, we take for v a derivative $D^\beta u$, β fixed, $|\beta| = k-1$, and choose for the weight functions $w_i(x)$ on the right-hand side of (2.218) the weight functions $w_\alpha(x)$, $|\alpha| = k$, appearing in the definition of the space $W^{k,p}(\Omega, w)$. More precisely, if we denote, for $\beta = (\beta_1, \beta_2, \ldots, \beta_n)$, $|\beta| = k-1$, by $\beta(i)$ the multiindex $\alpha = (\alpha_1, \alpha_2, \ldots, \alpha_N)$ with $\alpha_j = \beta_j$ for $i \neq j$ and $\alpha_i = \beta_i + 1$, then $|\alpha| = k$ and we choose $w_i(x) = w_{\beta(i)}(x)$. Then we can determine the weight function $\omega(x)$ on the left-hand side of (2.218) as $w_\beta(x)$ if $q = p$ (with $w_\beta(x)$ appearing in the definition of the space $W^{k,p}(\Omega, w)$) and as $\omega_\beta(x)$ if $q = q(\beta) \neq p$ (with $\omega_\beta(x)$ appearing in the definition of the space $H^{k-1,p}(\Omega, \omega)$, see (2.142)).

In this way we determine the weight functions $w_\beta(x)$ and $\omega_\beta(x)$ for $|\beta| = k-1$, and analogously we determine the weight functions appearing at the derivatives of order $k-2$, etc.

Example 2.11 Consider again an operator A whose principal part is of the form (2.184), i.e.

$$a_\alpha(x;\xi) = w_\alpha(x)|\xi_\alpha|^{p-2}\xi_\alpha, \quad |\alpha| = k$$

with $w_\alpha(x) = \nu(x)$. Let Ω be a plane domain, $\Omega = \{(x_1, x_2);\ 0 < x_1 < 1, 0 < x_2 < b\}$ with $0 < b < \infty$, and choose the weight function $\nu(x)$ in the form

$$\nu(x) = \nu(x_1, x_2) = x_2^\lambda, \quad \lambda \text{ a real number.}$$

In this case conditions (2.177) are obviously fulfilled. Let us consider, for simplicity, the Dirichlet problem. Then the basic space V is just $W_0^{k,p}(\Omega, \nu)$ and we can carry out all considerations for v from the dense set $C_0^\infty(\Omega)$.

The Hardy inequality

$$\left(\int_0^b |v(x_1, x_2)|^q x_2^\mu\,dx_2\right)^{\frac{1}{q}} \leq c\left(\int_0^b |\frac{\partial v}{\partial x_2}(x_1, x_2)|^p x_2^\lambda\,dx_2\right)^{\frac{1}{p}} \tag{2.220}$$

holds for all $v \in C_0^\infty(\Omega)$ with a constant $c > 0$ independent of v and *independent of* x_1 provided

$$\lambda \neq p-1 \quad \text{and} \quad \mu > \lambda \frac{q}{p} + \frac{q}{p} - q - 1 \tag{2.221}$$

(see Example 6.8 in [55]).

(i) Let us determine the weight functions $w_\beta(x)$, $|\beta| = k-1$. Take $q = p$ in (2.220), i.e.

$$\lambda \neq p-1, \quad \mu > \lambda - p; \tag{2.222}$$

it follows from (2.220) that

$$\begin{aligned}
\int_\Omega |v(x)|^p \omega(x)\,dx &= \int_0^1 \left(\int_0^b |v(x_1,x_2)|^p x_2^\mu \,dx_2 \right) dx_1 \\
&\leq c^p \left(\int_0^1 \left(\int_0^b \left| \frac{\partial v}{\partial x_2}(x_1,x_2) \right|^p x_2^\lambda \,dx_2 \right) dx_1 \right) \\
&\leq c^p \int_\Omega \left(\left| \frac{\partial v}{\partial x_1}(x) \right|^p v(x) + \left| \frac{\partial v}{\partial x_2}(x) \right|^p v(x) \right) dx\,.
\end{aligned} \tag{2.223}$$

This is inequality (2.218) with $\omega(x) = x_2^\mu$, $q = p$, which indicates that if we put $v = D^\beta u$, $|\beta| = k-1$, we can choose

$$w_\beta(x) = x_2^\mu\,.$$

Hence, we have

$$\begin{aligned}
D^\alpha u &\in L^p(\Omega, w_\alpha) \quad \text{with} \quad w_\alpha(x) = x_2^\lambda \quad \text{for} \quad |\alpha| = k\,, \\
D^\beta u &\in L^p(\Omega, w_\beta) \quad \text{with} \quad w_\beta(x) = x_2^\mu \quad \text{for} \quad |\beta| = k-1\,,
\end{aligned}$$

where λ and μ satisfy (2.222). Analogously we can show that

$$D^\gamma u \in L^p(\Omega, w_\gamma) \quad \text{with} \quad w_\gamma(x) = x_2^\delta \quad \text{for} \quad |\gamma| = k-2$$

provided

$$\mu \neq p-1, \quad \delta > \mu - p,$$

etc. By this procedure, we construct the whole family of weights w_α, i.e., the space

$$W^{k,p}(\Omega, w)$$

or more precisely, the space $W_0^{k,p}(\Omega, w)$. Let us emphasize that the imbedding corresponding to the inequality (2.220) is compact (see Example 7.10 (ii) in [55], where the case $\lambda < p-1$ is discussed; the case $\lambda > p-1$ follows similarly by using Example 6.8 (ii) in [55]). Hence, we simultaneously have the *compact imbedding*

$$V = W_0^{k,p}(\Omega, w) \hookrightarrow\hookrightarrow W_0^{k-1,p}(\Omega, w)\,.$$

(ii) If we liked to determine the space $H^{k-1,p}(\Omega,\omega)$ (satisfying assumption L), i.e. the weight functions ω_β for $q \neq p$, we have to distinguish two cases.

(a) The case $1 < q < p < \infty$. In this case we obtain again a Hardy type inequality similar to (2.223). Namely, we get

$$\begin{aligned}&\left(\int_\Omega |v(x)|^q \omega(x)\,dx\right)^{\frac{1}{q}}\\ &\quad \le c\left(\int_\Omega \left(\left|\frac{\partial v}{\partial x_1}(x)\right|^p + \left|\frac{\partial v}{\partial x_2}(x)\right|^p\right) \nu(x)\,dx\right)^{\frac{1}{p}}\end{aligned} \tag{2.224}$$

with $\omega(x) = x_2^\mu$. But now μ is determined by (2.221), i.e.

$$D^\beta u \in L^{q(\beta)}(\Omega,\omega_\beta) \quad \text{for} \quad |\beta| = k-1 \quad \text{with} \quad \omega_\beta(x) = x_2^\mu\,, \quad \mu > \lambda\frac{q}{p}+\frac{q}{p}-q-1\,.$$

Indeed, it follows from (2.220) that

$$\begin{aligned}&\int_0^1 \left(\int_0^b |v(x_1,x_2)|^q x_2^\mu\,dx_2\right) dx_1\\ &\quad \le c\int_0^1 \left(\int_0^b \left|\frac{\partial v}{\partial x_2}(x_1,x_2)\right|^p x_2^\lambda\,dx_2\right)^{\frac{q}{p}} dx_1\,.\end{aligned} \tag{2.225}$$

Since $q < p$, we can use the Hölder inequality for the outer integral with the exponent $r = \dfrac{p}{q} > 1$ obtaining

$$\begin{aligned}&\left(\int_0^1 \left(\int_0^b \left|\frac{\partial v}{\partial x_2}(x)\right|^p x_2^\lambda\,dx_2\right) dx_1\right)^{\frac{1}{r}} \left(\int_0^1 1^{r'}\,dx_1\right)^{\frac{1}{r'}}\\ &\quad = \left(\int_\Omega \left|\frac{\partial v}{\partial x_2}(x)\right|^p x_2^\lambda\,dx\right)^{\frac{q}{p}}\\ &\quad \le \left(\int_\Omega \left(\left|\frac{\partial v}{\partial x_1}(x)\right|^p + \left|\frac{\partial v}{\partial x_2}(x)\right|^p\right) x_2^\lambda\,dx\right)^{\frac{q}{p}}.\end{aligned} \tag{2.226}$$

Formula (2.224) now follows from (2.225) and (2.226).

Similarly we can determine $\omega_\beta(x)$ for $|\beta| = k-2, k-3$, etc.

(b) The case $1 < p < q < \infty$. In this case we cannot use the Hölder inequality. We have to use a more complicated procedure described in [55], Subsection 12.13. We will not go into details here. Let us only mention that we again obtain an inequality of the form (2.224) with $\omega(x) = x_2^\mu$ but now under the additional condition

$$\frac{1}{p} - \frac{1}{q} > \frac{1}{2}$$

and with

$$\mu > \lambda \frac{p}{q} .$$

(iii) Due to these investigations, we can consider lower order "coefficients" of the form

$$a_\beta(x; \xi) = x_2^\mu |\xi_\beta|^{q-2} \xi_\beta \quad \text{for} \quad |\beta| = k - 1$$

and similarly for $|\beta| < k - 1$, with μ depending on q and p. ◇

Remark 2.21 In Example 2.10, we investigated the *Dirichlet problem* on the plane domain $\Omega = (0, 1) \times (0, b)$ and considered the Hardy inequality (2.220) for functions $v \in C_0^\infty(\Omega)$. Let us point out that due to the fact that in (2.220) only the integral over $(0, b)$ for x_2 is of importance, the same results can be derived if we consider Dirichlet boundary conditions only *on the top and the bottom* of our rectangle Ω, i.e., for $(x_1, x_2) \in \partial\Omega$ with $x_2 = 0$ and $x_2 = b$.

Moreover, we can consider Dirichlet boundary conditions *only on the bottom*, $x_2 = 0$ (or only on the top, $x_2 = b$) but then we can consider only values λ such that $\lambda < p - 1$ (or $\lambda > p - 1$) instead of $\lambda \neq p - 1$.

Let us add one simple example illustrating the modified coercivity condition (2.171).

Example 2.12 Consider the Dirichlet problem for the second order equation of the type (2.113), where

$$(Au)(x) = -\sum_{i=1}^N \frac{\partial}{\partial x_i} \left(w_i(x) \left| \frac{\partial u}{\partial x_i} \right|^{p-2} \frac{\partial u}{\partial x_i} \right) + \omega(x) |u(x)|^{q-2} u(x)$$

(here we use the "usual" indices instead of multiindices). We have

$$\begin{aligned} a_i(x; \xi_0, \xi_1, \dots, \xi_N) &= w_i(x) |\xi_i|^{p-2} \xi_i \quad \text{for} \quad i = 1, 2, \dots, N, \\ a_0(x; \xi_0, \xi_1, \dots, \xi_N) &= \omega(x) |\xi_0|^{q-2} \xi_0 \end{aligned} \tag{2.227}$$

with $p, q > 1$. We use the space $V = W_0^{1,p}(\Omega, w)$ with $w = \{w_0, w_1, \dots, w_N\}$ where w_0 is chosen in such a way that the compact imbedding

$$W_0^{1,p}(\Omega, w) \hookrightarrow\hookrightarrow L^p(\Omega, w_0)$$

holds. The coercivity condition (2.170) has the form

$$\sum_{i=0}^N a_i(x; \xi_0, \xi_1, \dots, \xi_N) \xi_i \geq c_{16} \sum_{i=1}^N w_i(x) |\xi_i|^p + c_{17} w_0(x) |\xi_0|^p - c_{18} ,$$

i.e., for our special choice (2.227),

$$\sum_{i=0}^{N} w_i(x)|\xi_i|^p + \omega(x)|\xi_0|^q \geq c_{16} \sum_{i=1}^{N} w_i(x)|\xi_i|^p + c_{17} w_0(x)|\xi_0|^p - c_{18} \,. \quad (2.228)$$

Since there are no obvious connections between $\omega(x)$ and $w_0(x)$, q and p, we cannot guarantee that (2.228) is fulfilled for every $\xi \in \mathbb{R}^{N+1}$.

On the other hand, the modified coercivity condition (2.171) is fulfilled. Indeed, it has the form

$$\sum_{i=0}^{N} a_i(x; \xi_0, \xi_1, \dots, \xi_N)\xi_i \geq c_{16} \sum_{i=1}^{N} w_i(x)|\xi_i|^p + c_{17}\omega(x)|\xi_0|^q - c_{18}$$

and is obviously satisfied with

$$c_{16} = c_{17} = 1, \quad c_{18} \geq 0 \,.$$

To ensure the existence of a weak solution by means of Theorem 2.4, we have to suppose that the Hardy-type inequality

$$\left(\int_\Omega |u(x)|^q \omega(x)\, dx \right)^{\frac{1}{q}} \leq c \left(\sum_{i=1}^{N} \int_\Omega \left| \frac{\partial u}{\partial x_i}(x) \right|^p w_i(x)\, dx \right)^{\frac{1}{p}} \quad (2.229)$$

holds for every $u \in V = W_0^{1,p}(\Omega, w)$, that the corresponding compact imbedding

$$W_0^{1,p}(\Omega, w) \hookrightarrow\hookrightarrow L^q(\Omega, \omega)$$

holds, and further that $q > p$.

If we suppose that the inequality

$$\int_\Omega |u(x)|^p w_0(x)\, dx \leq c \sum_{i=1}^{N} \int_\Omega \left| \frac{\partial u}{\partial x_i}(x) \right|^p w_1(x)\, dx \quad (2.230)$$

holds, then we need not assume that $p < q$. Indeed, since the right-hand side in (2.229) is an *equivalent norm* in V, we have

$$\langle Tu, u \rangle = \sum_{i=1}^{N} \int_\Omega a_i(x; u(x), \nabla u(x)) \frac{\partial u}{\partial x_i}(x)\, dx + \int_\Omega a_0(x, u(x), \nabla u(x)) u(x)\, dx$$

$$= \sum_{i=1}^{N} \int_\Omega \left| \frac{\partial u}{\partial x_i}(x) \right|^p w_1(x)\, dx + \int_\Omega |u(x)|^q \omega(x)\, dx \geq c\|u\|_V^p + \|u\|_{q,\omega}^q \geq c\|u\|_V^p \,,$$

and hence

$$\lim_{\|u\|_V \to \infty} \frac{\langle Tu, u \rangle}{\|u\|_V} = \infty \, .$$

There are several results concerning the validity of inequalities (2.229) and (2.230), called also *weighted Friedrichs inequalities*. See, e.g., D.E. Edmunds, B. Opic [25] for the special case $w_i = w_1$ for $i = 2, \ldots, N$ with w_0, w_1 having singularities or degeneracies only on the boundary and satisfying certain (rather complicated) additional assumptions. $\diamond$

Chapter 3

The degenerated p-Laplacian on a bounded domain

3.1 Basic notation

In this chapter we will assume that $p > 1$ is a real number and Ω is a bounded open subset of $\mathbb{R}^N$. We will study the existence of the principal eigenvalue and the corresponding nonnegative (or positive) eigenfunction of the *nonhomogeneous degenerated quasilinear eigenvalue problem*

$$\begin{aligned} -\operatorname{div}(a(x,u)|\nabla u|^{p-2}\nabla u) &= \lambda b(x,u)|u|^{p-2}u \quad \text{in} \quad \Omega\,, \\ u &= 0 \quad \text{on} \quad \partial\Omega\,. \end{aligned} \tag{3.1}$$

Let us point out that the coefficient $a(x,u)$ in the principal part of (3.1) will depend in a general way on both x and u. For this reason we call the differential operator in the left-hand side of (3.1) the *degenerated p-Laplacian.* It is also worth noting that the problem (3.1) has a nonvariational structure which makes it different from the *homogeneous nondegenerated* eigenvalue problem for the p-Laplacian

$$\begin{aligned} -\operatorname{div}(|\nabla u|^{p-2}\nabla u) &= \lambda|u|^{p-2}u \quad \text{in} \quad \Omega\,, \\ u &= 0 \quad \text{on} \quad \partial\Omega\,, \end{aligned}$$

which has been extensively studied during the last ten years: see, e.g. A. Anane [5], G. Barles [7], T. Bhattacharya [8], J. García Azorero, I. Peral Alonso [28], P. Lindqvist [46], M. Otani, T. Teshima [17] and the book of P. Drábek [17] where also some other references can be found.

Let $\nu(x)$ be a weight function in Ω satisfying the conditions

$$\nu \in L^1_{\mathrm{loc}}(\Omega)\,, \quad \nu^{-\frac{1}{p-1}} \in L^1_{\mathrm{loc}}(\Omega)\,. \tag{3.2}$$

We will work in the special weighted Sobolev space $W^{1,p}(\nu,\Omega)$ which is the set of all real valued functions u defined in Ω for which

$$\|u\|_{1,p,\nu} = \left(\int_\Omega |u|^p\,dx + \int_\Omega \nu|\nabla u|^p\,dx\right)^{\frac{1}{p}} < \infty\,. \tag{3.3}$$

Since we deal with the Dirichlet problem we define also $X := W_0^{1,p}(\nu, \Omega)$ as the closure of $C_0^\infty(\Omega)$ in $W^{1,p}(\nu, \Omega)$ with respect to the norm $\|\cdot\|_{1,p,\nu}$. Due to (3.2), both spaces are well-defined uniformly convex Banach spaces (see Section 1.5).

We will also use the following results from Chapter 1, Example 1.3. Let $s \geq \dfrac{1}{p-1}$. Then the continuous imbedding

$$W^{1,p}(\nu, \Omega) \hookrightarrow W^{1,p_s}(\Omega) \tag{3.4}$$

holds provided

$$\nu^{-s} \in L^1(\Omega) \quad \text{and} \quad p_s = \frac{ps}{s+1}$$

(see (1.20)).

Assume that $s + 1 \leq ps < N(s+1)$. Then

$$X \hookrightarrow W_0^{1,p_s}(\Omega) \hookrightarrow L^{p_s^*}(\Omega), \tag{3.5}$$

where

$$1 \leq p_s^* = \frac{Np_s}{N - p_s} = \frac{Nps}{N(s+1) - ps}.$$

If $ps \geq N(s+1)$ then the imbeddings (3.5) hold with arbitrary p_s^*, $1 \leq p_s^* < \infty$. Moreover, the compact imbedding

$$X \hookrightarrow\hookrightarrow L^r(\Omega)$$

holds provided $1 \leq r < p_s^*$ (see (1.24)). It follows that $s > \dfrac{N}{p}$ implies $p_s^* > p$. In particular, we have

$$X \hookrightarrow\hookrightarrow L^{p+\eta}(\Omega) \tag{3.6}$$

for $0 \leq \eta < p_s^* - p$ provided

$$\nu^{-s} \in L^1(\Omega) \quad \text{and} \quad s \in \left(\frac{N}{p}, \infty\right) \cap \left[\frac{1}{p-1}, \infty\right). \tag{3.7}$$

It follows from the weighted Friedrichs inequality (1.28) that the norm

$$\|u\|_X = \left(\int_\Omega \nu |\nabla u|^p \, dx\right)^{\frac{1}{p}}$$

on the space X is equivalent to the norm $\|\cdot\|_{1,p,\nu}$ defined by (3.3).

3.2 Existence of the least eigenvalue of the homogeneous eigenvalue problem

We will first study the *degenerated* (and/or *singular*) homogeneous eigenvalue problem associated with (3.1), but now with a and b *independent* of u.

Let us suppose that ν is the weight function satisfying (3.2) and (3.7). Let $a(x)$, $b(x)$ be measurable functions satisfying

$$\frac{\nu(x)}{c_1} \leq a(x) \leq c_1 \nu(x), \tag{3.8}$$

$$b(x) \geq 0 \tag{3.9}$$

for a.e. $x \in \Omega$ with some constant $c_1 \geq 1$. We assume that either $b \in L^{\frac{q}{q-p}}(\Omega)$ with some q satisfying $p < q < p_s^*$ or $b \in L^\infty(\Omega)$ (we set $q = p$ in the latter case). Moreover, let

$$\text{meas}\{x \in \Omega;\ b(x) > 0\} > 0. \tag{3.10}$$

The proofs in this and the following section will be given for the case $q > p$, i.e. $b \in L^{\frac{q}{q-p}}(\Omega)$. They can be done in the same way also in the case $q = p$, i.e. $b \in L^\infty(\Omega)$.

Let us consider a *homogeneous eigenvalue problem*

$$\begin{aligned} -\operatorname{div}(a(x)|\nabla u|^{p-2}\nabla u) &= \lambda b(x)|u|^{p-2}u \quad \text{in} \quad \Omega\,, \\ u &= 0 \quad \text{on} \quad \partial\Omega\,. \end{aligned} \tag{3.11}$$

Observe that $a(x)$ is also a weight function, and due to (3.8) our space X coincides with $W_0^{1,p}(a, \Omega)$ and the norm $\|\cdot\|_X$ is equivalent to

$$\|u\|_a = \left(\int_\Omega a|\nabla u|^p\, dx\right)^{\frac{1}{p}}.$$

Definition 3.1 We will say that $\lambda \in \mathbb{R}$ is an *eigenvalue* of the eigenvalue problem (3.11) if there exists $u \in X$, $u \not\equiv 0$ such that

$$\int_\Omega a(x)|\nabla u|^{p-2}\nabla u \nabla\varphi\, dx = \lambda \int_\Omega b(x)|u|^{p-2}u\varphi\, dx \tag{3.12}$$

holds for any $\varphi \in X$. Then u is called an *eigenfunction* corresponding to the eigenvalue λ.

Lemma 3.1 *There exists the* least (*i.e. the* first *or* principal) *eigenvalue* $\lambda_1 > 0$ *and at least one corresponding eigenfunction* $u_1 \geq 0$ *a.e. in* Ω ($u_1 \not\equiv 0$) *of the eigenvalue problem* (3.11).

Proof. Set

$$\lambda_1 = \inf\left\{\int_\Omega a(x)|\nabla v|^p\, dx\right\}$$

the infimum being taken over all v such that $\int_\Omega b(x)|v|^p\, dx = 1$. We shall prove that λ_1 is the least eigenvalue of (3.11). The expression for λ_1 presented above will be referred to as its *variational characterization*. Obviously $\lambda_1 \geq 0$. Let $\{v_n\}_{n=1}^\infty$ be the minimizing sequence for λ_1, i.e.

$$\int_\Omega b(x)|v_n|^p\, dx = 1 \quad \text{and} \quad \int_\Omega a(x)|\nabla v_n|^p\, dx = \lambda_1 + \delta_n \tag{3.13}$$

with $\delta_n \to 0+$ for $n \to \infty$. It follows from (3.13) that $\|v_n\|_X \leq c_2$. The reflexivity of X yields the weak convergence $v_n \rightharpoonup u_1$ in X for some u_1 (at least for some subsequence of $\{v_n\}$). The compact imbedding $X \hookrightarrow\hookrightarrow L^q(\Omega)$ (see (3.6)) implies the strong convergence $v_n \to u_1$ in $L^q(\Omega)$. It follows from (3.9), (3.13), from the Minkowski and Hölder inequalities that

$$\begin{aligned}
1 &= \lim_{n\to\infty}\left(\int_\Omega b(x)|v_n|^p\, dx\right)^{\frac{1}{p}} \\
&\leq \lim_{n\to\infty}\left(\int_\Omega b(x)|v_n - u_1|^p\, dx\right)^{\frac{1}{p}} + \left(\int_\Omega b(x)|u_1|^p\, dx\right)^{\frac{1}{p}} \\
&\leq \lim_{n\to\infty}\left(\int_\Omega (b(x))^{\frac{q}{q-p}}\, dx\right)^{\frac{q-p}{pq}}\left(\int_\Omega |v_n - u_1|^q\, dx\right)^{\frac{1}{q}} + \left(\int_\Omega b(x)|u_1|^p\, dx\right)^{\frac{1}{p}} \\
&= \left(\int_\Omega b(x)|u_1|^p\, dx\right)^{\frac{1}{p}},
\end{aligned}$$

and analogously

$$\begin{aligned}
\left(\int_\Omega b(x)|u_1|^p\, dx\right)^{\frac{1}{p}} &\leq \lim_{n\to\infty}\left(\int_\Omega (b(x))^{\frac{q}{q-p}}\, dx\right)^{\frac{q-p}{pq}}\left(\int_\Omega |u_1 - v_n|^q\, dx\right)^{\frac{1}{q}} \\
&\quad + \lim_{n\to\infty}\left(\int_\Omega b(x)|v_n|^p\, dx\right)^{\frac{1}{p}} = 1\,.
\end{aligned}$$

Hence

$$\int_\Omega b(x)|u_1|^p\, dx = 1\,.$$

In particular, $u_1 \not\equiv 0$. The weak lower semicontinuity of the norm in X yields

$$\begin{aligned}
\lambda_1 &\leq \int_\Omega a(x)|\nabla u_1|^p\, dx = \|u_1\|_a^p \leq \liminf_{n\to\infty}\|v_n\|_a^p \\
&= \liminf_{n\to\infty}\int_\Omega a(x)|\nabla v_n|^p\, dx = \liminf_{n\to\infty}(\lambda_1 + \delta_n) = \lambda_1\,,
\end{aligned}$$

i.e.

$$\lambda_1 = \int_\Omega a(x)|\nabla u_1|^p \, dx \,. \tag{3.14}$$

It follows from (3.14) that $\lambda_1 > 0$ and it is easy to see that λ_1 is the least eigenvalue of (3.11) with the corresponding eigenfunction u_1.

Moreover, if u is an eigenfunction corresponding to λ_1 then $|u|$ is also an eigenfunction corresponding to λ_1. Hence we can suppose that $u_1 \geq 0$ a.e. in Ω. □

Remark 3.1 It follows from the proof of Lemma 3.1 that $v_n \rightharpoonup u_1$ in X and $\|v_n\|_X \to \|u_1\|_X$. The uniform convexity of X then implies the strong convergence $v_n \to u_1$ in X (see N. Dunford, J. T. Schwartz [24]).

Lemma 3.2 *Let $u \in X$, $u \geq 0$ a.e. in Ω, be the eigenfunction corresponding to the least eigenvalue $\lambda_1 > 0$ of the eigenvalue problem* (3.11). *Then $u \in L^\infty(\Omega)$.*

Proof. For $M > 0$ define

$$v_M(x) = \min\{u(x), M\} \,.$$

Let us choose $\varphi = v_M^{\kappa p+1}$ $(\kappa \geq 0)$ in

$$\int_\Omega a(x)|\nabla u|^{p-2}\nabla u \nabla\varphi \, dx = \lambda_1 \int_\Omega b(x)|u|^{p-2}u\varphi \, dx \,. \tag{3.15}$$

Obviously $\varphi \in X \cap L^\infty(\Omega)$. It follows from (3.15) that

$$(\kappa p + 1)\int_\Omega a(x)v_M^{\kappa p}|\nabla v_M|^p \, dx = \lambda_1 \int_\Omega b(x)u^{p-1}v_M^{\kappa p+1} \, dx \,. \tag{3.16}$$

Due to (3.8) and the imbedding $X \hookrightarrow L^{p_s^*}(\Omega)$ we have

$$\begin{aligned}(\kappa p + 1)\int_\Omega a(x)v_M^{\kappa p}|\nabla v_M|^p \, dx &\geq \frac{\kappa p + 1}{c_1}\int_\Omega \nu(x)v_M^{\kappa p}|\nabla v_M|^p \, dx \\ &= \frac{\kappa p + 1}{c_1(\kappa+1)^p}\int_\Omega \nu(x)|\nabla(v_M^{\kappa+1})|^p \, dx \\ &\geq c_3\frac{(\kappa p + 1)}{(\kappa+1)^p}\left(\int_\Omega (v_M^{\kappa+1})^{p_s^*} \, dx\right)^{\frac{p}{p_s^*}} .\end{aligned} \tag{3.17}$$

Hence it follows from (3.9), (3.15), (3.16), (3.17) and the Hölder inequality that

$$\begin{aligned}\left(\int_\Omega v_M^{(\kappa+1)p_s^*} \, dx\right)^{\frac{p}{p_s^*}} &\leq \frac{\lambda_1}{c_3}\frac{(\kappa+1)^p}{\kappa p + 1}\int_\Omega b(x)u^{p-1}v_M^{\kappa p+1} \, dx \\ &\leq \frac{\lambda_1}{c_3}\frac{(\kappa+1)^p}{\kappa p + 1}\left(\int_\Omega b(x)^{\frac{q}{q-p}} \, dx\right)^{\frac{q-p}{q}}\left(\int_\Omega u^{(\kappa+1)q} \, dx\right)^{\frac{p}{q}} .\end{aligned} \tag{3.18}$$

Due to the assumptions on $b(x)$ we obtain formally

$$\left(\int_\Omega v_M^{(\kappa+1)p_s^*}\,dx\right)^{\frac{p}{p_s^*}} \le c_4 \frac{(\kappa+1)^p}{\kappa p+1}\left(\int_\Omega u^{(\kappa+1)q}\,dx\right)^{\frac{p}{q}}, \tag{3.19}$$

i.e.

$$\|v_M\|_{(\kappa+1)p_s^*} \le c_5^{\frac{1}{\kappa+1}} \left(\frac{\kappa+1}{(\kappa p+1)^{\frac{1}{p}}}\right)^{\frac{1}{\kappa+1}} \|u\|_{(\kappa+1)q} \tag{3.20}$$

with $c_5 = c_4^{\frac{1}{p}}$. The expression (3.20) is a starting point for a bootstrap argument which plays an important role in L^∞-estimates throughout Chapters 3 and 4. Since $u \in L^{p_s^*}(\Omega)$, we can choose $\kappa := \kappa_1$ in (3.20) such that $(\kappa_1+1)q = p_s^*$, i.e. $\kappa_1 = \dfrac{p_s^*}{q} - 1$. Then we have

$$\|v_M\|_{(\kappa_1+1)p_s^*} \le c_5^{\frac{1}{\kappa_1+1}} \left(\frac{\kappa_1+1}{(\kappa_1 p+1)^{\frac{1}{p}}}\right)^{\frac{1}{\kappa_1+1}} \|u\|_{p_s^*} \tag{3.21}$$

for any $M > 0$. Due to $u(x) = \lim_{M\to\infty} v_M(x)$ for a.e. $x \in \Omega$, the Fatou lemma and (3.21) imply

$$\|u\|_{(\kappa_1+1)p_s^*} \le c_5^{\frac{1}{\kappa_1+1}} \left(\frac{\kappa_1+1}{(\kappa_1 p+1)^{\frac{1}{p}}}\right)^{\frac{1}{\kappa_1+1}} \|u\|_{p_s^*}. \tag{3.22}$$

Hence, we can choose $\kappa := \kappa_2$ in (3.20) such that $(\kappa_2+1)q = (\kappa_1+1)p_s^* = \dfrac{(p_s^*)^2}{q}$ and repeating the same argument we get

$$\|u\|_{(\kappa_2+1)p_s^*} \le c_5^{\frac{1}{\kappa_2+1}} \left(\frac{\kappa_2+1}{(\kappa_2 p+1)^{\frac{1}{p}}}\right)^{\frac{1}{\kappa_2+1}} \|u\|_{(\kappa_1+1)p_s^*}.$$

By induction we obtain

$$\|u\|_{(\kappa_n+1)p_s^*} \le c_5^{\frac{1}{\kappa_n+1}} \left(\frac{\kappa_n+1}{(\kappa_n p+1)^{\frac{1}{p}}}\right)^{\frac{1}{\kappa_n+1}} \|u\|_{(\kappa_{n-1}+1)p_s^*} \tag{3.23}$$

for any $n \in \mathbb{N}$, where $(\kappa_n+1) = \left(\dfrac{p_s^*}{q}\right)^n$. It follows from (3.22), (3.23) that

$$\|u\|_{(\kappa_n+1)p_s^*} \le c_5^{\sum_{k=1}^n \frac{1}{\kappa_k+1}} \left[\left(\frac{\kappa_1+1}{(\kappa_1 p+1)^{\frac{1}{p}}}\right)^{\frac{1}{\sqrt{\kappa_1+1}}}\right]^{\frac{1}{\sqrt{\kappa_1+1}}}.$$

$$\cdot\left[\left(\frac{\kappa_2+1}{(\kappa_2 p+1)^{\frac{1}{p}}}\right)^{\frac{1}{\sqrt{\kappa_2+1}}}\right]^{\frac{1}{\sqrt{\kappa_2+1}}}\cdots\left[\left(\frac{\kappa_n+1}{(\kappa_n p+1)^{\frac{1}{p}}}\right)^{\frac{1}{\sqrt{\kappa_n+1}}}\right]^{\frac{1}{\sqrt{\kappa_n+1}}}\|u\|_{p_s^*}\,.$$

Since $\left(\frac{y+1}{(yp+1)^{\frac{1}{p}}}\right)^{\frac{1}{\sqrt{y+1}}} > 1$ for $y > 0$ and $\lim\limits_{y\to\infty}\left(\frac{y+1}{(yp+1)^{\frac{1}{p}}}\right)^{\frac{1}{\sqrt{y+1}}} = 1$, there exists $c_6 > 1$ (independent of κ_n) such that

$$\|u\|_{(\kappa_n+1)p_s^*} \le c_5^{\sum\limits_{k=1}^{n}\frac{1}{\kappa_k+1}}\, c_6^{\sum\limits_{k=1}^{n}\frac{1}{\sqrt{\kappa_k+1}}}\,\|u\|_{p_s^*}\,. \tag{3.24}$$

However, $\sum\limits_{k=1}^{n}\frac{1}{\kappa_k+1} = \sum\limits_{k=1}^{n}\left(\frac{q}{p_s^*}\right)^k$, $\sum\limits_{k=1}^{n}\frac{1}{\sqrt{\kappa_k+1}} = \sum\limits_{k=1}^{n}\left(\sqrt{\frac{q}{p_s^*}}\right)^k$ and $\frac{q}{p_s^*} < \sqrt{\frac{q}{p_s^*}} < 1$. Hence it follows from (3.24) that there exists a constant $c_7 > 0$ such that we get

$$\|u\|_{r_n} \le c_7\|u\|_{p_s^*} \tag{3.25}$$

with $r_n = (\kappa_n+1)p_s^* \to \infty$ when $n \to \infty$. Let us assume that $\|u\|_\infty > c_7\|u\|_{p_s^*}$. Then there exists $\eta > 0$ and a set $\mathcal{A}$ of positive measure in Ω such that $u(x) \ge c_7\|u\|_{p_s^*} + \eta$ for $x \in \mathcal{A}$. It follows that

$$\begin{aligned}\liminf_{r_n\to\infty}\left(\int_\Omega |u(x)|^{r_n}\,dx\right)^{\frac{1}{r_n}} &\ge \liminf_{r_n\to\infty}\left(\int_{\mathcal{A}} |u(x)|^{r_n}\,dx\right)^{\frac{1}{r_n}}\\ &\ge \liminf_{r_n\to\infty}(c_7\|u\|_{p_s^*}+\eta)(\operatorname{meas}\mathcal{A})^{\frac{1}{r_n}}\\ &= c_7\|u\|_{p_s^*}+\eta\,,\end{aligned}$$

which contradicts (3.25). Hence

$$\|u\|_\infty \le c_7\|u\|_{p_s^*}$$

and the proof is complete. □

In the case of "classical" p-Laplacian (i.e. when $a(x) \equiv 1$) the first eigenvalue is simple as proved e.g. in A. Anane [5] and P. Lindqvist [46]. We will prove in Theorem 3.1 an analogous result also in our case. However, our situation is a bit more complicated due to the fact that we cannot apply the Harnack type inequality in order to prove that any eigenfunction corresponding to $\lambda_1 > 0$ is either positive or negative in Ω. Under our assumptions on the weight function $\nu(x)$ we cannot in general avoid that the eigenfunction $u_1 \ge 0$ from Lemma 3.1 vanishes inside Ω. Nevertheless, we have the following assertion.

Proposition 3.1 *There exists precisely one nonnegative eigenfunction u_1, $\|u_1\|_q = 1$, corresponding to the least eigenvalue $\lambda_1 > 0$ of the eigenvalue problem* (3.11).

Proof. Let the assertion be not true. We first prove that there are two distinct normed nonnegative eigenfunctions corresponding to $\lambda_1 > 0$ which vanish in the same set in Ω. Due to the variational characterization of λ_1 (see the beginning of the proof of Lemma 3.1) the function $u \in X$ is an eigenfunction corresponding to λ_1 if and only if

$$\begin{aligned} 0 &= \int_\Omega a(x)|\nabla u|^p\,dx - \lambda_1 \int_\Omega b(x)|u|^p\,dx \\ &= \inf_{v \in X} \left(\int_\Omega a(x)|\nabla v|^p\,dx - \lambda_1 \int_\Omega b(x)|v|^p\,dx \right). \end{aligned}$$

It follows that if $u_1, u_2 \in X$ are two eigenfunctions corresponding to λ_1 then also

$$v_1(x) = \max_{x\in\Omega}\{u_1(x), u_2(x)\}, \quad v_2(x) = \min_{x\in\Omega}\{u_1(x), u_2(x)\}$$

are eigenfunctions corresponding to λ_1 provided $v_2 \not\equiv 0$. Indeed, we have $v_1, v_2 \in X$ and

$$\begin{aligned} &\int_\Omega a(x)|\nabla v_1|^p\,dx - \lambda_1 \int_\Omega b(x)|v_1|^p\,dx + \int_\Omega a(x)|\nabla v_2|^p\,dx - \lambda_1 \int_\Omega b(x)|v_2|^p\,dx \\ &\quad = \int_\Omega a(x)|\nabla u_1|^p\,dx - \lambda_1 \int_\Omega b(x)|u_1|^p\,dx \\ &\qquad + \int_\Omega a(x)|\nabla u_2|^p\,dx - \lambda_1 \int_\Omega b(x)|u_2|^p\,dx. \end{aligned}$$

Hence

$$\begin{aligned} &\int_\Omega a(x)|\nabla v_1|^p\,dx - \lambda_1 \int_\Omega b(x)|v_1|^p\,dx \\ &\quad = \int_\Omega a(x)|\nabla v_2|^p\,dx - \lambda_1 \int_\Omega b(x)|v_2|^p\,dx = 0. \end{aligned}$$

Let $u_1 \geq 0$ and $u_2 \geq 0$ be two eigenfunctions corresponding to λ_1 such that $u_1 \not\equiv u_2$, $\min_{x\in\Omega}\{u_1(x), u_2(x)\} \not\equiv 0$ and

$$\|u_1\|_q = \|u_2\|_q = 1.$$

Denote $v_3(x) = k_1 v_2(x) = k_1 \min_{x\in\Omega}\{u_1(x), u_2(x)\}$, where $k_1 > 0$ is chosen in such a way that

$$\|v_3\|_q = 1.$$

Then $v_3 \in X$ is again an eigenfunction corresponding to λ_1 such that $v_3 \not\equiv u_1$. Moreover,

$$\{x \in \Omega;\ u_1(x) = 0\} \subseteq \{x \in \Omega;\ v_3(x) = 0\}.$$

Set $v_5(x) = k_2 v_4(x) = k_2 \max_{x\in\Omega}\{u_1(x), v_3(x)\}$, where $k_2 > 0$ is chosen such that

$$\|v_5\|_q = 1 .$$

Then $v_5 \in X$ is an eigenfunction corresponding to λ_1 such that $v_5 \not\equiv u_1$ and

$$\{x \in \Omega;\ v_5(x) = 0\} = \{x \in \Omega;\ u_1(x) = 0\} .$$

Let now $u_1 \geq 0$ and $u_2 \geq 0$ be two eigenfunctions corresponding to λ_1 such that $u_1 \not\equiv u_2$, $\|u_1\|_q = \|u_2\|_q = 1$ and

$$\min_{x\in\Omega}\{u_1(x), u_2(x)\} \equiv 0 .$$

Denote $\tilde{u}_1 = k_3 \max\{u_1(x), u_2(x)\}$, where $0 < k_3 < 1$ is chosen such that

$$\|\tilde{u}_1\|_q = 1 ,$$

and $\tilde{u}_2 = k_4 \max\{u_1(x), \tilde{u}_1(x)\}$, where $0 < k_4 < 1$ is such that

$$\|\tilde{u}_2\|_q = 1 .$$

Then $\tilde{u}_1$ and $\tilde{u}_2$ are eigenfunctions corresponding to λ_1 such that $\tilde{u}_1 \not\equiv \tilde{u}_2$ and

$$\{x \in \Omega;\ \tilde{u}_1 = 0\} = \{x \in \Omega;\ \tilde{u}_2 = 0\} .$$

We will prove now the assertion of the proposition via contradiction. Due to the argument presented above we may assume that $u \geq 0$ and $v \geq 0$ are two eigenfunctions corresponding to λ_1 such that

$$\|u\|_q = \|v\|_q = 1 , \quad u \not\equiv v , \tag{3.26}$$

and u, v vanish in Ω on the same set (almost everywhere). Then

$$\int_\Omega a(x)|\nabla u|^{p-2}\nabla u \nabla\varphi\, dx = \lambda_1 \int_\Omega b(x)|u|^{p-2}u\varphi\, dx \tag{3.27}$$

for any $\varphi \in X$, and

$$\int_\Omega a(x)|\nabla v|^{p-2}\nabla v \nabla\psi\, dx = \lambda_1 \int_\Omega b(x)|v|^{p-2}v\psi\, dx \tag{3.28}$$

for any $\psi \in X$. For $\varepsilon > 0$ set

$$u_\varepsilon = u + \varepsilon \quad \text{and} \quad v_\varepsilon = v + \varepsilon .$$

Substitute

$$\varphi = \frac{u_\varepsilon^p - v_\varepsilon^p}{u_\varepsilon^{p-1}}$$

into (3.27) and

$$\psi = \frac{v_\varepsilon^p - u_\varepsilon^p}{v_\varepsilon^{p-1}}$$

into (3.28). Since $\frac{u_\varepsilon}{v_\varepsilon}, \frac{v_\varepsilon}{u_\varepsilon} \in L^\infty(\Omega)$ by Lemma 3.2 and

$$\nabla\varphi = \left[1 + (p-1)\left(\frac{v_\varepsilon}{u_\varepsilon}\right)^p\right]\nabla u - p\left(\frac{v_\varepsilon}{u_\varepsilon}\right)^{p-1}\nabla v\,,$$

$$\nabla\psi = \left[1 + (p-1)\left(\frac{u_\varepsilon}{v_\varepsilon}\right)^p\right]\nabla v - p\left(\frac{u_\varepsilon}{v_\varepsilon}\right)^{p-1}\nabla u\,,$$

we have $\varphi, \psi \in X$. Adding (3.27) and (3.28) (with φ and ψ chosen above) we obtain

$$\begin{aligned}
&\int_\Omega a(x)\left\{\left[1 + (p-1)\left(\frac{v_\varepsilon}{u_\varepsilon}\right)^p\right]|\nabla u|^p + \left[1 + (p-1)\left(\frac{u_\varepsilon}{v_\varepsilon}\right)^p\right]|\nabla v|^p\right\} dx \\
&\quad - \int_\Omega a(x)\left\{p\left(\frac{v_\varepsilon}{u_\varepsilon}\right)^{p-1}|\nabla u|^{p-2}\nabla u\nabla v + p\left(\frac{u_\varepsilon}{v_\varepsilon}\right)^{p-1}|\nabla v|^{p-2}\nabla v\nabla u\right\} dx \\
&= \lambda_1\int_\Omega b(x)\left[\left(\frac{u}{u_\varepsilon}\right)^{p-1} - \left(\frac{v}{v_\varepsilon}\right)^{p-1}\right](u_\varepsilon^p - v_\varepsilon^p)\,dx\,.
\end{aligned}$$

Since $|\nabla \log u_\varepsilon| = \dfrac{|\nabla u|}{u_\varepsilon}$, the last equality is equivalent to

$$\begin{aligned}
&\int_\Omega a(x)(u_\varepsilon^p - v_\varepsilon^p)[|\nabla\log u_\varepsilon|^p - |\nabla\log v_\varepsilon|^p]\,dx \\
&\quad - \int_\Omega a(x)pv_\varepsilon^p|\nabla\log u_\varepsilon|^{p-2}\nabla\log u_\varepsilon(\nabla\log v_\varepsilon - \nabla\log u_\varepsilon)\,dx \\
&\quad - \int_\Omega a(x)pu_\varepsilon^p|\nabla\log v_\varepsilon|^{p-2}\nabla\log v_\varepsilon(\nabla\log u_\varepsilon - \nabla\log v_\varepsilon)\,dx \\
&= \lambda_1\int_\Omega b(x)\left[\left(\frac{u}{u_\varepsilon}\right)^{p-1} - \left(\frac{v}{v_\varepsilon}\right)^{p-1}\right](u_\varepsilon^p - v_\varepsilon^p)\,dx\,.
\end{aligned} \tag{3.29}$$

Let $p \geq 2$. We use Lemma 1.3 in order to estimate the left-hand side of (3.29) (we set first $\chi_1 = \nabla\log u_\varepsilon$, $\chi_2 = \nabla\log v_\varepsilon$ and then $\chi_1 = \nabla\log v_\varepsilon$, $\chi_2 = \nabla\log u_\varepsilon$). We obtain

$$\begin{aligned}
&\lambda_1 \int_\Omega b(x)\left[\left(\frac{u}{u_\varepsilon}\right)^{p-1} - \left(\frac{v}{v_\varepsilon}\right)^{p-1}\right](u_\varepsilon^p - v_\varepsilon^p)\,dx \\
&\quad \geq \frac{1}{2^{p-1}-1}\int_\Omega a(x)|\nabla \log u_\varepsilon - \nabla \log v_\varepsilon|^p (u_\varepsilon^p + v_\varepsilon^p)\,dx \\
&\quad = \frac{1}{2^{p-1}-1}\int_\Omega a(x)\left(\frac{1}{v_\varepsilon^p} + \frac{1}{u_\varepsilon^p}\right)|v_\varepsilon \nabla u - u_\varepsilon \nabla v|^p\,dx \geq 0\,.
\end{aligned} \tag{3.30}$$

Let $1 < p < 2$. We use Lemma 1.3 again in order to estimate the left-hand side of (3.29) (similarly as above) and we obtain

$$\begin{aligned}
&\lambda_1 \int_\Omega b(x)\left[\left(\frac{u}{u_\varepsilon}\right)^{p-1} - \left(\frac{v}{v_\varepsilon}\right)^{p-1}\right](u_\varepsilon^p - v_\varepsilon^p)\,dx \\
&\quad \geq \frac{3p(p-1)}{16}\int_\Omega a(x)\left(\frac{1}{u_\varepsilon^p} + \frac{1}{v_\varepsilon^p}\right)\frac{|v_\varepsilon \nabla u - u_\varepsilon \nabla v|^2}{(v_\varepsilon|\nabla u| + u_\varepsilon|\nabla v|)^{2-p}}\,dx \geq 0\,.
\end{aligned} \tag{3.31}$$

We have $u, v \in L^\infty(\Omega)$ (see Lemma 3.2) and

$$\frac{u}{u_\varepsilon} \to 1\,, \quad \frac{v}{v_\varepsilon} \to 1 \quad (\varepsilon \to 0+) \tag{3.32}$$

a.e. in Ω where $u > 0$ and $v > 0$, respectively;

$$\frac{u}{u_\varepsilon} = 0\,, \quad \frac{v}{v_\varepsilon} = 0 \quad (\text{for any} \quad \varepsilon > 0) \tag{3.33}$$

elsewhere (since u and v vanish on the same set in Ω). Hence it follows from (3.32), (3.33) and the Lebesgue theorem that for any p, $1 < p < \infty$,

$$\lambda_1 \int_\Omega b(x)\left[\left(\frac{u}{u_\varepsilon}\right)^{p-1} - \left(\frac{v}{v_\varepsilon}\right)^{p-1}\right](u_\varepsilon^p - v_\varepsilon^p)\,dx \to 0 \quad (\varepsilon \to 0+)\,.$$

It follows from (3.30), (3.31) and from the Fatou lemma that

$$|v\nabla u - u\nabla v| = 0 \quad \text{a.e. in} \quad \Omega$$

for any $1 < p < \infty$. Hence there exists a constant $k > 0$ such that $u = kv$ a.e. in Ω. But (3.26) yields $k = 1$, i.e. $u = v$ a.e. in Ω, which is a contradiction. □

The second part of the proof of Proposition 3.1 follows the lines of the proof of Lemma 3.1 in P. Lindqvist [46] for the nondegenerate case ($a(x) \equiv 1$ in Ω) but we have performed it here in detail for the reader's convenience.

Theorem 3.1 *The least eigenvalue* $\lambda_1 > 0$ *of the eigenvalue problem* (3.11) *is simple (i.e. there exists precisely one pair of normed eigenfunctions corresponding to* λ_1*). These eigenfunctions do not change the sign in* Ω*.*

Proof. Let u be the eigenfunction of (3.11) associated with $\lambda_1 > 0$ and changing sign in Ω. Then (3.12) holds with $\lambda = \lambda_1$ and with test functions $\varphi = u^+$, $\varphi = u^-$. Hence, we get

$$\int_\Omega a(x)|\nabla u^+|^p\,dx - \lambda_1 \int_\Omega b(x)|u^+|^p\,dx = 0\,,$$

$$\int_\Omega a(x)|\nabla u^-|^p\,dx - \lambda_1 \int_\Omega b(x)|u^-|^p\,dx = 0\,.$$

It follows from the variational characterization of λ_1 (see the beginning of the proof of Lemma 3.1) that both $\dfrac{u^+}{\|u^+\|_q}$ and $\dfrac{u^-}{\|u^-\|_q}$ are nonnegative eigenfunctions associated with $\lambda_1 > 0$. Proposition 3.1 then implies that $\dfrac{u^+}{\|u^+\|_q} = \dfrac{u^-}{\|u^-\|_q}$ a.e. in Ω, which is a contradiction. The assertion now follows from Proposition 3.1. □

Remark 3.2 As already mentioned before Proposition 3.1 we cannot apply Harnack-type inequality and prove that the first eigenfunction u is either positive or negative in Ω. This is the case when e.g. $a(x) \equiv 1$ (see A. Anane [5], P. Lindqvist [46]) or if the weight ν is of *special type.* Let us assume that for any compact subset $\Omega' \subset \Omega$ there exist positive constants $k_1(\Omega')$ and $k_2(\Omega')$ such that

$$0 < k_1(\Omega') \le \nu(x) \le k_2(\Omega') \quad \text{for a.e. } x \in \Omega'\,. \tag{3.34}$$

Then we can apply Theorem 1.9 and Remark 1.3 on Ω' to get that u_1 is either positive or negative in Ω'. The inequalities (3.34) allow also to apply the regularity result (see Theorem 1.11) to prove that $u_1 \in C^{1,\alpha}(\Omega')$ with a certain $\alpha = \alpha(\Omega') \in (0, 1)$ if a and b are smooth enough. Since Ω' was arbitrary it follows from these considerations that we can choose $u_1 > 0$ in Ω and u_1 locally in $C^{1,\alpha}(\overline{\Omega})$. Let us note that the inequalities (3.34) are fulfilled if e.g. ν is a weight of the type *power of the distance from the boundary.*

3.3 Existence of the least eigenvalue of the nonhomogeneous eigenvalue problem

In this section we will consider the *nonhomogeneous eigenvalue problem*

$$\begin{aligned} -\operatorname{div}(a(x,u)|\nabla u|^{p-2}\nabla u) &= \lambda b(x,u)|u|^{p-2}u \quad \text{in} \quad \Omega\,, \\ u &= 0 \quad \text{on} \quad \partial\Omega\,. \end{aligned} \tag{3.35}$$

Let $g\colon [0,\infty) \to [1,\infty)$ be a nondecreasing function, $c > 0$ a constant and let $b(x)$ satisfy the same assumptions as in Section 3.2. We assume that $a(x,s)$, $b(x,s) \in \mathrm{CAR}$ and

$$\frac{\nu(x)}{c_1} \le a(x,s) \le c_1 g(|s|)\nu(x)\,, \tag{3.36}$$

$$0 \le b(x,s) \le b(x) + c|s|^{q-p} \tag{3.37}$$

hold for a.e. $x \in \Omega$ and for all $s \in \mathbb{R}$.

Moreover, assume that

$$\operatorname{meas}\{x \in \Omega;\ b(x,v(x)) > 0\} > 0 \tag{3.38}$$

for any $v \in L^q(\Omega)$. (Note that the condition (3.38) is fulfilled e.g. if $b(x,s) > 0$ for a.e. $x \in \Omega$ and for all $s \neq 0$.)

Definition 3.2 We will say that $\lambda \in \mathbb{R}$ is an *eigenvalue* of the eigenvalue problem (3.35) if there exists $u \in X$, $u \not\equiv 0$, such that

$$\int_\Omega a(x,u(x))|\nabla u|^{p-2}\nabla u\nabla\varphi\,dx = \lambda\int_\Omega b(x,u(x))|u|^{p-2}u\varphi\,dx \tag{3.39}$$

holds for any $\varphi \in X$. Then u is called an *eigenfunction* corresponding to the eigenvalue λ.

Proposition 3.2 *Let $u \in L^\infty(\Omega)$, q as at the beginning of Section 3.2, let $\|u\|_q = R > 0$, $u \ge 0$ be any eigenfunction of (3.35) corresponding to the eigenvalue λ. Then there exists $d(R) > 0$ (independent of q) such that $\|u\|_\infty \le d(R)$.*

Proof. Choose $\varphi = u^{\kappa p+1}$ in (3.39) with $\kappa \ge 0$. We obtain

$$(\kappa p+1)\int_\Omega a(x,u(x))u^{\kappa p}|\nabla u|^p\,dx = \lambda\int_\Omega b(x,u(x))u^{(\kappa+1)p}\,dx\,,$$

i.e.

$$\frac{\kappa p+1}{(\kappa+1)^p}\int_\Omega a(x,u(x))|\nabla(u^{\kappa+1})|^p\,dx = \lambda\int_\Omega b(x,u(x))u^{(\kappa+1)p}\,dx\,. \tag{3.40}$$

Now, the proof follows the lines of that of Lemma 3.2 using the assumptions (3.36), (3.37), (3.38) instead of (3.8), (3.9), (3.10). □

We will now define a problem equivalent to (3.35). Let $R > 0$ and $d = d(R) > 0$ be as in Proposition 3.2. We define

$$\tilde{a}(x,s) = \begin{cases} a(x,s) & \text{for } x \in \Omega,\ |s| \le d(R), \\ a(x,d(R)) & \text{for } x \in \Omega,\ s > d(R), \\ a(x,-d(R)) & \text{for } x \in \Omega,\ s < -d(R). \end{cases} \tag{3.41}$$

Let us consider the nonhomogeneous eigenvalue problem

$$\begin{aligned} -\operatorname{div}(\tilde{a}(x,u)|\nabla u|^{p-2}\nabla u) &= \lambda b(x,u)|u|^{p-2}u \quad \text{in} \quad \Omega\,, \\ u &= 0 \quad \text{on} \quad \partial\Omega\,. \end{aligned} \tag{3.42}$$

Then it follows from Proposition 3.2 that $u \in X \cap L^\infty(\Omega)$, $\|u\|_q = R$, $u \ge 0$ is an eigenfunction of (3.42) *if and only if* it is an eigenfunction of (3.35).

We will apply the Schauder fixed point theorem. The original idea comes from L. Boccardo [9]. For a given $v \in L^q(\Omega)$ set $a_v(x) = \tilde{a}(x, v(x))$, $b_v(x) = b(x, v(x))$. It follows from (3.36), (3.37), (3.38) and (3.41) that $a_v(x)$ and $b_v(x)$ fulfil (3.8), (3.9), (3.10) for any fixed $v \in L^q(\Omega)$, $v \ne 0$. Let us consider the homogeneous eigenvalue problem

$$\begin{aligned} -\operatorname{div}(a_v(x)|\nabla u|^{p-2}\nabla u) &= \lambda b_v(x)|u|^{p-2}u \quad \text{in} \quad \Omega\,, \\ u &= 0 \quad \text{on} \quad \partial\Omega \end{aligned} \tag{3.43}$$

for any fixed $v \in L^q(\Omega)$. Due to the results of Section 3.2 there exists the *least* eigenvalue $\lambda_v > 0$ of (3.43) and *precisely one* corresponding eigenfunction u_v such that $u_v \ge 0$ a.e. in Ω, $u_v \in L^\infty(\Omega)$ and $\|u_v\|_q = R$. Hence we can define the *operator*

$$\mathcal{S}\colon L^q(\Omega) \to L^q(\Omega)$$

which associates to $v \in L^q(\Omega)$ the first nonnegative eigenfunction u_v of (3.43) such that $\|u_v\|_q = R$.

Let us assume for a moment that $\mathcal{S}$ is a *compact operator* (see Proposition 3.3 below). Since it maps the ball $B_R = \{u \in L^q(\Omega)\,,\ \|u\|_q \le R\}$ into itself it follows from the Schauder fixed point theorem that $\mathcal{S}$ has a *fixed point* $u \in B_R$. Hence there exists $\lambda_u > 0$ such that

$$\begin{aligned} -\operatorname{div}(a_u(x)|\nabla u|^{p-2}\nabla u) &= \lambda_u b_u(x)|u|^{p-2}u \quad \text{in} \quad \Omega\,, \\ u &= 0 \quad \text{on} \quad \partial\Omega\,, \end{aligned}$$

and it follows from the above considerations that $\lambda_u > 0$ is the least eigenvalue of (3.35) and $u \in L^\infty(\Omega)$, $u \ge 0$ a.e. in Ω, is the corresponding eigenfunction satisfying $\|u\|_q = R$.

The *main result of this section* is the following assertion.

Theorem 3.2 *Let the assumptions from the beginning of this section be fulfilled. Then for a given real number $R > 0$ there exists the least eigenvalue $\lambda > 0$ and the corresponding eigenfunction $u \in X \cap L^{\infty}(\Omega)$ of the nonhomogeneous eigenvalue problem* (3.35) *such that $u \geq 0$ a.e. in Ω and $\|u\|_q = R$.*

This theorem does not guarantee the simplicity of λ and also the existence of changing sign eigenfunctions is not excluded. Theorem 3.2 follows from the considerations presented above if we prove the compactness of the operator $\mathcal{S}$.

For this purpose, let us define the Nemytskij operators

$$G_1 : u \mapsto |u|^{p-2}u\,, \quad G_2 : u \mapsto |u|^p\,, \quad G_3 : u \mapsto b(x, u(x))\,.$$

Then G_i is a bounded and continuous operator from $L^q(\Omega)$ into $L^{\frac{q}{p-1}}(\Omega)$ for $i = 1$, from $L^q(\Omega)$ into $L^{\frac{q}{p}}(\Omega)$ for $i = 2$, and from $L^q(\Omega)$ into $L^{\frac{q}{q-p}}(\Omega)$ for $i = 3$ (see Chapter 1). The Nemytskij operator

$$G_4 : (u, z_1, \dots, z_n) \mapsto \tilde{a}(x, u(x))(z_1^2(x) + \dots + z_n^2(x))^{\frac{p-1}{2}}$$

is bounded and continuous from

$$L^q(\Omega) \times L^p(\Omega, \nu) \times \dots \times L^p(\Omega, \nu) \quad \text{into} \quad L^{\frac{p}{p-1}}(\Omega, \nu^{-\frac{1}{p-1}})$$

(see Section 1.3). Let us now present two technical lemmas. The proof of the first lemma follows the lines of J.-L. Lions [48], Chapter 2, where an analogous assertion in a nonweighted space is proved.

Lemma 3.3 *Let $J : X \to X^*$ be an operator defined by*

$$\langle J(u), \varphi \rangle = \int_{\Omega} a(x) |\nabla u|^{p-2} \nabla u \nabla \varphi \, dx$$

for any $u, \varphi \in X$ (here $\langle \cdot, \cdot \rangle$ denotes the duality between X^* and X, and $a(x)$ is as in Section 3.2*). Then J is surjective and $J^{-1} : X^* \to X$ is bounded and continuous.*

Proof. The operator J is bounded, strictly monotone, continuous and coercive. Then it follows from the Browder theorem (see, e.g., S. Fučík, A. Kufner [27]) that J is surjective. It follows from the Hölder inequality that

$$\langle J(v) - J(u), v - u \rangle \geq (\|v\|_a^{p-1} - \|u\|_a^{p-1})(\|v\|_a - \|u\|_a) \tag{3.44}$$

for any $u, v \in X$. The boundedness of J^{-1} follows immediately from (3.44). The continuity of J^{-1} will be proved via contradiction. Let us suppose that J^{-1} is not continuous. Then there exists a sequence $\{f_n\}$ such that $f_n \to f$ in X^* and

$\|J^{-1}(f_n) - J^{-1}(f)\|_a \geq \delta$ for some $\delta > 0$. Denote $u_n = J^{-1}(f_n)$, $u = J^{-1}(f)$. It follows from (3.44) that

$$\|f_n\|_{X^*}\|u_n\|_a \geq \langle f_n, u_n\rangle = \langle J(u_n), u_n\rangle \geq \|u_n\|_a^p ,$$

i.e.

$$\|u_n\|_a^{p-1} \leq \|f_n\|_{X^*}$$

($\|\cdot\|_{X^*}$ denotes the norm in the dual space X^*). Then $\{u_n\}$ is bounded in X and we can assume that there exists $\tilde{u} \in X$ such that $u_n \rightharpoonup \tilde{u}$ in X. Hence we have

$$\langle J(u_n)-J(\tilde{u}), u_n-\tilde{u}\rangle = \langle J(u_n)-J(u), u_n-\tilde{u}\rangle+\langle J(u)-J(\tilde{u}), u_n-\tilde{u}\rangle \to 0 \quad (3.45)$$

since $J(u_n) \to J(u)$ in X^*. It follows from (3.44) (where we set $v = u_n$, $u = \tilde{u}$), (3.45) that $\|u_n\|_a \to \|\tilde{u}\|_a$. The uniform convexity of X equipped with the norm $\|\cdot\|_a$ implies $u_n \to \tilde{u}$ in X. This convergence together with the convergence $J(u_n) \to J(u)$ in X^* implies $\tilde{u} = u$, which is a contradiction. The continuity of J^{-1} is proved. □

Lemma 3.4 *Let $z, z_n \in X$ satisfy*

$$\int_\Omega a_v(x)|\nabla z|^{p-2}\nabla z\nabla\varphi\, dx = \langle f, \varphi\rangle ,$$

$$\int_\Omega a_{v_n}(x)|\nabla z_n|^{p-2}\nabla z_n\nabla\psi\, dx = \langle f_n, \psi\rangle$$

for any $\varphi, \psi \in X$, and let $v_n \to v$ in $L^q(\Omega)$, $f_n \to f$ in X^*. Then $z_n \to z$ in X.

Proof. Define operators $J, J_n\colon X \to X^*$ by

$$\langle J(u), \varphi\rangle = \int_\Omega a_v(x)|\nabla u|^{p-2}\nabla u\nabla\varphi\, dx ,$$

$$\langle J_n(u), \psi\rangle = \int_\Omega a_{v_n}(x)|\nabla u|^{p-2}\nabla u\nabla\psi\, dx$$

for any $\varphi, \psi, u \in X$. Hence $J(z) = f$ and $J_n(z_n) = f_n$.

Let $n \in \mathbb{N}$ be fixed. Consider the equation

$$J_n(u) = h .$$

It implies

$$\int_\Omega a_{v_n}(x)|\nabla u|^p\, dx = \langle h, u\rangle ,$$

$$\|u\|_X^p \leq c_2\|h\|_{X^*}\|u\|_X , \qquad \|J_n^{-1}(h)\|_X \leq c_2^{\frac{1}{p-1}}\|h\|_{X^*}^{\frac{1}{p-1}} \quad (3.46)$$

for any $h \in X^*$, where $c_2 > 0$ is independent of n and h.

Analogously

$$\|J^{-1}(h)\|_X \le c_2^{\frac{1}{p-1}} \|h\|_{X^*}^{\frac{1}{p-1}} \tag{3.47}$$

(cf. Lemma 3.3). Applying Lemma 3.3 to $a(x) := a_v(x)$ we obtain continuity of J^{-1}.

Assume that $\{u_n\}$ is a sequence satisfying $u_n \to z$ in X. It follows from the continuity of the Nemytskij operator G_4 that

$$\begin{aligned}
&\|J_n(u_n) - J(u_n)\|_{X^*} \\
&\quad = \sup_{\|\varphi\|_X \le 1} |\langle J_n(u_n) - J(u_n), \varphi\rangle| \\
&\quad = \sup_{\|\varphi\|_X \le 1} \left| \int_\Omega (a_{v_n}(x) - a_v(x)) |\nabla u_n|^{p-2} \nabla u_n \nabla\varphi \, dx \right| \\
&\quad \le \sup_{\|\varphi\|_X \le 1} \left| \int_\Omega [a_{v_n}(x) |\nabla u_n|^{p-2} \nabla u_n - a_v(x) |\nabla z|^{p-2} \nabla z] \nabla\varphi \, dx \right| \\
&\qquad + \sup_{\|\varphi\|_X \le 1} \left| \int_\Omega [a_v(x) |\nabla z|^{p-2} \nabla z - a_v(x) |\nabla u_n|^{p-2} \nabla u_n] \nabla\varphi \, dx \right| \\
&\quad \le \sup_{\|\varphi\|_X \le 1} \left(\int_\Omega \nu(x)^{-\frac{1}{p-1}} |a_{v_n}(x) |\nabla u_n|^{p-2} \nabla u_n \right. \\
&\qquad \left. - a_v(x) |\nabla z|^{p-2} \nabla z|^{\frac{p}{p-1}} \, dx \right)^{\frac{p-1}{p}} \left(\int_\Omega \nu(x) |\nabla\varphi|^p \, dx \right)^{\frac{1}{p}} \\
&\qquad + \sup_{\|\varphi\|_X \le 1} \left(\int_\Omega \nu(x)^{-\frac{1}{p-1}} |a_v(x) |\nabla z|^{p-2} \nabla z \right. \\
&\qquad \left. - a_v(x) |\nabla u_n|^{p-2} \nabla u_n|^{\frac{p}{p-1}} \, dx \right)^{\frac{p-1}{p}} \left(\int_\Omega \nu(x) |\nabla\varphi|^p \, dx \right)^{\frac{1}{p}} \to 0
\end{aligned} \tag{3.48}$$

for $n \to \infty$.

Set $u_n = J^{-1}(f_n)$. Then the assumptions of Lemma 3.4 and continuity of J^{-1} imply

$$u_n \to z \quad \text{in} \quad X. \tag{3.49}$$

The relations (3.46)–(3.49) and continuity of J^{-1} now yield

$$\begin{aligned}
\|z_n - z\|_X &\le \|J_n^{-1}(f_n) - J^{-1}(f_n)\|_X + \|J^{-1}(f_n) - J^{-1}(f)\|_X \\
&\le \|J_n^{-1}(J_n - J) J^{-1}(f_n)\|_X + \|J^{-1}(f_n) - J^{-1}(f)\|_X \\
&\le c_2^{\frac{1}{p-1}} \|J_n(u_n) - J(u_n)\|_{X^*}^{\frac{1}{p-1}} + \|J^{-1}(f_n) - J^{-1}(f)\|_X \to 0
\end{aligned}$$

for $n \to \infty$, which completes the proof. □

The following assertion actually completes the proof of Theorem 3.2.

Proposition 3.3 *The operator $\mathcal{S}: L^q(\Omega) \to L^q(\Omega)$ introduced on page 123 is compact.*

Proof. We prove that $\mathcal{S}$ is a continuous operator from $L^q(\Omega)$ into X. The assertion then follows from the compact imbedding $X \hookrightarrow\hookrightarrow L^q(\Omega)$ (see Section 3.2). Let $u_{v_n} = \mathcal{S}(v_n)$, $u_v = \mathcal{S}(v)$. Suppose to the contrary that $v_n \to v$ in $L^q(\Omega)$ and

$$\|u_{v_n} - u_v\|_X \geq \delta \tag{3.50}$$

for some $\delta > 0$. We have

$$\int_\Omega a_v(x)|\nabla u_v|^{p-2}\nabla u_v \nabla\varphi\, dx = \lambda_v \int_\Omega b_v(x)|u_v|^{p-2}u_v\varphi\, dx\,, \tag{3.51}$$

$$\int_\Omega a_{v_n}(x)|\nabla u_{v_n}|^{p-2}\nabla u_{v_n} \nabla\psi\, dx = \lambda_{v_n} \int_\Omega b_{v_n}(x)|u_{v_n}|^{p-2}u_{v_n}\psi\, dx \tag{3.52}$$

for any $\varphi, \psi \in X$. It follows from Lemma 3.3 that for any $v_n \in L^q(\Omega)$ there exists $z_n \in X$ such that

$$\int_\Omega a_{v_n}(x)|\nabla z_n|^{p-2}\nabla z_n \nabla\varphi\, dx = \lambda_v \int_\Omega b_v(x)|u_v|^{p-2}u_v\varphi\, dx \tag{3.53}$$

for any $\varphi \in X$. Lemma 3.4 yields $z_n \to u_v$ in X (and hence also in $L^q(\Omega)$). Applying the Hölder inequality, (3.37) and the Minkowski inequality, we obtain

$$\begin{aligned}
&\left|\int_\Omega b(x, v(x))|u_v|^{p-2}u_v(z_n - u_v)\, dx\right| \\
&\leq \left(\int_\Omega (b(x, v(x)))^{\frac{q}{q-1}} |u_v|^{\frac{q(p-1)}{q-1}}\, dx\right)^{\frac{q-1}{q}} \left(\int_\Omega |z_n - u_v|^q\, dx\right)^{\frac{1}{q}} \\
&\leq \left(\int_\Omega (b(x, v(x)))^{\frac{q}{q-p}}\, dx\right)^{\frac{q-p}{q}} \left(\int_\Omega |u_v|^q\, dx\right)^{\frac{p-1}{q}} \left(\int_\Omega |z_n - u_v|^q\, dx\right)^{\frac{1}{q}} \\
&\leq \left[\left(\int_\Omega b(x)^{\frac{q}{q-p}}\, dx\right)^{\frac{q-p}{q}} + c\left(\int_\Omega |v(x)|^q\, dx\right)^{\frac{q-p}{q}}\right] \\
&\quad \cdot \left(\int_\Omega |u_v|^q\, dx\right)^{\frac{p-1}{q}} \left(\int_\Omega |z_n - u_v|^q\, dx\right)^{\frac{1}{q}} \to 0
\end{aligned} \tag{3.54}$$

for $n \to \infty$. Applying the Hölder inequality, (3.37), the Minkowski inequality and the continuity of the Nemytskij operators G_2, G_3 we obtain

$$\begin{aligned}
&\left|\int_\Omega [b(x, v_n(x))|z_n|^p - b(x, v(x))|u_v|^p]\,dx\right| \\
&\quad\le \left|\int_\Omega b(x, v_n(x))[|z_n|^p - |u_v|^p]\,dx\right| \\
&\qquad + \left|\int_\Omega [b(x, v_n(x)) - b(x, v(x))]|u_v|^p]\,dx\right| \\
&\quad\le \left[\left(\int_\Omega b(x)^{\frac{q}{q-p}}\,dx\right)^{\frac{q-p}{q}} + c\left(\int_\Omega |v_n(x)|^q\,dx\right)^{\frac{q-p}{q}}\right] \\
&\qquad \cdot \left(\int_\Omega ||z_n|^p - |u_v|^p|^{\frac{q}{p}}\,dx\right)^{\frac{p}{q}} \\
&\qquad + \left(\int_\Omega |b(x, v_n(x)) - b(x, v(x))|^{\frac{q}{q-p}}\,dx\right)^{\frac{q-p}{q}} \left(\int_\Omega |u_v|^q\,dx\right)^{\frac{p}{q}} \to 0
\end{aligned} \tag{3.55}$$

for $n \to \infty$. It follows from the variational characterization of λ_{v_n}, (3.51)–(3.55) that

$$\begin{aligned}
\lambda_{v_n} \le \frac{\int_\Omega a_{v_n}(x)|\nabla z_n|^p\,dx}{\int_\Omega b_{v_n}(x)|z_n|^p\,dx} &= \frac{\lambda_v \int_\Omega b_v(x)|u_v|^{p-2}u_v z_n\,dx}{\int_\Omega b_{v_n}(x)|z_n|^p\,dx} \\
&\to \lambda_v \frac{\int_\Omega b_v(x)|u_v|^p\,dx}{\int_\Omega b_v(x)|u_v|^p\,dx} = \lambda_v\,.
\end{aligned}$$

Hence

$$\limsup_{n\to\infty} \lambda_{v_n} \le \lambda_v\,. \tag{3.56}$$

Applying the Hölder inequality, the Minkowski inequality and the assumptions (3.36), (3.37) we obtain from (3.52) (with $\psi = u_{v_n}$) that

$$\begin{aligned}
\frac{1}{c_1}\|u_{v_n}\|_X^p &\le \int_\Omega a_{v_n}(x)|\nabla u_{v_n}|^p\,dx = \lambda_{v_n}\int_\Omega b_{v_n}(x)|u_{v_n}|^p\,dx \\
&\le \lambda_{v_n}\left[\left(\int_\Omega |b(x)|^{\frac{q}{q-p}}\,dx\right)^{\frac{q-p}{q}} + c\left(\int_\Omega |v_n(x)|^q\,dx\right)^{\frac{q-p}{q}}\right] \\
&\quad \cdot \left(\int_\Omega |u_{v_n}|^q\,dx\right)^{\frac{p}{q}}.
\end{aligned} \tag{3.57}$$

It follows from the assumption $\|u_{v_n}\|_q = R$, from $v_n \to v$ in $L^q(\Omega)$ and from (3.57) that

$$\|u_{v_n}\|_X \le c_3 \tag{3.58}$$

for any $n \in \mathbb{N}$. Due to (3.58) we have

$$u_{v_n} \rightharpoonup u \quad \text{in} \quad X \tag{3.59}$$

(at least for a subsequence) with some $u \in X$ and hence $u_{v_n} \to u$ in $L^q(\Omega)$. The Hölder inequality, the Minkowski inequality, (3.37) and the continuity of the Nemytskij operators G_1 and G_3 imply

$$\begin{aligned}
&\left|\int_\Omega (b(x, v_n(x))|u_{v_n}|^{p-2}u_{v_n} - b(x, v(x))|u|^{p-2}u)\varphi\, dx\right| \\
&\quad \le \left|\int_\Omega (b(x, v_n(x)) - b(x, v(x)))|u_{v_n}|^{p-2}u_{v_n}\varphi\, dx\right| \\
&\qquad + \left|\int_\Omega b(x, v(x))(|u_{v_n}|^{p-2}u_{v_n} - |u|^{p-2}u)\varphi\, dx\right| \\
&\quad \le \left(\int_\Omega |b(x, v_n(x)) - b(x, v(x))|^{\frac{q}{q-p}}\, dx\right)^{\frac{q-p}{q}} \\
&\qquad \cdot \left(\int_\Omega |u_{v_n}|^q\, dx\right)^{\frac{p-1}{q}} \left(\int_\Omega |\varphi|^q\, dx\right)^{\frac{1}{q}} \\
&\qquad + \left[\left(\int_\Omega |b(x)|^{\frac{q}{q-p}}\, dx\right)^{\frac{q-p}{q}} + c\left(\int_\Omega |v(x)|^q\, dx\right)^{\frac{q-p}{q}}\right] \\
&\qquad \cdot \left(\int_\Omega ||u_{v_n}|^{p-2}u_{v_n} - |u|^{p-2}u|^{\frac{q}{p-1}}\, dx\right)^{\frac{p-1}{q}} \left(\int_\Omega |\varphi|^q\, dx\right)^{\frac{1}{q}} \to 0
\end{aligned} \tag{3.60}$$

for any $\varphi \in X$. Passing to suitable subsequences we can assume that

$$\lambda_{v_n} \to \lambda \in [0, \lambda_v] \tag{3.61}$$

(see (3.56)).

Let $\bar{u} \in X$ be the unique solution of

$$\int_\Omega a_v(x)|\nabla\bar{u}|^{p-2}\nabla\bar{u}\nabla\varphi\, dx = \lambda\int_\Omega b_v(x)|\bar{u}|^{p-2}\bar{u}\varphi\, dx \tag{3.62}$$

for any $\varphi \in X$ (Lemma 3.3 guarantees the existence of $\bar{u}$). It follows from (3.52), (3.60)–(3.62) and from Lemma 3.4 that

$$u_{v_n} \to \bar{u} \quad \text{in} \quad X\,. \tag{3.63}$$

Now (3.59), (3.63) imply $u = \bar{u}$ and $u_{v_n} \to u$ in X. Hence we have

$$\lambda_v \ge \lambda = \frac{\int_\Omega a_v(x)|\nabla u|^p\, dx}{\int_\Omega b_v(x)|u|^p\, dx} \ge \inf_{\substack{\tilde{u}\neq 0\\ \tilde{u}\in X}} \frac{\int_\Omega a_v(x)|\nabla\tilde{u}|^p\, dx}{\int_\Omega b_v(x)|\tilde{u}|^p\, dx} = \frac{\int_\Omega a_v(x)|\nabla u_v|^p\, dx}{\int_\Omega b_v(x)|u_v|^p\, dx} = \lambda_v.$$

It follows that $\lambda = \lambda_v$ and $u = u_v$ (note that in Section 3.2 we proved that $u_v \geq 0$ satisfying $\|u_v\|_q = R$ is unique).

In particular, this means that

$$u_{v_n} \to u_v \quad \text{in} \quad X\,,$$

which contradicts (3.50). This completes the proof of Proposition 3.3. □

Remark 3.3 Similarly as we have mentioned in Section 3.2, the proofs work in the same way with $L^\infty(\Omega)$ instead of $L^{\frac{q}{q-p}}(\Omega)$ in the case $q = p$. Hence we obtain the following *special version* of Theorem 3.2.

Theorem 3.3 *Let* (3.36)–(3.38) *be fulfilled with* $b \in L^\infty(\Omega)$ *and* $q = p$. *Then for a given real number* $R > 0$ *there exists the least eigenvalue* $\lambda > 0$ *and the corresponding eigenfunction* $u \in X \cap L^\infty(\Omega)$ *of* (3.35) *such that* $u \geq 0$ *a.e. in* Ω *and* $\|u\|_p = R$.

Remark 3.4 Since the eigenvalue problem (3.43) is homogeneous, we can define the operator $\tilde{S}\colon L^q(\Omega) \to L^q(\Omega)$ which associates to $v \in L^q(\Omega)$ the first nonpositive eigenfunction $-u_v$ of (3.43) such that $\|-u_v\|_q = R$. It is clear from the above considerations that $\tilde{S}$ has the *same properties* as S. Hence repeating the same argument we can prove the following dual version of Theorem 3.2.

Theorem 3.4 *Let the assumptions of Theorem* 3.2 *be fulfilled. Then for a given real number* $R > 0$ *there exists the least eigenvalue* $\tilde{\lambda} > 0$ *and the corresponding eigenfunction* $\tilde{u} \in X \cap L^\infty(\Omega)$ *of the nonhomogeneous eigenvalue problem* (3.35) *such that* $\tilde{u} \leq 0$ *a.e. in* Ω *and* $\|\tilde{u}\|_q = R$.

Remark 3.5 Let λ and $\tilde{\lambda}$ be the least eigenvalues guaranteed by Theorems 3.2 and 3.4, respectively, for a given fixed $R > 0$. Then $\lambda \neq \tilde{\lambda}$ is possible due to the fact that the eigenvalue problem (3.35) is in general not homogeneous. In particular, dealing with the nonhomogeneous problem we have to understand the notion of the least eigenvalue in a different way than in the homogeneous case. Its value depends not only on $R > 0$ but also on the sign of the corresponding eigenfunction (i.e. we have to distinguish between the least eigenvalue corresponding to the positive eigenfunction and the least eigenvalue corresponding to the negative eigenfunction).

Remark 3.6 If the weight function ν is of "the power type" in the sense of Remark 3.2 then the eigenfunction in Theorem 3.2 (and 3.3) is *positive* in Ω and *locally* in $C^{1,\alpha}(\overline{\Omega})$ (cf. Section 3.2).

Example 3.1 Let Ω be a bounded domain in $\mathbb{R}^N$, $p > 1$, let $\nu(x)$ be positive and measurable in Ω satisfying $\nu \in L^1_{\text{loc}}(\Omega)$, $\nu^{-s} \in L^1(\Omega)$ for $s > \max\left\{\frac{N}{p}, \frac{1}{p-1}\right\}$. Consider the eigenvalue problem

$$\begin{aligned} -\operatorname{div}(\nu(x)e^{u^2}|\nabla u|^{p-2}\nabla u) = \lambda |u|^{p-2}u \quad \text{in} \quad \Omega\,, \\ u = 0 \quad \text{on} \quad \partial\Omega\,. \end{aligned} \tag{3.64}$$

In this case we have

$$a(x,s) = \nu(x)e^{s^2}, \quad b(x,s) \equiv 1, \quad g(s) = e^{s^2}$$

for a.e. $x \in \Omega$ and for all $s \in \mathbb{R}$.

It follows from Theorem 3.3 that *for any given real number $R > 0$ there exists the least eigenvalue $\lambda > 0$ and the corresponding eigenfunction $u \in X \cap L^\infty(\Omega)$ of* (3.64) *such that $u \geq 0$ a.e. in Ω and $\|u\|_p = R$.*

Analogously we can apply Theorem 3.4 to get the dual assertion. ◇

Example 3.2 Let us consider the plane domain $\Omega = (-1,1) \times (-1,1)$. For $x = (x_1, x_2) \in \Omega$ set

$$\nu(x) = \begin{cases} 1, & x_1 \leq 0, \\ x_2^\eta (1-x_1)^\gamma, & x_1 > 0, x_2 > 0, \\ |x_2|^\mu (1-x_1)^\gamma, & x_1 > 0, x_2 < 0 \end{cases}$$

with η, μ, γ real numbers. Consider the eigenvalue problem

$$\begin{aligned} -\operatorname{div}(\nu(x)(1+u^4)|\nabla u|^2 \nabla u) = \lambda u^9 \quad \text{in} \quad \Omega, \\ u = 0 \quad \text{on} \quad \partial\Omega. \end{aligned} \tag{3.65}$$

In this case we have $p = 4$,

$$a(x,s) = \nu(x)(1+s^4), \quad b(x,s) = s^6, \quad g(s) = 1 + s^4$$

for a.e. $x \in \Omega$ and for all $s \in \mathbb{R}$. Thus the principal part of the differential operator has a *degeneration* (or *singularity*) which is concentrated on a part Γ of the boundary $\partial\Omega$,

$$\Gamma = \{x = (x_1, x_2)\,;\ x_1 = 1\,,\ x_2 \in (-1,1)\},$$

as well as on a segment S in the interior of Ω,

$$S = \{x = (x_1, x_2)\,;\ x_1 \in (0,1)\,,\ x_2 = 0\}.$$

Condition (3.2) indicates that we have to choose η and μ from the interval $(-1, 3)$ with no condition on γ. Let us assume that

$$\eta, \mu \in \left(-1, \frac{4}{3}\right), \qquad \gamma \in \left(-\infty, \frac{4}{3}\right). \tag{3.66}$$

It follows from (3.66) that $\nu^{-1} \in L^{\frac{3}{4}}(\Omega)$ and $p_s^* = 12$ (see Section 3.2). Hence the growth condition (3.37) is fulfilled e.g. with $q = 10$. Applying Theorem 3.2 and Remark 3.6 we have the following assertion.

Let us assume (3.66). *Then for a given real number* $R > 0$ *there exists the least eigenvalue* $\lambda > 0$ *and the corresponding eigenfunction* $u \in W_0^{1,4}(\nu, \Omega) \cap L^\infty(\Omega)$ *of* (3.65) *such that* $u > 0$ *in* Ω *and* $\|u\|_{10} = R$. *Moreover,* u *is locally in* $C^{1,\alpha}(\overline{\Omega})$.

Note that for η, μ and γ *positive* we have *degenerations* of the same extent at Γ and S. On the other hand, the *singularity* can occur in a limited extent at S (for η or μ *negative*, but bigger than -1), but big enough at Γ (for any $\gamma < 0$). Note also that applying Theorem 3.4 and Remark 3.6 we get a dual assertion concerning the existence of negative eigenfunction. ◇

3.4 Maximum principle for degenerated (singular) equations

In this and the following section we will assume, for simplicity, that $q = p$, i.e. (3.36)–(3.38) are fulfilled with $b \in L^\infty(\Omega)$ (cf. Remark 3.3). Due to Theorem 3.3 we can associate to every $R > 0$ the least eigenvalue λ_R of (3.35) and so we have a function

$$R \mapsto \lambda_R$$

mapping $(0, \infty)$ into $(0, \infty)$. Using the assumptions (3.36) and (3.37) we prove an *estimate* of λ_R *from below* uniformly with respect to $R > 0$. Set

$$\lambda^\# = \frac{1}{c_1 c_2 \|b(x)\|_\infty},$$

where c_1 is from (3.36) and $c_2 > 0$ is the p-th power of the constant of the imbedding $X \hookrightarrow L^p(\Omega)$ i.e.

$$\|u\|_p^p \le c_2 \|u\|_X^p \tag{3.67}$$

for any $u \in X$. Let us assume that $v \in X$ is such that both integrals

$$\int_\Omega a(x, v(x))|\nabla v|^p\,dx \quad \text{and} \quad \int_\Omega b(x, v(x))|v|^p\,dx$$

are finite.

Lemma 3.5 *Let* $\lambda \le \lambda^\#$. *Then for any* $v \in X$ *we have*

$$\int_\Omega a(x, v(x))|\nabla v|^p\,dx - \lambda \int_\Omega b(x, v(x))|v|^p\,dx \ge 0. \tag{3.68}$$

Proof. The assertion is clear if $\lambda \le 0$. Let $\lambda > 0$. It follows from (3.36) that

$$\int_\Omega a(x, v(x))|\nabla v|^p \, dx \ge \frac{1}{c_1} \int_\Omega v(x)|\nabla v|^p \, dx = \frac{1}{c_1} \|v\|_X^p . \tag{3.69}$$

On the other hand, applying (3.37) and (3.67) we obtain

$$\int_\Omega b(x, v(x))|v|^p \, dx \le \|b(x)\|_\infty \int_\Omega |v|^p \, dx \le c_2 \|b(x)\|_\infty \|v\|_X^p . \tag{3.70}$$

Combining (3.69) and (3.70) we get

$$\begin{aligned} &\int_\Omega a(x, v(x))|\nabla v|^p \, dx - \lambda \int_\Omega b(x, v(x))|v|^p \, dx \\ &\qquad \ge \left(\frac{1}{c_1} - \lambda c_2 \|b(x)\|_\infty \right) \|v\|_X^p . \end{aligned} \tag{3.71}$$

Since $\lambda \le \lambda^\#$ implies $\frac{1}{c_1} - \lambda c_2 \|b(x)\|_\infty \ge 0$, the assertion follows from (3.71). □

Lemma 3.6 *Let* $\lambda^* = \inf_{R>0} \lambda_R$. *Then* $\lambda^* \ge \lambda^\#$.

Proof. Let $R > 0$ be arbitrary. Let λ_R be an eigenvalue and $u_R \in X \cap L^\infty(\Omega)$, $u_R \ge 0$, $\|u_R\|_p = R$ be the corresponding eigenfunction of (3.35). Choosing $\varphi = u_R$ as a test function in (3.39) we obtain

$$\int_\Omega a(x, u_R(x))|\nabla u_R|^p \, dx = \lambda_R \int_\Omega b(x, u_R(x))|u_R|^p \, dx . \tag{3.72}$$

The estimates (3.69) and (3.70) with v replaced by u_R yield

$$\frac{1}{c_1} \le \lambda_R c_2 \|b(x)\|_\infty , \quad \text{i.e.} \quad \lambda_R \ge \frac{1}{c_1 c_2 \|b(x)\|_\infty} = \lambda^\# .$$

Since $R > 0$ is arbitrary, we have $\lambda^* \ge \lambda^\#$. □

Let us consider BVP

$$\begin{aligned} -\operatorname{div}(a(x, u)|\nabla u|^{p-2} \nabla u) &= \vartheta b(x, u)|u|^{p-2} u + h \quad \text{in} \quad \Omega , \\ u = 0 \quad &\text{on} \quad \partial\Omega , \end{aligned} \tag{3.73}$$

with a real parameter ϑ and with $h \in L^{p'}(\Omega)$, $\frac{1}{p} + \frac{1}{p'} = 1$.

Definition 3.3 We will say that BVP (3.73) satisfies *the maximum principle* if $h \ge 0$ a.e. in Ω implies $u \ge 0$ a.e. in Ω for all possible weak solutions $u \in X$ of (3.73).

Lemma 3.7 (Sufficient condition) *Let $b(x,s) > 0$ for a.e. $x \in \Omega$ and for all $s \in \mathbb{R}$. If*

$$\vartheta < \lambda^{\#} \tag{3.74}$$

then BVP (3.73) satisfies the maximum principle.

Proof. Let $u \in X$ be the weak solution of (3.73) corresponding to $h \in L^{p'}(\Omega)$, $h \geq 0$ a.e. in Ω. Then

$$\int_\Omega a(x,u)|\nabla u|^{p-2}\nabla u \nabla \varphi \, dx = \vartheta \int_\Omega b(x,u)|u|^{p-2}u\varphi \, dx + \int_\Omega h\varphi \, dx \tag{3.75}$$

holds for any $\varphi \in X$. Choose $\varphi = u^- = \max\{-u(x), 0\}$ as a test function in (3.75). We obtain

$$-\int_\Omega a(x,u)|\nabla u^-|^p \, dx = -\vartheta \int_\Omega b(x,u)|u^-|^p \, dx + \int_\Omega hu^- \, dx \,. \tag{3.76}$$

It follows from (3.76) and from $h \geq 0$ a.e. in Ω that

$$\int_\Omega a(x,-u^-)|\nabla u^-|^p \, dx - \vartheta \int_\Omega b(x,-u^-)|u^-|^p \, dx \leq 0 \,. \tag{3.77}$$

On the other hand, it follows from Lemma 3.5 (see (3.68)) that

$$\int_\Omega a(x,-u^-)|\nabla u^-|^p \, dx - \lambda^{\#} \int_\Omega b(x,-u^-)|u^-|^p \, dx \geq 0 \,. \tag{3.78}$$

The inequalities (3.77) and (3.78) imply

$$(\lambda^{\#} - \vartheta) \int_\Omega b(x,-u^-)|u^-|^p \, dx \leq 0 \,.$$

Due to (3.74) and the assumption (3.37), (3.38) we get $u^- = 0$ a.e. in Ω, i.e. $u \geq 0$ a.e. in Ω. Hence BVP (3.73) satisfies the maximum principle. □

Lemma 3.8 (Necessary condition) *Let both $a(x,s)$ and $b(x,s)$ be even in s, i.e.*

$$a(x,s) = a(x,-s) \quad \text{and} \quad b(x,s) = b(x,-s)$$

for all $s \in \mathbb{R}$ and for a.e. $x \in \Omega$. Let BVP (3.73) satisfy the maximum principle. Then

$$\vartheta < \lambda_R$$

for any $R > 0$, i.e. in particular $\vartheta < \lambda^*$.

Proof. We will proceed via contradiction. Let us assume that there exists $R > 0$ such that $\vartheta \geq \lambda_R$ and (3.73) satisfies the maximum principle. Let u_R be the corresponding nonnegative eigenfunction of (3.35) satisfying $\|u_R\|_p = R$ (see Theorem 3.3). Set

$$h(x) = (\vartheta - \lambda_R) b(x, u_R(x)) |u_R(x)|^{p-2} u_R(x) . \tag{3.79}$$

Then $h \geq 0$ a.e. in Ω. Since both $a(x, s)$ and $b(x, s)$ are even in s, the function $-u_R$ is the weak solution of BVP (3.73) with h given by (3.79). But $u_R \not\equiv 0$ and $-u_R \leq 0$ a.e. in Ω. This contradicts the fact that BVP (3.73) satisfies the maximum principle. $\square$

Remark 3.7 Let us consider the p-Laplacian, i.e., take

$$a(x, s) \equiv 1 \quad \text{and} \quad b(x, s) \equiv 1 . \tag{3.80}$$

Then obviously

$$\lambda_R = \lambda^* = \lambda^\# = \lambda_1 \quad \text{for} \quad R > 0 ,$$

where $\lambda_1 > 0$ is the first eigenvalue of the p-Laplacian (see P. Drábek [17] and the references therein). The functions $a(x, s)$, $b(x, s)$ given by (3.80) satisfy the assumptions of Lemmas 3.7 and 3.8. It follows from here that the BVP

$$-\operatorname{div}(|\nabla u|^{p-2} \nabla u) = \vartheta |u|^{p-2} u + h \quad \text{in} \quad \Omega ,$$
$$u = 0 \quad \text{on} \quad \partial\Omega ,$$

satisfies the maximum principle *if and only if*

$$\vartheta < \lambda_1$$

(cf. J. Fleckinger, J. Hernández, de F. Thélin [26]).

3.5 Positive solutions of degenerated (singular) BVP

In this section we will prove the existence result for the BVP

$$\begin{aligned} -\operatorname{div}(a(x, u) |\nabla u|^{p-2} \nabla u) &= \lambda b(x, u) |u|^{p-2} u + f(x, u) \quad \text{in} \quad \Omega , \\ u &= 0 \quad \text{on} \quad \partial\Omega . \end{aligned} \tag{3.81}$$

We will assume that a and b are as in Section 3.4 and $f(x, s) \in \mathrm{CAR}$ satisfies the assumption

$$|f(x, s)| \leq \sigma(x) + \rho(x) |s|^{p-1} \tag{3.82}$$

for a.e. $x \in \Omega$ and for all $s \in \mathbb{R}$, where $\rho, \sigma \in L^\infty(\Omega)$, $\rho, \sigma \geq 0$.

Definition 3.4 We will say that $u \in X$ is a *weak solution* of BVP (3.81) if

$$\begin{aligned}\int_\Omega & a(x,u(x))|\nabla u|^{p-2}\nabla u \nabla\varphi\,dx \\ &= \lambda\int_\Omega b(x,u(x))|u|^{p-2}u\varphi\,dx + \int_\Omega f(x,u(x))\varphi\,dx\end{aligned} \tag{3.83}$$

holds for any $\varphi \in X$.

Theorem 3.5 *Let us assume* (3.7), (3.36), (3.37) (*with* $q = p, b \in L^\infty(\Omega)$, $c = 0$), (3.82) *and*

$$\lambda < \frac{1 - c_1c_2\|\rho\|_\infty}{c_1c_2\|b\|_\infty} (= \lambda^\#(1 - c_1c_2\|\rho\|_\infty)), \tag{3.84}$$

where c_1 *and* c_2 *are as in Section* 3.4. *Then BVP* (3.81) *has at least one weak solution* $u \in X \cap L^\infty(\Omega)$.

Proof. We will prove the assertion in three steps. In the first step we prove an apriori estimate in $L^r(\Omega)$ for any $1 \le r \le \infty$ for any possible weak solution of BVP (3.81). In the second step we find a suitable operator representation of BVP (3.81). In the third step we apply the degree theory in order to prove the existence of a weak solution.

Step 1 (Apriori estimate in $L^r(\Omega)$, $1 \le r \le \infty$) Let us suppose that $u \in X$ is a weak solution of BVP (3.81). For $\varphi \in X$ let us set formally

$$\begin{aligned}&\langle T(u), \varphi\rangle \\ &= \int_\Omega a(x,u)|\nabla u|^{p-2}\nabla u\nabla\varphi\,dx - \lambda\int_\Omega b(x,u)|u|^{p-2}u\varphi\,dx - \int_\Omega f(x,u)\varphi\,dx\,.\end{aligned}$$

Choosing $\varphi = u$ as a test function in (3.83) we obtain due to (3.36), (3.37), (3.67), (3.68) and (3.82) that

$$\begin{aligned}0 = \langle T(u), u\rangle &= \frac{\lambda}{\lambda^\#}\left(\int_\Omega a(x,u)|\nabla u|^p\,dx - \lambda^\#\int_\Omega b(x,u)|u|^p\,dx\right) \\ &\qquad + \frac{\lambda^\# - \lambda}{\lambda^\#}\int_\Omega a(x,u)|\nabla u|^p\,dx - \int_\Omega f(x,u)u\,dx \\ &\ge \frac{\lambda^\# - \lambda}{c_1\lambda^\#}\|u\|_X^p - c_2\|\rho\|_\infty\|u\|_X^p - \|\sigma\|_\infty\|u\|_1 \\ &\ge \frac{\lambda^\#(1 - c_1c_2\|\rho\|_\infty) - \lambda}{c_1\lambda^\#}\|u\|_X^p - c_2^{\frac{1}{p}}(\text{meas}\,\Omega)^{\frac{p-1}{p}}\|\sigma\|_\infty\|u\|_X\,.\end{aligned} \tag{3.85}$$

It follows from (3.84) and (3.85) that $\|u\|_X \le c_3$ for any possible weak solution u of BVP (3.81). By the imbedding (3.5) we have $\|u\|_{p_s^*} \le c_4$. In particular, we also have

$$\|u^+\|_{p_s^*} \le c_4\,. \tag{3.86}$$

For a given real $M > 0$ set

$$\begin{aligned} v_M(x) &= \min\{u(x), M\} && \text{on} \quad \Omega(u > 0) := \{x \in \Omega;\ u(x) > 0\}, \\ v_M(x) &= 0 && \text{on} \quad \{x \in \Omega,\ u(x) < 0\}. \end{aligned}$$

For a real $\kappa \geq 0$ set $\varphi = v_M^{\kappa p+1} \in X \cap L^\infty(\Omega)$ and choose this φ as a test function in (3.83). We obtain

$$\begin{aligned} &(\kappa p + 1) \int_{\Omega(u>0)} a(x,u) v_M^{\kappa p} |\nabla v_M|^p \, dx \\ &\quad = \lambda \int_{\Omega(u>0)} b(x,u) u^{p-1} v_M^{\kappa p+1} \, dx + \int_{\Omega(u>0)} f(x,u) v_M^{\kappa p+1} \, dx . \end{aligned} \tag{3.87}$$

The left-hand side of (3.87) is estimated by using (3.5), (3.36) and (3.67):

$$\begin{aligned} &(\kappa p + 1) \int_{\Omega(u>0)} a(x,u) v_M^{\kappa p} |\nabla v_M|^p \, dx \\ &\quad \geq \frac{\kappa p + 1}{c_1} \int_{\Omega(u>0)} \nu(x) v_M^{\kappa p} |\nabla v_M|^p \, dx \\ &\quad = \frac{\kappa p + 1}{c_1 (\kappa + 1)^p} \int_{\Omega(u>0)} \nu(x) |\nabla (v_M^{\kappa+1})|^p \, dx \\ &\quad \geq \frac{\kappa p + 1}{c_1 (c_2')^p (\kappa + 1)^p} \left(\int_{\Omega(u>0)} (v_M^{\kappa+1})^{p_s^*} \, dx \right)^{\frac{p}{p_s^*}} . \end{aligned} \tag{3.88}$$

The right-hand side of (3.87) is estimated by using (3.37), (3.82) and the Hölder inequality:

$$\begin{aligned} &\lambda \int_{\Omega(u>0)} b(x,u) u^{p-1} v_M^{\kappa p+1} \, dx + \int_{\Omega(u>0)} f(x,u) v_M^{\kappa p+1} \, dx \\ &\quad \leq \lambda \|b\|_\infty \int_{\Omega(u>0)} u^{(\kappa+1)p} \, dx + \|\rho\|_\infty \int_{\Omega(u>0)} u^{(\kappa+1)p} \, dx \\ &\qquad + \|\sigma\|_\infty (\operatorname{meas} \Omega)^{\frac{p-1}{(\kappa+1)p}} \left(\int_{\Omega(u>0)} u^{(\kappa+1)p} \, dx \right)^{\frac{\kappa p+1}{(\kappa+1)p}} . \end{aligned} \tag{3.89}$$

We have either

$$\left(\int_{\Omega(u>0)} u^{(\kappa+1)p} \, dx \right)^{\frac{\kappa p+1}{(\kappa+1)p}} \leq 1$$

or

$$\left(\int_{\Omega(u>0)} u^{(\kappa+1)p} \, dx \right)^{\frac{\kappa p+1}{(\kappa+1)p}} \leq \int_{\Omega(u>0)} u^{(\kappa+1)p} \, dx$$

due to the fact that $\dfrac{\kappa p+1}{(\kappa+1)p} < 1$ for any $\kappa > 0$. Then it follows from (3.87), (3.88) and (3.89) that

$$\left(\int_{\Omega(u>0)} v_M^{(\kappa+1)p_s^*}\,dx\right)^{\frac{p}{p_s^*}} \le c_5\frac{(\kappa+1)^p}{\kappa p+1}\left(\int_{\Omega(u>0)} u^{(\kappa+1)p}\,dx + c_5'\right)$$

with some $c_5, c_5' > 0$.

Using the bootstrap argument (similarly as in the proof of Lemma 3.2) starting with $(\kappa_1+1)p = p_s^*$ we get

$$\|u^+\|_{(\kappa+1)p_s^*} < c(\kappa)$$

for any finite number $\kappa > 0$. To get the uniform estimate with respect to κ we have to argue as follows. If there is a sequence $\kappa_n \to \infty$ such that

$$\int_{\Omega(u>0)} u^{(\kappa_n+1)p}\,dx \le 1\,,$$

we get $\|u^+\|_{(\kappa_n+1)p_s^*} \le 1$ and hence

$$\|u^+\|_\infty \le 1$$

(cf. the proof of Lemma 3.2). In the opposite case there is $\kappa_0 > 0$ such that

$$\int_{\Omega(u>0)} u^{(\kappa+1)p}\,dx > 1$$

for any $\kappa \ge \kappa_0$. Then we derive directly from (3.89) that

$$\left(\int_{\Omega(u>0)} v_M^{(\kappa+1)p_s^*}\,dx\right)^{\frac{p}{p_s^*}} \le \tilde{c}_5\frac{(\kappa+1)^p}{\kappa p+1}\int_{\Omega(u>0)} u^{(\kappa+1)p}\,dx$$

holds for any $\kappa \ge \kappa_0$. Then repeating the iterations from the proof of Lemma 3.2 (but starting with the exponent $(\kappa_1+1)p = (\kappa_0+1)p_s^*$ instead of p_s^*) we derive the uniform estimate

$$\|u^+\|_{(\kappa_n+1)p_s^*} \le \tilde{c}_6\|u^+\|_{(\kappa_0+1)p_s^*}$$

for $\kappa_n \to \infty$. This implies again

$$\|u^+\|_\infty \le c_6\,.$$

Similarly we can handle u^- and hence finally, due to (3.86), we have an apriori estimate in $L^\infty(\Omega)$ of any possible weak solution of BVP (3.81):

$$\|u\|_\infty \le d \tag{3.90}$$

with some $d > 0$ (which is independent of u).

Step 2 (Operator representation of BVP (3.81)) Let us define the function

$$\tilde{a}(x,s) = \begin{cases} a(x,s) & \text{for } x \in \Omega,\ |s| \le d, \\ a(x,d) & \text{for } x \in \Omega,\ s > d, \\ a(x,-d) & \text{for } x \in \Omega,\ s < -d, \end{cases} \tag{3.91}$$

where $d > 0$ is the real number from (3.90). Then due to Step 1 the function $u \in X$ is a weak solution of BVP (3.81) if and only if u satisfies (3.90) and it is a weak solution of BVP

$$\begin{aligned} -\operatorname{div}(\tilde{a}(x,u)|\nabla u|^{p-2}\nabla u) &= \lambda b(x,u)|u|^{p-2}u + f(x,u) \quad \text{in} \quad \Omega, \\ u &= 0 \quad \text{on} \quad \partial\Omega. \end{aligned} \tag{3.92}$$

Define operators

$$A, B, F\colon X \to X^*$$

in the following way:

$$\langle A(u), \varphi\rangle = \int_\Omega \tilde{a}(x,u)|\nabla u|^{p-2}\nabla u \nabla\varphi\, dx,$$

$$\langle B(u), \varphi\rangle = \int_\Omega b(x,u)|u|^{p-2}u\varphi\, dx,$$

$$\langle F(u), \varphi\rangle = \int_\Omega f(x,u)\varphi\, dx$$

for any $u, \varphi \in X$. Then (3.36), (3.37), (3.82) and (3.91) guarantee that A, B and F are well defined operators. Set

$$\tilde{T}(u) = A(u) - \lambda B(u) - F(u)$$

for any $u \in X$. Then $u \in X$ is a weak solution of BVP (3.92) if and only if u is a solution of the operator equation

$$\tilde{T}(u) = 0. \tag{3.93}$$

Step 3 (Application of the degree theory) According to Section 1.3 the Nemytskij operator

$$G\colon (u, z_1, \dots, z_n) \to \tilde{a}(x,u)(z_1^2 + \cdots + z_n^2)^{\frac{p-1}{2}}$$

from $L^p(\Omega) \times L^p(\Omega, \nu) \times \cdots \times L^p(\Omega, \nu)$ into $L^{\frac{p}{p-1}}(\Omega, \nu^{-\frac{1}{p-1}})$ is continuous. Then for $u_n \to u_0$ in X we have

$$\begin{aligned} &\sup_{\|\varphi\|_X \le 1} \int_\Omega (\tilde{a}(x,u_n)|\nabla u_n|^{p-2}\nabla u_n - \tilde{a}(x,u_0)|\nabla u_0|^{p-2}\nabla u_0)\nabla\varphi\, dx \\ &\quad \le \sup_{\|\varphi\|_X \le 1} \Bigg(\int_\Omega \nu^{-\frac{1}{p-1}} |\tilde{a}(x,u_n)|\nabla u_n|^{p-2}\nabla u_n \\ &\qquad - \tilde{a}(x,u_0)|\nabla u_0|^{p-2}\nabla u_0|^{\frac{p}{p-1}}\, dx \Bigg)^{\frac{p-1}{p}} \left(\int_\Omega \nu |\nabla\varphi|^p\, dx \right)^{\frac{1}{p}} \to 0 \end{aligned}$$

for any fixed $\varphi \in X$. Hence A is a continuous operator.

We have also the following estimates for $u_n \rightharpoonup u_0$ in X:

$$\begin{aligned}
&\sup_{\|\varphi\|_X \le 1} |\langle B(u_n) - B(u_0), \varphi\rangle| \\
&\qquad = \sup_{\|\varphi\|_X \le 1} \left| \int_\Omega (b(x,u_n)|u_n|^{p-2}u_n - b(x,u_0)|u_0|^{p-2}u_0)\varphi \, dx \right| \\
&\qquad \le \sup_{\|\varphi\|_X \le 1} \left(\int_\Omega |b(x,u_n)|u_n|^{p-2}u_n - b(x,u_0)|u_0|^{p-2}u_0|^{\frac{p}{p-1}} \, dx \right)^{\frac{p-1}{p}} \\
&\qquad\qquad \cdot \left(\int_\Omega |\varphi|^p \, dx \right)^{\frac{1}{p}} \to 0
\end{aligned}$$

and similarly

$$\begin{aligned}
\sup_{\|\varphi\|_X \le 1} |\langle F(u_n) - F(u_0), \varphi\rangle| &= \sup_{\|\varphi\|_X \le 1} \left| \int_\Omega (f(x,u_n) - f(x,u_0))\varphi \, dx \right| \\
&\le \sup_{\|\varphi\|_X \le 1} \left(\int_\Omega |f(x,u_n) - f(x,u_0)|^{\frac{p}{p-1}} \, dx \right)^{\frac{p-1}{p}} \left(\int_\Omega |\varphi|^p \, dx \right)^{\frac{1}{p}} \to 0
\end{aligned}$$

due to the continuity of the Nemytskij operators

$$u \mapsto b(\cdot, u)|u|^{p-2}u$$

and

$$u \mapsto f(\cdot, u)$$

from $L^p(\Omega)$ into $L^{p'}(\Omega)$ and due to the compact imbedding $X \hookrightarrow\hookrightarrow L^p(\Omega)$. Hence B and F are compact operators. This implies that $\tilde{T}$ is a continuous and hence a demicontinuous operator. Let us prove that it satisfies condition $\alpha(X)$ (see Definition 1.1). Due to the compactness of B and F it is sufficient to prove that A satisfies condition $\alpha(X)$ (see Lemma 1.2). Let $u_n \rightharpoonup u_0$ in X, and

$$\limsup_{n\to\infty} \langle A(u_n), u_n - u_0\rangle \le 0 \,. \tag{3.94}$$

Then

$$\limsup_{n\to\infty} \langle A(u_n), u_n - u_0\rangle = \limsup_{n\to\infty} \langle A(u_n) - A(u_0), u_n - u_0\rangle \tag{3.95}$$

and

$$\begin{aligned}
&\langle A(u_n) - A(u_0), u_n - u_0\rangle \\
&\qquad = \int_\Omega (\tilde{a}(x,u_n) - \tilde{a}(x,u_0))|\nabla u_0|^{p-2}\nabla u_0(\nabla u_n - \nabla u_0) \, dx \\
&\qquad\quad + \int_\Omega \tilde{a}(x,u_n)(|\nabla u_n|^{p-2}\nabla u_n - |\nabla u_0|^{p-2}\nabla u_0)(\nabla u_n - \nabla u_0) \, dx \,.
\end{aligned} \tag{3.96}$$

The first integral approaches zero due to the continuity of the Nemytskij operator G and the compactness of the imbedding $X \hookrightarrow\hookrightarrow L^p(\Omega)$:

$$\begin{aligned}
&\left|\int_\Omega (\tilde{a}(x,u_n) - \tilde{a}(x,u_0))|\nabla u_0|^{p-2}\nabla u_0(\nabla u_n - \nabla u_0)\,dx\right| \\
&\le \Big(\int_\Omega \nu^{-\frac{1}{p-1}} |\tilde{a}(x,u_n)|\nabla u_0|^{p-2}\nabla u_0 \\
&\qquad - \tilde{a}(x,u_0)|\nabla u_0|^{p-2}\nabla u_0|^{\frac{p}{p-1}}\,dx\Big)^{\frac{p-1}{p}} \Big(\int_\Omega \nu|\nabla u_n - \nabla u_0|^p\,dx\Big)^{\frac{1}{p}} \\
&\le c_7\|G(u_n,\nabla u_0) - G(u_0,\nabla u_0)\|_{L^{\frac{p}{p-1}}(\Omega,\nu^{-\frac{1}{p-1}})} \to 0\,.
\end{aligned} \tag{3.97}$$

The second integral in (3.96) is estimated from below as follows:

$$\begin{aligned}
&\int_\Omega \tilde{a}(x,u_n)(|\nabla u_n|^{p-2}\nabla u_n - |\nabla u_0|^{p-2}\nabla u_0)(\nabla u_n - \nabla u_0)\,dx \\
&\quad\ge \frac{1}{c_1}\int_\Omega \nu(x)(|\nabla u_n|^{p-2}\nabla u_n - |\nabla u_0|^{p-2}\nabla u_0)(\nabla u_n - \nabla u_0)\,dx \\
&\quad\ge \frac{1}{c_1}\left[\Big(\int_\Omega \nu|\nabla u_n|^p\,dx\Big)^{\frac{p-1}{p}} - \Big(\int_\Omega \nu|\nabla u_0|^p\,dx\Big)^{\frac{p-1}{p}}\right] \\
&\qquad\cdot\left[\Big(\int_\Omega \nu|\nabla u_n|^p\,dx\Big)^{\frac{1}{p}} - \Big(\int_\Omega \nu|\nabla u_0|^p\,dx\Big)^{\frac{1}{p}}\right] \\
&\quad= \frac{1}{c_1}\Big(\|u_n\|_X^{p-1} - \|u_0\|_X^{p-1}\Big)\Big(\|u_n\|_X - \|u_0\|_X\Big)\,.
\end{aligned} \tag{3.98}$$

It follows from (3.94)–(3.98) that

$$0 \ge \limsup_{n\to\infty}(\|u_n\|_X^{p-1} - \|u_0\|_X^{p-1})(\|u_n\|_X - \|u_0\|_X) \ge 0\,. \tag{3.99}$$

We obtain from (3.99) that

$$\|u_n\|_X \to \|u_0\|_X$$

for $n \to \infty$.

The uniform convexity of X then implies that u_n converge strongly to u_0 in X. Thus the condition $\alpha(X)$ is satisfied.

Similarly as in (3.85) we prove that

$$\langle \tilde{T}(u), u\rangle > 0 \tag{3.100}$$

for any $u \in \partial B_R(0) = \{u \in X;\ \|u\|_X = R\}$ with $R > 0$ sufficiently large (it is sufficient to verify that $\tilde{a}(x,s)$ satisfies the same bound from below as $a(x,s)$). Hence the degree $\mathrm{Deg}[\tilde{T}; B_R(0), 0]$ is well defined (see Section 1.7) and due to (3.100) we have

$$\mathrm{Deg}[\tilde{T}; B_R(0), 0] = 1 \tag{3.101}$$

(see Theorem 1.6). The basic property of the degree and (3.101) yield that there is $u \in X$, $\|u\|_X < R$ satisfying (3.93). Hence u is a weak solution of the BVP (3.81). Due to the discussions in Steps 1 and 2 we have $u \in L^\infty(\Omega)$. □

Theorem 3.6 *Let us assume the same as in Theorem* 3.5 *and, moreover, suppose that* $b(x,s) > 0$, $f(x,s) \geq 0$ *for a.e.* $x \in \Omega$ *and for all* $s \in \mathbb{R}$. *Then BVP* (3.81) *has at least one weak solution* $u \in X \cap L^\infty(\Omega)$ *satisfying* $u \geq 0$ *a.e. in* Ω.

Proof. It follows from Theorem 3.5 that BVP (3.81) has at least one weak solution $u \in X \cap L^\infty(\Omega)$. Let us apply Lemma 3.7 where we put $\vartheta = \lambda$ and $h(x) = f(x, u(x))$ in (3.73). We immediately obtain $u(x) \geq 0$ for a.e. $x \in \Omega$. □

Example 3.3 Let us consider the BVP

$$\begin{aligned} -\operatorname{div}(\nu(x)e^{u^2}|\nabla u|^2 \nabla u) &= \frac{1}{c\pi^2} u^3 \arctan^2(u) + \frac{1}{3c} u^3 \quad \text{in} \quad \Omega\,, \\ u &= 0 \quad \text{on} \quad \partial\Omega\,, \end{aligned} \tag{3.102}$$

where $\nu(x)$ is the weight function satisfying (3.7), c is the 4-th power of the constant of the imbedding $W_0^{1,4}(\nu, \Omega) \hookrightarrow L^4(\Omega)$. In this case we have $p = 4$, $\lambda = \dfrac{1}{c\pi^2}$, $a(x,s) = \nu(x)e^{s^2}$, $b(x,s) = \arctan^2(s)$, $f(x,s) = \dfrac{1}{3c}s^3$, $b(x) = \dfrac{\pi^2}{4}$, $\rho(x) \equiv \dfrac{1}{3c}$, $\sigma(x) \equiv 0$. It is possible to show that the hypotheses of Theorem 3.5 are satisfied. Hence BVP (3.102) has at least one weak solution $u \in W_0^{1,4}(\nu, \Omega) \cap L^\infty(\Omega)$. ◇

Example 3.4 Let $\Omega = (0, \pi) \times (0, \pi)$ be a plane domain, $x = (x_1, x_2) \in \mathbb{R}^2$. Define

$$\nu(x) = \left|\frac{x_1}{\pi}\right|^\mu \left|\frac{x_2}{\pi}\right|^\eta, \quad \mu, \eta \in \mathbb{R}, \quad x \in \mathbb{R}^2$$

and consider the BVP

$$\begin{aligned} -\operatorname{div}(\nu(x)(1 + u^4)\nabla u) &= \left(\frac{u}{2\pi} + 1\right) \operatorname{arccotan}(u) \quad \text{in} \quad \Omega\,, \\ u &= 0 \quad \text{on} \quad \partial\Omega\,. \end{aligned} \tag{3.103}$$

In this case we have $N = p = 2$, $\lambda = 1$, $a(x,s) = \nu(x)(1 + s^4)$, $b(x,s) = \dfrac{1}{2\pi} \operatorname{arccotan}(s)$, $f(x,s) = \operatorname{arccotan}(s)$, $\rho(x) \equiv 0$, $\sigma(x) \equiv \pi$. We will assume

$$\mu < 0, \quad \eta < 0\,. \tag{3.104}$$

Then (3.104) guarantees the validity of (3.7). Since $\lambda_1 = 1$ is the first eigenvalue of the homogeneous Dirichlet problem for the Laplace operator on $\Omega = (0, \pi) \times (0, \pi)$, the condition (3.104) implies

$$\int_0^\pi \int_0^\pi u^2\, dx_1\, dx_2 \le \int_0^\pi \int_0^\pi |\nabla u|^2\, dx_1\, dx_2 \le \int_0^\pi \int_0^\pi \nu(x_1, x_2)|\nabla u|^2\, dx_1\, dx_2$$

for any $u \in W_0^1(\nu, \Omega)$. Hence we can put $c = 1$ for the imbedding constant $W_0^{1,2}(\nu, \Omega) \hookrightarrow L^2(\Omega)$. Thus $\lambda^\# \ge 2$ and the assumptions of Theorem 3.6 are fulfilled. Hence BVP (3.103) has at least one weak solution $u \in W_0^{1,2}(\nu, \Omega) \cap L^\infty(\Omega)$ such that $u \ge 0$ a.e. in Ω. Arguing as in Remark 3.2 we have even $u > 0$ in Ω in this case. ◇

3.6 Bifurcation from the least eigenvalue

In this section we will study the existence of nonnegative solutions of the BVP

$$\begin{aligned} -\operatorname{div}(a(x)|\nabla u|^{p-2}\nabla u) &= \lambda b(x)|u|^{p-2}u + f(\lambda, x, u) \quad in \quad \Omega, \\ u &= 0 \quad \text{on} \quad \partial\Omega \end{aligned} \tag{3.105}$$

and of the associated perturbed BVP

$$\begin{aligned} -\operatorname{div}(a(x,u)|\nabla u|^{p-2}\nabla u) &= \lambda b(x,u)|u|^{p-2}u + f(\lambda, x, u) \quad \text{in} \quad \Omega, \\ u &= 0 \quad \text{on} \quad \partial\Omega. \end{aligned} \tag{3.106}$$

Both problems are regarded as bifurcation problems with the bifurcation parameter $\lambda \in \mathbb{R}$.

Let us start with the investigation of BVP (3.105). Assume that $a = a(x), b = b(x)$ satisfy (3.8)–(3.10), where we assume everywhere $q = p$, i.e. $b \in L^\infty(\Omega)$. The principal eigenvalue λ_1 of the homogeneous eigenvalue problem (3.11) will play a very important role in our considerations and, in addition to Theorem 3.1, we will need also its following properties.

Lemma 3.9 *The least eigenvalue $\lambda_1 > 0$ of (3.11) is isolated and every eigenfunction corresponding to the eigenvalue $0 < \lambda_0 \ne \lambda_1$ changes sign in Ω.*

Proof. Assume first that $u_0 \ge 0$ is the eigenfunction associated with the eigenvalue λ_0 of (3.11) and $\lambda_0 \ne \lambda_1$. Then $\lambda_0 > \lambda_1$ by the variational characterization of the principal eigenvalue $\lambda_1 > 0$ (see the proof of Lemma 3.1) and

$$\int_\Omega a(x)|\nabla u_1|^{p-2}\nabla u_1 \nabla\varphi\, dx = \lambda_1 \int_\Omega b(x)|u_1|^{p-2}u_1\varphi\, dx\,, \tag{3.107}$$

$$\int_\Omega a(x)|\nabla u_0|^{p-2}\nabla u_0 \nabla\psi\, dx = \lambda_0 \int_\Omega b(x)|u_0|^{p-2}u_0\psi\, dx \tag{3.108}$$

for any $\varphi, \psi \in X$, where u_1 is a nonnegative eigenfunction associated with λ_1. For $\varepsilon > 0$ set

$$u_{1\varepsilon} = u_1 + \varepsilon \quad \text{and} \quad u_{0\varepsilon} = u_0 + \varepsilon$$

and similarly as in the proof of Proposition 3.1 substitute

$$\varphi = \frac{u_{1\varepsilon}^p - u_{0\varepsilon}^p}{u_{1\varepsilon}^{p-1}}$$

into (3.107) and

$$\psi = \frac{u_{0\varepsilon}^p - u_{1\varepsilon}^p}{u_{0\varepsilon}^{p-1}}$$

into (3.108). Note that $u_0 \in L^\infty(\Omega)$ since the proof of Lemma 3.2 can proceed in the same way with λ_0 instead of λ_1. Thus we can repeat the procedure from the proof of Proposition 3.1. Adding (3.107) and (3.108) with φ and ψ chosen above we obtain the integral identity with the same left-hand side as in the proof of Proposition 3.1, which was proved to be nonnegative. Hence also the right-hand side is nonnegative, i.e. in this case we have

$$\int_\Omega b(x)\left[\lambda_0\left(\frac{u_0}{u_{0\varepsilon}}\right)^{p-1} - \lambda_1\left(\frac{u_1}{u_{1\varepsilon}}\right)^{p-1}\right](u_{0\varepsilon}^p - u_{1\varepsilon}^p)\, dx \geq 0\,.$$

Letting $\varepsilon \to 0$, we get

$$(\lambda_0 - \lambda_1)\int_\Omega b(x)(u_0^p - u_1^p)\, dx \geq 0\,. \tag{3.109}$$

Renormalizing u_0 so that the last integral is negative we derive a contradiction. Hence u_0 must change sign.

Denote

$$\Omega_0^- = \{x \in \Omega;\ u_0(x) < 0\}\,.$$

Then it follows from (3.108), where we put $\psi = u_0^-$, that

$$\int_\Omega a(x)|\nabla u_0^-|^p\, dx = \lambda_0 \int_\Omega b(x)|u_0^-|^p\, dx\,,$$

i.e.

$$\frac{1}{c_1}\|u_0^-\|_X^p \leq \lambda_0\|b\|_\infty\left(\int_{\Omega_0^-}|u_0^-|^{p_s^*}\, dx\right)^{\frac{p}{p_s^*}}(\text{meas}\,\Omega_0^-)^{\frac{p_s^*-p}{p_s^*}}\,. \tag{3.110}$$

Due to the imbedding (3.5) it follows from here that

$$\operatorname{meas} \Omega_0^- \geq \text{const} > 0\,. \tag{3.111}$$

It follows from the estimate (3.110) that meas Ω_0^- is uniformly separated from zero if λ_0 varies in some neighbourhood of λ_1.

Suppose, now, that there exists a sequence of eigenpairs (λ_n, u_n) of (3.11) with $\lambda_n \to \lambda_1$. Then $\lambda_n > \lambda_1$ and without loss of generality we may assume that $\|u_n\|_X = 1$ and $u_n \rightharpoonup \tilde{u}$ in X for some $\tilde{u} \in X$. By simplicity of λ_1 (see Theorem 3.1) we get from (3.12) that either $\tilde{u} = u_1$ or $\tilde{u} = -u_1$. Assume further that $u_n \rightharpoonup u_1 \geq 0$ in Ω. Due to the compact imbedding $X \hookrightarrow\hookrightarrow L^p(\Omega)$ we have $u_n \to u_1$ in $L^p(\Omega)$ and so by the Egorov theorem (see e.g. E. Hewitt, K. Stromberg [33]) u_n converge uniformly to u_1 in Ω with the exception of a set with arbitrarily small measure. But this contradicts (3.111) with the subscript 0 replaced by n — notice that u_0 and λ_0 can be replaced by u_n and λ_n, respectively in the first part of the proof (cf. A. Anane [5]). Thus we have proved that λ_1 is isolated. This proves the lemma. □

The following assumptions are made concerning f:

(f1) $f = f(\lambda, x, s) \colon \mathbb{R} \times \Omega \times \mathbb{R} \to \mathbb{R}$, $f \in \text{CAR}$ (here we assume that $f(\cdot, x, \cdot)$ is continuous for a.e. $x \in \Omega$ and $f(\lambda, \cdot, s)$ is measurable for all $(\lambda, s) \in \mathbb{R}^2$);

(f2) there is $p < \gamma + 1 < p_s^*$ and there are functions $\sigma, \rho \in L^\infty(\Omega)$ and a non-negative continuous function $c(\lambda)$ on $\mathbb{R}$ such that $|f(\lambda, x, s)| \leq c(\lambda)(\sigma(x) + \rho(x)|s|^\gamma)$ holds for a.e. $x \in \Omega$ and for all $s, \lambda \in \mathbb{R}$;

(f3) the limit

$$\lim_{s \to 0} \frac{f(\lambda, x, s)}{|s|^{p-2}s} = 0 \tag{3.112}$$

exists uniformly for a.e. $x \in \Omega$ and λ in a bounded interval.

Let us define operators $J, S, F_\lambda \colon X \to X^*$ by

$$\langle J(u), \varphi \rangle = \int_\Omega a(x) |\nabla u|^{p-2} \nabla u \nabla \varphi \, dx\,,$$

$$\langle S(u), \varphi \rangle = \int_\Omega b(x) |u|^{p-2} u \varphi \, dx\,,$$

$$\langle F_\lambda(u), \varphi \rangle = \int_\Omega f(\lambda, x, u) \varphi \, dx$$

for any $u, \varphi \in X$. The following lemma is then a consequence of our assumptions concerning a, b, f and of the compact imbedding (3.6).

Lemma 3.10 *The operators J, S, F_λ are well defined, J and S are $(p-1)$-homogeneous, J is continuous and S, F_λ are compact.*

Proof. The assertions concerning J and S are obvious. Let $u_n \rightharpoonup u_0$ in X. Then due to (3.6), $u_n \to u_0$ in $L^{p+\eta_1}(\Omega)$ for $\eta_1 = \gamma + 1 - p$ and for any $\varphi \in X$ we have $\varphi \in L^{p+\eta_1}(\Omega)$. The growth condition (f2) implies that the Nemytskij operator

$$u \mapsto f(\lambda, \cdot, u(\cdot))$$

is a continuous map from $L^{p+\eta_1}(\Omega)$ into $L^{\frac{\gamma+1}{\gamma}}(\Omega)$. Then we have

$$\begin{aligned}|\langle F_\lambda(u_n) - F_\lambda(u_0), \varphi\rangle| &= \left|\int_\Omega (f(\lambda, x, u_n) - f(\lambda, x, u_0))\varphi\, dx\right| \\ &\le \left(\int_\Omega |f(\lambda, x, u_n) - f(\lambda, x, u_0)|^{\frac{\gamma+1}{\gamma}}\, dx\right)^{\frac{\gamma}{\gamma+1}} \left(\int_\Omega |\varphi|^{\gamma+1}\, dx\right)^{\frac{1}{\gamma+1}} \to 0\,.\end{aligned}$$

Hence F_λ is compact. □

The operator F_λ has, moreover, the following property.

Lemma 3.11 *The operator F_λ satisfies*

$$\lim_{\|u\|_X \to 0} \frac{\|F_\lambda(u)\|_{X^*}}{\|u\|_X^{p-1}} = 0 \tag{3.113}$$

uniformly for λ in a bounded subset of $\mathbb{R}$.

Proof. By definition of the norm in X^* we have

$$\begin{aligned}\lim_{\|u\|_X \to 0} \frac{\|F_\lambda(u)\|_{X^*}}{\|u\|_X^{p-1}} &= \lim_{\|u\|_X \to 0} \sup_{\|v\|_X \le 1} \frac{1}{\|u\|_X^{p-1}} |\langle F_\lambda(u), v\rangle| \\ &= \lim_{\|u\|_X \to 0} \sup_{\|v\|_X \le 1} \frac{1}{\|u\|_X^{p-1}} \left|\int_\Omega f(\lambda, x, u) v\, dx\right| \\ &\le \lim_{\|u\|_X \to 0} \sup_{\|v\|_X \le 1} \int_\Omega \frac{|f(\lambda, x, u)|}{|u|^{p-1}} |\tilde{u}|^{p-1} |v|\, dx\,,\end{aligned} \tag{3.114}$$

where $\tilde{u} = \dfrac{u}{\|u\|_X}$. Define, for $\delta > 0$, the set

$$\Omega_\delta(u) = \{x \in \Omega;\ |u(x)|^{p-1} \ge \delta\}\,.$$

Note that meas $\Omega_\delta(u) \to 0$ as $\|u\|_X \to 0$. Then due to (3.112), for any given $\varepsilon > 0$ there exists $\delta > 0$ such that

$$\frac{|f(\lambda, x, u)|}{|u|^{p-1}} \le \varepsilon$$

uniformly for $|u|^{p-1} < \delta$. Thus we can split the last integral in (3.114) into integrals on $\Omega \setminus \Omega_\delta(u)$ and $\Omega_\delta(u)$, respectively. For these integrals we have the following estimates:

$$\int_{\Omega\setminus\Omega_\delta(u)} \frac{|f(\lambda,x,u)|}{|u|^{p-1}}|\tilde{u}|^{p-1}|v|\,dx$$

$$\le \varepsilon \int_{\Omega\setminus\Omega_\delta(u)} |\tilde{u}|^{p-1}|v|\,dx \le \varepsilon\left(\int_\Omega |\tilde{u}|^p\,dx\right)^{\frac{1}{p'}}\left(\int_\Omega |v|^p\,dx\right)^{\frac{1}{p}} \le \varepsilon \,\mathrm{const}\,,$$

$$\int_{\Omega_\delta(u)} \frac{|f(\lambda,x,u)|}{|u|^{p-1}}|\tilde{u}|^{p-1}|v|\,dx$$

$$\le \int_{\Omega_\delta(u)} \frac{c(\lambda)\sigma(x)}{|u|^{p-1}}|\tilde{u}|^{p-1}|v|\,dx + \frac{c(\lambda)}{\|u\|_X^{p-1}}\int_{\Omega_\delta(u)} \rho(x)|u|^\gamma|v|\,dx$$

$$\le \frac{c_2}{\delta}\left(\int_{\Omega_\delta(u)} |\tilde{u}|^{p_s^*}\,dx\right)^{\frac{p-1}{p_s^*}}\left(\int_{\Omega_\delta(u)} (\sigma(x)|v|)^{\frac{p_s^*}{p_s^*-(p-1)}}\,dx\right)^{\frac{p_s^*-(p-1)}{p_s^*}}$$

$$+ \frac{c_2}{\|u\|_X^{p-1}}\left(\int_{\Omega_\delta(u)} |u|^{p_s^*}\,dx\right)^{\frac{\gamma}{p_s^*}}\left(\int_{\Omega_\delta(u)} (\rho(x)|v|)^{\frac{p_s^*}{p_s^*-\gamma}}\,dx\right)^{\frac{p_s^*-\gamma}{p_s^*}}$$

$$\le \frac{c_2}{\delta}\left(\int_{\Omega_\delta(u)} |\tilde{u}|^{p_s^*}\,dx\right)^{\frac{p-1}{p_s^*}}\left(\int_{\Omega_\delta(u)} |v|^{p_s^*}\,dx\right)^{\frac{1}{p_s^*}}\left(\int_{\Omega_\delta(u)} \sigma(x)^{\frac{p_s^*}{p_s^*-p}}\,dx\right)^{\frac{p_s^*-p}{p_s^*}}$$

$$+ \frac{c_2}{\|u\|_X^{p-1}}\left(\int_{\Omega_\delta(u)} |u|^{p_s^*}\,dx\right)^{\frac{\gamma}{p_s^*}}\left(\int_{\Omega_\delta(u)} |v|^{p_s^*}\,dx\right)^{\frac{1}{p_s^*}}\left(\int_{\Omega_\delta(u)} \rho(x)^{\gamma_s}\,dx\right)^{\frac{1}{\gamma_s}}$$

$$\le c_3\|\tilde{u}\|_X^{p-1}\|v\|_X\left(\int_{\Omega_\delta(u)} \sigma(x)^{\frac{Ns}{ps-N}}\,dx\right)^{\frac{ps-N}{Ns}}$$

$$+ c_4\|\tilde{u}\|_X^{\gamma-p+1}\|v\|_X\left(\int_{\Omega_\delta(u)} \rho(x)^{\gamma_s}\,dx\right)^{\frac{1}{\gamma_s}} \to 0\,,$$

where $\gamma_s = \dfrac{p_s^*}{p_s^* - (\gamma+1)}$, since meas $\Omega_\delta(u) \to 0$ and $\sigma, \rho \in L^\infty(\Omega)$. Thus (3.113) follows and the proof is complete. □

Remark 3.8 Let us call reader's attention to the fact that in order to establish the above estimates the following weaker assumptions on $\sigma(x)$ and $\rho(x)$ in (f2) suffice:

$$\sigma \in L^{\frac{Ns}{ps-N}}(\Omega)\,, \qquad \rho \in L^{\gamma_s}(\Omega)\,,$$

of course with a bit more complicated technique in the estimates.

Now let us give the definition of a solution of (3.105).

Definition 3.5 We say that $\lambda \in \mathbb{R}$ and $u \in X$ *solve* (3.105) *weakly* if

$$J(u) - \lambda S(u) - F_\lambda(u) = 0 \quad \text{in} \quad X^* . \tag{3.115}$$

Let us define the space $E := \mathbb{R} \times X$ equipped with the norm

$$\|(\lambda, u)\|_E = (|\lambda|^2 + \|u\|_X^2)^{\frac{1}{2}}, \qquad (\lambda, u) \in E . \tag{3.116}$$

Then we can define a bifurcation point of (3.105) as follows:

Definition 3.6 Let C be a connected set in E with respect to the topology induced by the norm (3.116) and $C \subset \{(\lambda, u) \in E;\ (\lambda, u)$ solves (3.105) weakly, $u \neq 0\}$. Then C is called a *continuum of nontrivial solutions* of (3.105). We say that $\lambda_0 \in \mathbb{R}$ is a *global bifurcation point* of (3.105) (in the sense of Rabinowitz) if there is a continuum of nontrivial solutions C of (3.105) such that $(\lambda_0, 0) \in \overline{C}$ (closure in E) and C is either unbounded in E or there is an eigenvalue $\hat{\lambda}$ of (3.11) such that $\hat{\lambda} \neq \lambda_0$ and $(\hat{\lambda}, 0) \in \overline{C}$.

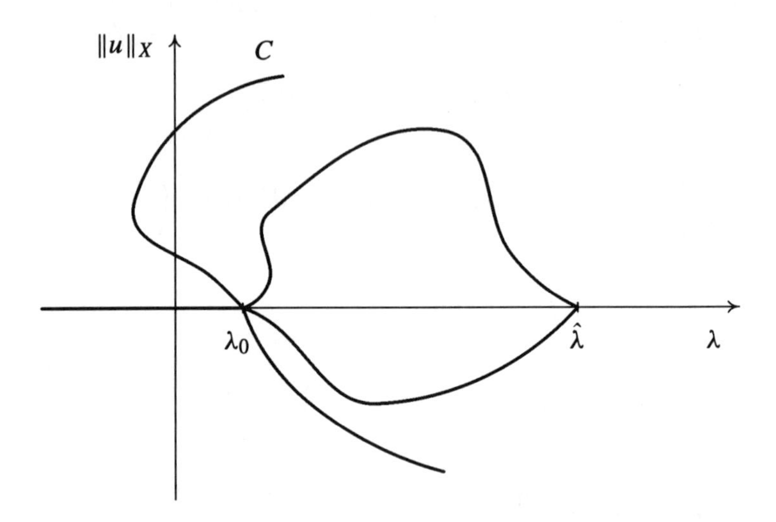

Figure 3.1: Global bifurcation point in the sense of Rabinowitz.

We will apply the topological degree theory for the operator given on the left-hand side of (3.115). Due to Lemmas 1.2 and 3.10 it is sufficient to prove the following assertion.

Lemma 3.12 *The operator $J: X \to X^*$ satisfies condition $\alpha(X)$.*

Proof. Let us assume $u_n \rightharpoonup u_0$ in X and

$$\lim_{n\to\infty} \sup\langle J(u_n), u_n - u_0\rangle \le 0 .$$

Then similarly as in Step 3 of the proof of Theorem 3.5 we have

$$\begin{aligned} 0 &\ge \lim_{n\to\infty} \sup\langle J(u_n) - J(u_0), u_n - u_0\rangle \\ &= \limsup_{n\to\infty} \int_\Omega a(x)(|\nabla u_n|^{p-2}\nabla u_n - |\nabla u_0|^{p-2}\nabla u_0)(\nabla u_n - \nabla u_0)\, dx \\ &\ge \frac{1}{c_1} \limsup_{n\to\infty} \int_\Omega \nu(x)(|\nabla u_n|^{p-2}\nabla u_n - |\nabla u_0|^{p-2}\nabla u_0)(\nabla u_n - \nabla u_0)\, dx \\ &\ge \frac{1}{c_1} \limsup_{n\to\infty}(\|u_n\|_X^{p-1} - \|u_0\|_X^{p-1})(\|u_n\|_X - \|u_0\|_X) \end{aligned}$$

and hence

$$\|u_n\|_X \to \|u_0\|_X .$$

The uniform convexity of X then implies that $u_n \to u_0$ in X. □

The following assertion is the bifurcation result for BVP (3.105).

Theorem 3.7 *Let us suppose that $a(x)$, $b(x)$ and f satisfy hypotheses mentioned above. Then the least eigenvalue $\lambda_1 > 0$ of* (3.11) *is a global bifurcation point of BVP* (3.105).

Proof. The proof consists in three steps. The first step is standard. The second step uses a procedure similar to that developed in P. Drábek [16, 17] and it is the main and new part of this proof. The third step is only a variation of the proof of Rabinowitz's Theorem 1.3 (see P. H. Rabinowitz [57]) and is therefore omitted.

Step 1 First consider operators $N_\lambda, \tilde{N}_\lambda : X \to X^*$ defined by

$$N_\lambda(u) := J(u) - \lambda S(u) - F_\lambda(u), \quad \tilde{N}_\lambda(u) := J(u) - \lambda S(u) .$$

It follows from the variational characterization of λ_1 that for $\lambda \in (0, \lambda_1)$ and any $u \in X$ with $\|u\|_X \ne 0$, we have

$$\langle \tilde{N}_\lambda(u), u\rangle > 0 .$$

Then the degree

$$\mathrm{Deg}[\tilde{N}_\lambda; B_r(0), 0] \tag{3.117}$$

is well defined for any $\lambda \in (0, \lambda_1)$ and any ball $B_r(0) \subset X$. Applying Theorem 1.6 we get

$$\mathrm{Deg}[\tilde{N}_\lambda; B_r(0), 0] = 1 , \quad \lambda \in (0, \lambda_1) . \tag{3.118}$$

Step 2 According to Lemma 3.9 there exists a $\delta > 0$ such that the interval $(\lambda_1, \lambda_1 + \delta)$ does not contain any eigenvalue of the problem (3.11). Hence the degree (3.117) is well defined also for $\lambda \in (\lambda_1, \lambda_1 + \delta)$. We are going to evaluate $\operatorname{Ind}(\tilde{N}_\lambda, 0)$ for $\lambda \in (\lambda_1, \lambda_1 + \delta)$.

Fix a number $K > 0$ and define a function $\psi : \mathbb{R} \to \mathbb{R}$ by

$$\psi(t) = \begin{cases} 0, & \text{for } t \leq K, \\ \dfrac{2\delta}{\lambda_1}(t - 2K), & \text{for } t \geq 3K, \end{cases}$$

and $\psi(t)$ is continuously differentiable, positive and strictly convex in $(K, 3K)$.

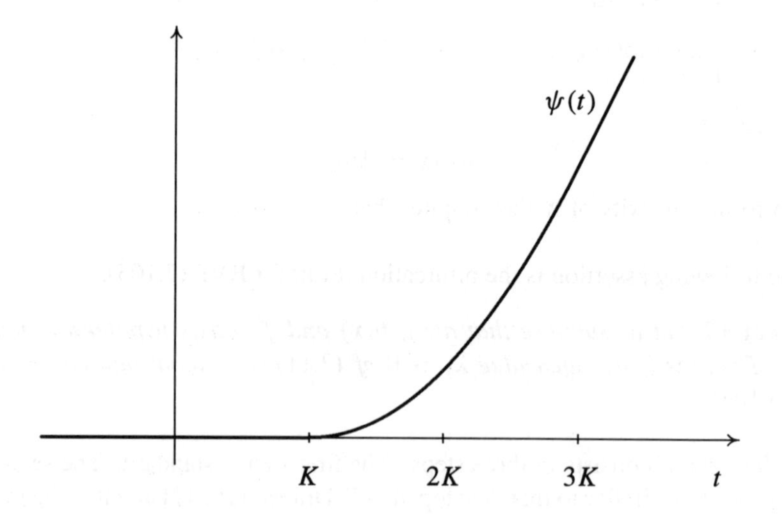

Figure 3.2: The graph of function ψ.

We define a functional

$$\Psi_\lambda(u) = \frac{1}{p}\langle J(u), u\rangle - \frac{\lambda}{p}\langle S(u), u\rangle + \psi\left(\frac{1}{p}\langle J(u), u\rangle\right).$$

Then Ψ_λ is continuously Fréchet differentiable and its critical point $u_0 \in X$ corresponds to a solution of the equation

$$J(u_0) - \frac{\lambda}{1 + \psi'\left(\frac{1}{p}\langle J(u_0), u_0\rangle\right)} S(u_0) = 0.$$

However, since $\lambda \in (\lambda_1, \lambda_1 + \delta)$, the only nontrivial critical points of Ψ'_λ occur if

$$\psi'\left(\frac{1}{p}\langle J(u_0), u_0\rangle\right) = \frac{\lambda}{\lambda_1} - 1. \tag{3.119}$$

Due to the definition of ψ we then have

$$\frac{1}{p}\langle J(u_0), u_0\rangle \in (K, 3K)$$

and due to (3.119) and the simplicity of λ_1, either $u_0 = -u_1$ or $u_0 = u_1$, where $u_1 \geq 0$ is the principal eigenfunction (which need not to be normed by 1 in general). So, for $\lambda \in (\lambda_1, \lambda_1 + \delta)$ the derivative Ψ'_λ has precisely three isolated critical points:

$$-u_1, 0, u_1 .$$

The functional Ψ_λ is weakly lower semicontinuous. Indeed, assume that $u_n \rightharpoonup \tilde{u}_0$ in X. Then

$$\langle S(u_n), u_n\rangle \to \langle S(\tilde{u}_0), \tilde{u}_0\rangle \tag{3.120}$$

due to the compactness of S (see Lemma 3.10), and then

$$\liminf_{n\to\infty}\left[\frac{1}{p}\langle J(u_n), u_n\rangle - \frac{\lambda}{p}\langle S(u_n), u_n\rangle + \psi\left(\frac{1}{p}\langle J(u_n), u_n\rangle\right)\right] \geq \Psi_\lambda(\tilde{u}_0)$$

by the facts that $\liminf\limits_{n\to\infty} \|u_n\|_a \geq \|\tilde{u}_0\|_a$, (3.120) holds, and that ψ is nondecreasing.

Observe that Ψ is coercive, i.e.

$$\lim_{\|u\|_X\to\infty} \Psi_\lambda(u) = \infty . \tag{3.121}$$

Indeed, we have

$$\begin{aligned}\Psi_\lambda(u) = \frac{1}{p}\langle J(u), u\rangle &- \frac{\lambda_1}{p}\langle S(u), u\rangle \\ &+ \frac{\lambda_1 - \lambda}{p}\langle S(u), u\rangle + \psi\left(\frac{1}{p}\langle J(u), u\rangle\right) .\end{aligned} \tag{3.122}$$

We also have

$$\langle J(u), u\rangle - \lambda_1\langle S(u), u\rangle \geq 0 \tag{3.123}$$

for any $u \in X$ (cf. the variational characterization of λ_1 in Lemma 3.1). It follows from (3.123) that

$$\begin{aligned}&\frac{\lambda_1 - \lambda}{p}\langle S(u), u\rangle + \psi\left(\frac{1}{p}\langle J(u), u\rangle\right) \\ &\quad\geq \frac{\lambda_1 - \lambda}{p\lambda_1}\langle J(u), u\rangle + \psi\left(\frac{1}{p}\langle J(u), u\rangle\right) \\ &\quad\geq -\frac{\delta}{p\lambda_1}\langle J(u), u\rangle + \frac{2\delta}{\lambda_1}\left(\frac{1}{p}\langle J(u), u\rangle - 2K\right) \to \infty\end{aligned} \tag{3.124}$$

for $\|u\|_X \to \infty$ due to the definition of ψ. Now, (3.121) follows from (3.122)–(3.124).

Since Ψ_λ is an even functional, there are precisely two points at which the minimum of Ψ_λ is achieved, say $-u_1, u_1$, while the origin 0 is obviously an isolated critical point of "saddle type". In virtue of Theorem 1.8 we have

$$\operatorname{Ind}(\Psi'_\lambda, -u_1) = \operatorname{Ind}(\Psi'_\lambda, u_1) = 1 . \tag{3.125}$$

Simultaneously, we have

$$\langle \Psi'(u), u\rangle > 0 \tag{3.126}$$

for any $u \in X$, $\|u\|_X = \kappa$ with $\kappa > 0$ large enough. Theorem 1.6 and (3.126) then imply

$$\operatorname{Deg}[\Psi'_\lambda; B_\kappa(0), 0] = 1 . \tag{3.127}$$

Let us verify (3.126). We use the estimate

$$\begin{aligned}
\langle \Psi'_\lambda(u), u\rangle &= \langle J(u), u\rangle - \lambda\langle S(u), u\rangle + \psi'\left(\frac{1}{p}\langle J(u), u\rangle\right)\langle J(u), u\rangle \\
&= \langle J(u), u\rangle - \lambda_1\langle S(u), u\rangle \\
&\quad + \psi'\left(\frac{1}{p}\langle J(u), u\rangle\right)\left(\langle J(u), u\rangle - \frac{\lambda - \lambda_1}{\psi'\left(\frac{1}{p}\langle J(u), u\rangle\right)}\langle S(u), u\rangle\right) \\
&\geq \frac{2\delta}{\lambda_1}\left(\langle J(u), u\rangle - \frac{\lambda_1}{2}\langle S(u), u\rangle\right) \to \infty \quad \text{for} \quad \|u\|_X \to \infty .
\end{aligned}$$

We have used again the variational characterization of λ_1 and the definition of ψ.

Let us choose, now, $\kappa > 0$ so large that $\pm u_1 \in B_\kappa(0)$. By the additivity property of the topological degree (see Theorem 1.7) and (3.125) and (3.127), we have

$$\operatorname{Ind}(\Psi'_\lambda, 0) = -1 . \tag{3.128}$$

We further have, by the definition of ψ,

$$\operatorname{Deg}[\tilde{N}_\lambda; B_r(0), 0] = \operatorname{Ind}(\Psi'_\lambda, 0) \tag{3.129}$$

for $r > 0$ small enough. We then conclude from (3.118), (3.128) and (3.129) that

$$\begin{aligned}
\operatorname{Ind}(\tilde{N}_\lambda, 0) &= 1 , & \lambda &\in (0, \lambda_1) , \\
\operatorname{Ind}(\tilde{N}_\lambda, 0) &= -1 , & \lambda &\in (\lambda_1, \lambda_1 + \delta) .
\end{aligned} \tag{3.130}$$

It follows from (3.113) and from the homotopy invariance property of the degree (see Theorem 1.5 (iii)) that for $r > 0$ small enough,

$$\operatorname{Deg}[N_\lambda; B_r(0), 0] = \operatorname{Deg}[\tilde{N}_\lambda; B_r(0), 0]$$

for $\lambda \in (0, \lambda_1 + \delta) \setminus \{\lambda_1\}$. We then deduce from (3.130) that

$$\begin{aligned}
\operatorname{Ind}(N_\lambda, 0) &= 1 , & \lambda &\in (0, \lambda_1) , \\
\operatorname{Ind}(N_\lambda, 0) &= -1 , & \lambda &\in (\lambda_1, \lambda_1 + \delta) .
\end{aligned}$$

Step 3 Following the proof of Theorem 1.3 in P. H. Rabinowitz [57] we can prove the conclusion of this theorem. □

Now, we will study the perturbed problem (3.106). Assume that $a(x,s)$ and $b(x,s)$ satisfy conditions (3.36)–(3.38) with $q = p$, $b \in L^\infty(\Omega)$, $c = 0$ (cf. Sections 3.4 and 3.5) and the function $f(\lambda, x, s)$ satisfies conditions (f1)–(f3) formulated in this section (see page 145). Let $d > 0$ be a given number and define a function $\tilde{a}(x,s)$ as in Step 2 of the proof of Theorem 3.5 (see Section 3.5). Then the operators $A, B \colon X \to X^*$ defined there make sense, A is continuous and B is compact. Assume that $\lim_{s\to 0} a(x,s) = a(x)$, $\lim_{s\to 0} b(x,s) = b(x)$ uniformly for a.e. $x \in \Omega$, i.e. $a(x,0) = a(x)$, $b(x,0) = b(x)$. Then $a(x)$ and $b(x)$ satisfy (3.8), (3.9) and (3.10) with $q = p$.

The following technical lemma will play an important role in the proof of the bifurcation result for the perturbed problem (3.106).

Lemma 3.13 *Let $\tilde{a}$ be defined by* (3.91), *with a given $d > 0$, $\varphi \in X$, $t_n \in [0,1]$, $\|u_n\|_X \to 0$, $\|u_n\|_X \neq 0$. Then*

$$\lim_{n\to\infty} \frac{1}{\|u_n\|_X^{p-1}} \int_\Omega (\tilde{a}(x, t_n u_n) - a(x)) |\nabla u_n|^{p-2} \nabla u_n \nabla\varphi \, dx = 0 . \tag{3.131}$$

Moreover, if $\tilde{u}_n \rightharpoonup u_0$ in X for $\tilde{u}_n = \dfrac{u_n}{\|u_n\|_X}$ then

$$\lim_{n\to\infty} \sup_{\|\varphi\|_X \le 1} \left| \int_\Omega (b(x, t_n u_n)|\tilde{u}_n|^{p-2}\tilde{u}_n - b(x)|u_0|^{p-2}u_0)\varphi \, dx \right| = 0 . \tag{3.132}$$

Proof. Let $\varphi \in X$ be fixed and denote $\tilde{u}_n = \dfrac{u_n}{\|u_n\|_X}$. Then we estimate

$$\begin{aligned} &\frac{1}{\|u_n\|_X^{p-1}} \left| \int_\Omega (\tilde{a}(x, t_n u_n) - a(x)) |\nabla u_n|^{p-2} \nabla u_n \nabla\varphi \, dx \right| \\ &\qquad = \left| \int_\Omega (\tilde{a}(x, t_n u_n) - a(x)) |\nabla \tilde{u}_n|^{p-2} \nabla \tilde{u}_n \nabla\varphi \, dx \right| \end{aligned} \tag{3.133}$$

and

$$\begin{aligned} &\left| \int_\Omega (b(x, t_n u_n)|\tilde{u}_n|^{p-2}\tilde{u}_n - b(x)|u_0|^{p-2}u_0)\varphi \, dx \right| \\ &\quad \le \left(\int_\Omega |b(x, t_n u_n)|\tilde{u}_n|^{p-2}\tilde{u}_n - b(x)|u_0|^{p-2}u_0|^{\frac{p}{p-1}} \, dx \right)^{\frac{p-1}{p}} \\ &\qquad \cdot \left(\int_\Omega |\varphi|^p \, dx \right)^{\frac{1}{p}} . \end{aligned} \tag{3.134}$$

Given $\varepsilon > 0$, due to (3.6) and the Egorov theorem there exists a set $\mathcal{A}_1 \subset \Omega$ with meas $\mathcal{A}_1 < \frac{\varepsilon}{2}$ such that $t_n u_n$ converge to 0 uniformly on $\Omega \setminus \mathcal{A}_1$. Due to the boundedness of Ω and to (3.2) there exists $\mathcal{A}_2 \subset \Omega$, meas $\mathcal{A}_2 < \frac{\varepsilon}{2}$ such that $\frac{1}{\delta} \geq \nu(x) \geq \delta > 0$ for a.e. $x \in \Omega \setminus \mathcal{A}_2$ and for some $\delta > 0$ (depending in general on $\varepsilon > 0$). Then we have for $\mathcal{A} = \mathcal{A}_1 \cup \mathcal{A}_2$:

$$\begin{aligned}
&\left|\int_\Omega (\tilde{a}(x, t_n u_n) - a(x))|\nabla \tilde{u}_n|^{p-2}\nabla \tilde{u}_n \nabla \varphi \, dx\right| \\
&\quad \leq \int_{\Omega\setminus\mathcal{A}} \left| \frac{|\tilde{a}(x, t_n u_n) - a(x)|}{\nu} \nu^{\frac{1}{p'}} |\nabla \tilde{u}_n|^{p-2}\nabla \tilde{u}_n \nu^{\frac{1}{p}} \nabla\varphi\right| dx \\
&\qquad + \int_{\mathcal{A}} c_5 \left|\nu^{\frac{1}{p'}} |\nabla \tilde{u}_n|^{p-2}\nabla \tilde{u}_n \nu^{\frac{1}{p}} \nabla\varphi\right| dx \\
&\quad \leq \frac{1}{\delta} \|\tilde{a}(x, t_n u_n) - a(x)\|_{L^\infty(\Omega\setminus\mathcal{A})} \left(\int_\Omega |\nabla \tilde{u}_n|^p \nu \, dx\right)^{\frac{1}{p'}} \left(\int_\Omega |\nabla \varphi|^p \nu \, dx\right)^{\frac{1}{p}} \\
&\qquad + c_5 \left(\int_\Omega |\nabla \tilde{u}_n|^p \nu \, dx\right)^{\frac{1}{p'}} \left(\int_{\mathcal{A}} |\nabla \varphi|^p \nu \, dx\right)^{\frac{1}{p}} .
\end{aligned}$$

Now, the first term approaches 0 due to the uniform convergence $t_n u_n \to 0$ in $\Omega \setminus \mathcal{A}$ while the second term is less than $c_6\varepsilon$ for some $c_6 > 0$ independent of ε. Hence (3.133) is arbitrarily small for n large enough and (3.131) follows.

Due to the compact imbedding (3.6) (with $\eta = 0$), we can assume $\tilde{u}_n \to u_0$ in $L^p(\Omega)$. Then we have

$$\left(\int_\Omega |b(x, t_n u_n)|\tilde{u}_n|^{p-2}\tilde{u}_n - b(x)|u_0|^{p-2}u_0|^{\frac{p}{p-1}} \, dx\right)^{\frac{p-1}{p}} \to 0$$

as $n \to \infty$. This follows from the continuity of the Nemytskij operator from $L^\infty(\Omega) \times L^p(\Omega)$ into $L^{p'}(\Omega)$ which is defined by

$$(v, u) \mapsto b(x, v)|u|^{p-2}u \, .$$

Hence (3.134) is arbitrarily small as $n \to \infty$ independently of $\|\varphi\|_X \leq 1$. □

Let us define the notion of a weak solution of the perturbed problem (3.106).

Definition 3.7 We say that $\lambda \in \mathbb{R}$ and $u \in X$ *solve* (3.106) *weakly* if

$$\begin{aligned}
&\int_\Omega a(x, u)|\nabla u|^{p-2}\nabla u \nabla\varphi \, dx \\
&\qquad = \lambda \int_\Omega b(x, u)|u|^{p-2}u\varphi \, dx + \int_\Omega f(\lambda, x, u)\varphi \, dx
\end{aligned} \tag{3.135}$$

holds for any $\varphi \in X$.

We have the following L^∞-estimate for any possible weak solution of (3.106).

Lemma 3.14 *Let a, b and f satisfy the above assumptions and let $\lambda \in \Lambda$, where Λ is a bounded set. Then there is a positive function $\hat{c} = \hat{c}(t)$ bounded on bounded sets such that*

$$\|u\|_\infty \le \hat{c}(\|u\|_X) \tag{3.136}$$

for any weak solution u of (3.106).

Proof. Let u be a weak solution of (3.106). Then we define $v_M(x)$ and $\varphi(x)$ in the same way as in the proof of Theorem 3.5, Step 1. Choosing this φ as a test function in (3.135) we obtain

$$\begin{aligned}(\kappa p+1)&\int_{\Omega(u>0)} a(x,u)v_M^{\kappa p}|\nabla v_M|^p\,dx\\ &=\lambda\int_{\Omega(u>0)} b(x,u)u^{p-1}v_M^{\kappa p+1}\,dx+\int_{\Omega(u>0)} f(\lambda,x,u)v_M^{\kappa p+1}\,dx\,.\end{aligned} \tag{3.137}$$

We have the same estimate for the left-hand side from below as in the proof of Theorem 3.5:

$$\begin{aligned}(\kappa p+1)&\int_{\Omega(u>0)} a(x,u)v_M^{\kappa p}|\nabla v_M|^p\,dx\\ &\ge \frac{\kappa p+1}{c_1(c_2')^p(\kappa+1)^p}\left(\int_{\Omega(u>0)}(v_M^{\kappa+1})^{p_s^*}\,dx\right)^{\frac{p}{p_s^*}}.\end{aligned} \tag{3.138}$$

Also the first integral on the right-hand side of (3.137) is estimated from above in the same way:

$$\lambda\int_{\Omega(u>0)} b(x,u)u^{p-1}v_M^{\kappa p+1}\,dx \le \lambda\|b\|_\infty\int_{\Omega(u>0)} u^{(\kappa+1)p}\,dx\,. \tag{3.139}$$

Concerning the second integral on the right-hand side of (3.137) we have the following estimate:

$$\begin{aligned}&\int_{\Omega(u>0)} f(\lambda,x,u)v_M^{\kappa p+1}\,dx\\ &\quad\le \text{const}\left(\|\sigma\|_\infty\int_{\Omega(u>0)} u^{\kappa p+1}\,dx+\|\rho\|_\infty\int_{\Omega(u>0)} u^{\kappa p+1+\gamma}\,dx\right).\end{aligned} \tag{3.140}$$

Let $q = \dfrac{pp_s^*}{p_s^*-(\gamma+1-p)}$. Then we have $p<q<p_s^*$ and

$$p_s^* = (\gamma+1-p)\frac{q}{q-p}\,.$$

The integrals on the right-hand sides of (3.139) and (3.140) are estimated as follows:

$$\int_{\Omega(u>0)} u^{(\kappa+1)p}\,dx \le (\text{meas }\Omega)^{\frac{q-p}{q}} \left(\int_{\Omega(u>0)} u^{(\kappa+1)q}\,dx\right)^{\frac{p}{q}},$$

$$\int_{\Omega(u>0)} u^{\kappa p+1}\,dx \le (\text{meas }\Omega)^{1-\frac{\kappa p+1}{(\kappa+1)q}} \left(\int_{\Omega(u>0)} u^{(\kappa+1)q}\,dx\right)^{\frac{\kappa p+1}{(\kappa+1)q}},$$

$$\begin{aligned}\int_{\Omega(u>0)} u^{\kappa p+1+\gamma}\,dx &= \int_{\Omega(u>0)} u^{(\kappa+1)p}u^{\gamma+1-p}\,dx\\ &\le \left(\int_{\Omega(u>0)} u^{(\gamma+1-p)\frac{q}{q-p}}\,dx\right)^{\frac{q-p}{q}} \left(\int_{\Omega(u>0)} u^{(\kappa+1)q}\,dx\right)^{\frac{p}{q}}\\ &= \left(\int_{\Omega(u>0)} u^{p_s^*}\,dx\right)^{\frac{q-p}{q}} \left(\int_{\Omega(u>0)} u^{(\kappa+1)q}\,dx\right)^{\frac{p}{q}}.\end{aligned}$$

These estimates together with (3.138)–(3.140) yield

$$\begin{aligned}\left(\int_{\Omega(u>0)} v_M^{(\kappa+1)p_s^*}\,dx\right)^{\frac{p}{p_s^*}} \le c_1(c_2')^p\frac{(\kappa+1)^p}{\kappa p+1}\Bigg[c_7\left(\int_{\Omega(u>0)} u^{(\kappa+1)q}\,dx\right)^{\frac{p}{q}}\\ + c_8\left(\int_{\Omega(u>0)} u^{(\kappa+1)q}\,dx\right)^{\frac{\kappa p+1}{(\kappa+1)q}}\\ + c_9\left(\int_{\Omega(u>0)} u^{p_s^*}\,dx\right)^{\frac{q-p}{q}}\left(\int_{\Omega(u>0)} u^{(\kappa+1)q}\,dx\right)^{\frac{p}{q}}\Bigg].\end{aligned} \tag{3.141}$$

Since $\dfrac{\kappa p+1}{(\kappa+1)q} \le \dfrac{p}{q}$ for any $\kappa > 0$, we have either

$$\left(\int_{\Omega(u>0)} u^{(\kappa+1)q}\,dx\right)^{\frac{\kappa p+1}{(\kappa+1)q}} \le 1$$

or

$$\left(\int_{\Omega(u>0)} u^{(\kappa+1)q}\,dx\right)^{\frac{\kappa p+1}{(\kappa+1)q}} \le \left(\int_{\Omega(u>0)} u^{(\kappa+1)q}\,dx\right)^{\frac{p}{q}}.$$

Thus it follows from (3.141) that

$$\begin{aligned}&\left(\int_{\Omega(u>0)} v_M^{(\kappa+1)p_s^*}\,dx\right)^{\frac{1}{(\kappa+1)p_s^*}}\\ &\le c_{10}^{\frac{1}{\kappa+1}}\left(\frac{(\kappa+1)}{(\kappa p+1)^{\frac{1}{p}}}\right)^{\frac{1}{\kappa+1}}\left[c_{11}+(c_{12}\|u\|_X^{\frac{p_s^*(q-p)}{q}}+c_{13})\left(\int_{\Omega(u>0)} u^{(\kappa+1)q}\,dx\right)^{\frac{p}{q}}\right]^{\frac{1}{(\kappa+1)p}}\end{aligned}$$

$$\leq \left[c_{14} + c_{15} \left(\int_{\Omega(u>0)} u^{(\kappa+1)q} \, dx \right)^{\frac{1}{(\kappa+1)q}} \right].$$

Now, we can conclude the proof in the same way as in the proof of Theorem 3.5, Step 1. □

Let us define the continuum of nontrivial solutions C of (3.106) in the same way as in Definition 3.6 for problem (3.105). However, in the case of the more general problem (3.106) we shall work with a weaker notion of the bifurcation point.

Definition 3.8 We say that $\lambda_0 \in \mathbb{R}$ is *a local bifurcation point* of (3.106) if there is a continuum of nontrivial solutions C of (3.106) such that $(\lambda_0, 0) \in \overline{C}$.

The following assertion is the bifurcation result for BVP (3.106).

Theorem 3.8 *Let us suppose that $a(x, s)$, $b(x, s)$ and f satisfy hypotheses mentioned above. Then the least eigenvalue $\lambda_1 > 0$ of (3.11) is a local bifurcation point (in the sense of Definition 3.8) of BVP (3.106).*

Proof. Due to Lemma 3.14, we can choose $d > 0$ large enough so that in a small neighbourhood B of $(\lambda_1, 0)$ in the space E, any weak solution (λ, u) of BVP (3.106) satisfies $\|u\|_\infty < d$. Then $\tilde{a}(x, u(x)) = a(x, u(x))$ for a.e. $x \in \Omega$ and (3.135) is equivalent to

$$\tilde{T}_\lambda(u) := A(u) - \lambda B(u) - F_\lambda(u) = 0. \tag{3.142}$$

Let us recall that A satisfies condition $\alpha(X)$ and B, F_λ are compact operators. So the degrees mentioned below are well defined and we will prove that for any $\varepsilon > 0$ small enough, there exists $\delta > 0$ such that

$$\mathrm{Deg}[\tilde{T}_\lambda; B_\delta(0), 0] = \mathrm{Deg}[J - \lambda S; B_\delta(0), 0] \tag{3.143}$$

for $\lambda = \lambda_1 \pm \varepsilon$. Define the homotopy

$$\mathcal{H}(t, \lambda, u) = A(t, u) - \lambda B(t, u) - t F_\lambda(u)$$

where $A(t, u)$, $B(t, u)$ are defined by

$$\langle A(t, u), \varphi \rangle = \int_\Omega \tilde{a}(x, tu) |\nabla u|^{p-2} \nabla u \nabla \varphi \, dx,$$

$$\langle B(t, u), \varphi \rangle = \int_\Omega b(x, tu) |u|^{p-2} u \varphi \, dx$$

for any $u, \varphi \in X$. Then

$$\mathcal{H}(1, \lambda, u) = \tilde{T}_\lambda(u), \quad \mathcal{H}(0, \lambda, u) = J(u) - \lambda S(u).$$

Hence it is sufficient to prove that for $\delta > 0$ and $\varepsilon > 0$ mentioned above and $\lambda = \lambda_1 \pm \varepsilon$, $\mathcal{H}(t, \lambda, u)$ is an admissible homotopy between $\tilde{T}_\lambda$ and $J - \lambda S$ on $B_\delta(0)$. Suppose this is not true. Then there exist $u_n \in X$ with $\|u_n\|_X \to 0$, $\|u_n\|_X \neq 0$, $t_n \in [0, 1]$, such that

$$\mathcal{H}(t_n, \lambda, u_n) = 0 .$$

That is, for any $\varphi \in X$ we have

$$\langle A(t_n, u_n), \varphi \rangle - \lambda \langle B(t_n, u_n), \varphi \rangle - t_n \langle F_\lambda(u_n), \varphi \rangle = 0 . \tag{3.144}$$

Dividing (3.144) by $\|u_n\|_X^{p-1}$ and denoting $\tilde{u}_n = \dfrac{u_n}{\|u_n\|_X}$ and

$$\langle \tilde{A}(t_n, \tilde{u}_n), \varphi \rangle = \int_\Omega \tilde{a}(x, t_n u_n) |\nabla \tilde{u}_n|^{p-2} \nabla \tilde{u}_n \nabla \varphi \, dx ,$$

$$\langle \tilde{B}(t_n, \tilde{u}_n), \varphi \rangle = \int_\Omega b(x, t_n u_n) |\tilde{u}_n|^{p-2} \tilde{u}_n \varphi \, dx ,$$

we obtain for all $\varphi \in X$ the identity

$$\langle \tilde{A}(t_n, \tilde{u}_n), \varphi \rangle - \lambda \langle \tilde{B}(t_n, \tilde{u}_n), \varphi \rangle - t_n \frac{\langle F_\lambda(u_n), \varphi \rangle}{\|u_n\|_X^{p-1}} = 0 .$$

By Lemma 3.11 we have

$$\sup_{\|\varphi\|_X \le 1} \frac{|\langle F_\lambda(u_n), \varphi \rangle|}{\|u_n\|_X^{p-1}} \to 0 .$$

We thus conclude

$$\sup_{\|\varphi\|_X \le 1} \left| \langle \tilde{A}(t_n, \tilde{u}_n) - \lambda \tilde{B}(t_n, \tilde{u}_n), \varphi \rangle \right| \to 0 ,$$

i.e.

$$\tilde{A}(t_n, \tilde{u}_n) - \lambda \tilde{B}(t_n, \tilde{u}_n) \to 0 \quad \text{in} \quad X^* . \tag{3.145}$$

Since $\tilde{u}_n$ is bounded in X, we can assume that $\tilde{u}_n \rightharpoonup u_0$ in X for some u_0. Lemma 3.13 then implies

$$\tilde{B}(t_n, \tilde{u}_n) \to S(u_0) \quad \text{in} \quad X^* . \tag{3.146}$$

We thus infer that

$$\tilde{A}(t_n, \tilde{u}_n) \to \lambda S(u_0) \tag{3.147}$$

in X^*. It follows from this strong convergence and from $\tilde{u}_n \rightharpoonup u_0$ in X that

$$\lim_{n \to \infty} \langle \tilde{A}(t_n, \tilde{u}_n), \tilde{u}_n - u_0 \rangle = 0 .$$

This implies

$$0 = \lim_{n\to\infty} \langle \tilde{A}(t_n, \tilde{u}_n) - J(u_0), \tilde{u}_n - u_0 \rangle$$
$$= \lim_{n\to\infty} \int_\Omega (\tilde{a}(x, t_n u_n) - a(x))|\nabla u_0|^{p-2}\nabla u_0(\nabla \tilde{u}_n - \nabla u_0)\, dx$$
$$+ \lim_{n\to\infty} \int_\Omega \tilde{a}(x, t_n u_n)(|\nabla \tilde{u}_n|^{p-2}\nabla \tilde{u}_n - |\nabla u_0|^{p-2}\nabla u_0)(\nabla \tilde{u}_n - \nabla u_0)\, dx\,.$$

Now, the same estimates as in (3.97) and (3.98) imply $\|\tilde{u}_n\|_X \to \|u_0\|_X$ and hence $\tilde{u}_n \to u_0$ in X. In particular, $\|u_0\|_X = 1$, and it follows from (3.145)–(3.147) that

$$J(u_0) - \lambda S(u_0) = 0\,,$$

which contradicts the fact that λ_1 is an isolated eigenvalue and $\lambda \neq \lambda_1$ belongs to its small neighbourhood. Thus $\mathcal{H}$ is an admissible homotopy and (3.143) follows.

By Step 1 of the proof of Theorem 3.7 we have

$$\operatorname{Deg}[\tilde{T}_\lambda; B_\delta(0), 0] = \pm 1$$

for $\lambda = \lambda_1 \pm \varepsilon$, which means that λ_1 is a local bifurcation point of the operator equation (3.142) by the classical bifurcation result (cf. Step 3 in the proof of Theorem 3.7). □

Example 3.5 Let Ω be a bounded domain in $\mathbb{R}^N$, $p > 1$, let $\nu(x)$ be positive and measurable in Ω satisfying $\nu \in L^1_{loc}(\Omega)$, $\nu^{-s} \in L^1(\Omega)$ for $s > \max\left\{\frac{N}{p}, \frac{1}{p-1}\right\}$ and $f(\lambda, x, s) = |s|^\gamma$ with $\gamma \in (p-1, p_s^* - 1)$. Consider the BVP

$$\begin{aligned} -\operatorname{div}(\nu(x)|\nabla u|^{p-2}\nabla u) &= \lambda |u|^{p-2}u + |u|^\gamma \quad \text{in} \quad \Omega\,, \\ u &= 0 \quad \text{on} \quad \partial\Omega\,. \end{aligned} \tag{3.148}$$

Then it follows from Theorem 3.7 that the principal eigenvalue $\lambda_1 > 0$ of the eigenvalue problem

$$\begin{aligned} -\operatorname{div}(\nu(x)|\nabla u|^{p-2}\nabla u) &= \lambda |u|^{p-2}u \quad \text{in} \quad \Omega\,, \\ u &= 0 \quad \text{on} \quad \partial\Omega \end{aligned} \tag{3.149}$$

is a global bifurcation point of BVP (3.148) (in the sense of Rabinowitz). ◇

Example 3.6 Let ν and f be as in Example 3.5. Consider the BVP

$$\begin{aligned} -\operatorname{div}(\nu(x)e^{u^2}|\nabla u|^{p-2}\nabla u) &= \lambda e^{-u^2}|u|^{p-2}u + |u|^\gamma \quad \text{in} \quad \Omega\,, \\ u &= 0 \quad \text{on} \quad \partial\Omega\,. \end{aligned} \tag{3.150}$$

Then it follows from Theorem 3.8 that the principal eigenvalue $\lambda_1 > 0$ of the eigenvalue problem (3.149) is a local bifurcation point of BVP (3.150) (in the sense of Definition 3.8). ◇

Chapter 4
The p-Laplacian in $\mathbb{R}^N$

4.1 Nonlinear eigenvalue problem

In this chapter we will study nonlinear problems containing the p-Laplacian in $\mathbb{R}^N$. We start with a nonlinear eigenvalue problem which can be regarded as the generalized *Emden–Fowler equation,* i.e. the equation

$$-\Delta u = \lambda f(x, u) \quad \text{in} \quad \mathbb{R}^N ,$$

which has been studied recently in E. S. Noussair, C. A. Swanson [54] and W. Rother [58]. Namely, we consider the nonlinear eigenvalue problem

$$\begin{cases} -\operatorname{div}(a(x)|\nabla u|^{p-2}\nabla u) = \lambda f(x, u) \quad \text{in} \quad \mathbb{R}^N , \\ u > 0 \quad \text{in} \quad \mathbb{R}^N , \qquad \lim\limits_{|x|\to\infty} u(x) = 0 , \end{cases} \tag{4.1}$$

for a general p, $1 < p < N$, $\lambda > 0$ which was recently investigated in P. Drábek [15] with functions $a = a(x)$, $f = f(x, s)$ satisfying the following conditions:

(a) $a \in L^\infty(\mathbb{R}^N)$ and there exists a constant $a_0 > 0$ such that

$$a(x) \geq a_0 \quad \text{for a.e.} \quad x \in \mathbb{R}^N ; \tag{4.2}$$

(f1) $f(x, s)\colon \mathbb{R}^N \times [0, \infty) \to [0, \infty)$, $f \in \text{CAR}$;

(f2) there exist an open set Ω in $\mathbb{R}^N$ and a constant $\delta_0 > 0$ such that $f(x, \delta_0) > 0$ holds for a.e. $x \in \Omega$;

(f3) there are a nonnegative function $\rho = \rho(x)$ and a constant δ, $0 < \delta < \infty$ such that

$$0 \leq f(x, s) \leq \rho(x)s^\gamma \quad \text{for a.e. } x \in \mathbb{R}^N \quad \text{and all } s \in [0, \infty) , \tag{4.3}$$

where $p < \gamma + 1 < p^* = \dfrac{Np}{N-p}$ and $\rho \in L^{\gamma_1}(\mathbb{R}^N) \cap L^{\gamma_1+\delta}(\mathbb{R}^N)$ with

$$\gamma_1 = \frac{p^*}{p^* - (\gamma + 1)} = \frac{Np}{Np - (\gamma + 1)(N - p)} .$$

Let us extend f by zero for $s < 0$ and denote

$$F(x, s) = \int_0^s f(x, \tau)\, d\tau .$$

Then it follows from (f2) and (f3) that

$$F(x, s) > 0 \tag{4.4}$$

holds a.e. in $\Omega \times (\delta_0, \infty)$ and

$$0 \leq F(x, s) \leq \frac{1}{\gamma + 1}\rho(x)(s^+)^{\gamma+1} \tag{4.5}$$

holds a.e. in $\mathbb{R}^N \times \mathbb{R}$, where $s^+ = \max\{s, 0\}$.

For this section, the basic function space will be

$$X = \{u \in L^{p^*}(\mathbb{R}^N);\ |\nabla u| \in L^p(\mathbb{R}^N)\} .$$

Due to the Sobolev inequality (1.12) and (4.2) there exists a constant $c_1 > 0$ such that

$$\|u\|_{p^*} \leq c_1 \left(\int_{\mathbb{R}^N} a(x)|\nabla u|^p\, dx \right)^{\frac{1}{p}} \tag{4.6}$$

holds for all $u \in X$. Hence X equipped with the norm

$$\|u\|_X = \left(\int_{\mathbb{R}^N} a(x)|\nabla u|^p\, dx \right)^{\frac{1}{p}}$$

is a uniformly convex Banach space.

Definition 4.1 We will say that the equation in (4.1) holds in the *weak sense* for some $u \in X$ and $\lambda > 0$ if the integral identity

$$\int_{\mathbb{R}^N} a(x)|\nabla u|^{p-2}\nabla u \nabla\varphi\, dx = \lambda \int_{\mathbb{R}^N} f(x, u)\varphi\, dx \tag{4.7}$$

is fulfilled for any $\varphi \in X$.

Note that it follows from the assumptions about a and f that both integrals in (4.7) are well-defined if $u, \varphi \in X$. Note also that finding $u \in X$ for which (4.1) holds in the weak sense we are able to show that the additional requirements in (4.1), namely $u > 0$ in $\mathbb{R}^N$ and $\lim_{|x|\to\infty} u(x) = 0$, are satisfied automatically (as consequences of Theorems 1.9 and 1.10 — see the proof of Theorems 4.1 and 4.2 for the details).

The following two estimates will play the key role in the proof of existence of $u \in X$ and $\lambda > 0$ satisfying (4.1) in the weak sense.

Lemma 4.1 *For any $K \in [0, \infty)$ and for any $u \in X$, we have*

$$\int_{|x|\geq K} F(x, u(x))\,dx \leq \frac{1}{\gamma+1}\left(\int_{|x|\geq K} |\rho(x)|^{\gamma_1}\,dx\right)^{\frac{1}{\gamma_1}} (c_1\|u\|_X)^{\gamma+1}. \tag{4.8}$$

Proof. It follows from (4.5), (4.6) and the Hölder inequality that

$$\begin{aligned}\int_{|x|\geq K} F(x, u(x))\,dx &\leq \frac{1}{\gamma+1}\int_{|x|\geq K} \rho(x)|u(x)|^{\gamma+1}\,dx \\ &\leq \frac{1}{\gamma+1}\left(\int_{|x|\geq K} |\rho(x)|^{\gamma_1}\,dx\right)^{\frac{1}{\gamma_1}}\left(\int_{|x|\geq K} |u(x)|^{(\gamma+1)\gamma_1'}\,dx\right)^{\frac{1}{\gamma_1'}} \\ &\leq \frac{1}{\gamma+1}\left(\int_{|x|\geq K} |\rho(x)|^{\gamma_1}\,dx\right)^{\frac{1}{\gamma_1}}\left(\int_{\mathbb{R}^N} |u(x)|^{p^*}\,dx\right)^{\frac{\gamma+1}{p^*}} \\ &\leq \frac{1}{\gamma+1}\left(\int_{|x|\geq K} |\rho(x)|^{\gamma_1}\,dx\right)^{\frac{1}{\gamma_1}} (c_1\|u\|_X)^{\gamma+1}.\end{aligned}$$

□

Lemma 4.2 *For a.e. $x \in \mathbb{R}^N$ and every $s \in \mathbb{R}$ the following estimate holds:*

$$F(x, s) \leq \frac{1}{\gamma+1}\left(\frac{\rho(x)^{\gamma_1+\delta}}{\gamma_1+\delta} + \frac{(s^+)^{(\gamma+1)(\gamma_1+\delta)'}}{(\gamma_1+\delta)'}\right). \tag{4.9}$$

Proof. Let us consider the Young inequality in the following form:

$$ab \leq \frac{a^\varepsilon}{\varepsilon} + \frac{b^{\varepsilon'}}{\varepsilon'}, \quad \frac{1}{\varepsilon} + \frac{1}{\varepsilon'} = 1. \tag{4.10}$$

If we set $a = \rho(x)$, $b = (s^+)^{\gamma+1}$, $\varepsilon = \gamma_1 + \delta$ then the estimate (4.9) follows from (4.5) and (4.10). □

Remark 4.1 Elementary calculation yields

$$(\gamma+1)(\gamma_1+\delta)' < p^*. \tag{4.11}$$

Indeed, $\gamma_1' = \dfrac{p^*}{\gamma+1}$ and $(\gamma_1+\delta)' < \gamma_1'$. Hence (4.11) follows.

We have the following existence result.

Theorem 4.1 *Let us suppose $1 < p < N$, (a), (f1), (f2) and (f3). Then there exist a constant $\lambda > 0$ and a nonnegative function $u \in X$, $u \not\equiv 0$, such that the equation* (4.1) *holds in the weak sense. Moreover,*

$$u \in L^r(\mathbb{R}^N) \quad \textit{for all} \quad r \in [p^*, \infty]$$

and the function u decays uniformly as $|x| \to \infty$.

Proof. The proof of this theorem will be performed in three steps. In the first step we prove the existence of $\lambda > 0$ and $u \in X$, $u \geq 0$, $u \not\equiv 0$ in $\mathbb{R}^N$, which satisfy (4.1) in the weak sense. In the second step we prove that $u \in L^r(\mathbb{R}^N)$, $p^* \leq r \leq \infty$. Finally, in the third step, we prove that u is positive in $\mathbb{R}^N$ and decays uniformly at infinity.

Step 1 Let us consider the functional

$$\Phi(u) = \frac{\int_{\mathbb{R}^N} F(x, u(x))\, dx}{\|u\|_X^p + \|u\|_X^{p^*}}$$

defined for all $u \in X$, $u \not\equiv 0$. Choose $K = 0$ in Lemma 4.1. Then it follows from (4.8) and $p < \gamma + 1 < p^*$ that there exists a real constant $c_2 > 0$ such that

$$\Phi(u) \leq c_2 \tag{4.12}$$

holds for all $0 \not\equiv u \in X$. We conclude from (4.12) that

$$S = \sup_{0 \not\equiv u \in X} \Phi(u)$$

is a finite number.

Let us choose a function $\varphi_0 \in C_0^\infty(\mathbb{R}^N)$ such that

$$\operatorname{supp} \varphi_0 \subset \Omega \quad \text{and} \quad \sup_{x \in \mathbb{R}^N} \varphi_0(x) > \delta_0$$

(for Ω and δ_0 see (f2)). Then due to (4.4) we have

$$S > c_3 := \frac{1}{2} \Phi(\varphi_0) > 0 \,. \tag{4.13}$$

Let $\{u_n\}_{n=1}^\infty \subset X$, $u_n \not\equiv 0$, be the maximizing sequence for Φ, i.e.

$$\lim_{n \to \infty} \Phi(u_n) = S \,.$$

Due to (4.13), we may assume without restriction that

$$\Phi(u_n) \geq c_3 \tag{4.14}$$

holds for all $n \in \mathbb{N}$. Since $\Phi(u_n^+) \geq \Phi(u_n)$, we may further assume that $u_n \geq 0$ in $\mathbb{R}^N$. It follows from Lemma 4.1 and (4.14) that there is a constant $c_4 > 0$ such that for any $n \in \mathbb{N}$ we have

$$\|u_n\|_X^p + \|u_n\|_X^{p^*} \leq c_4 \|u_n\|_X^{\gamma+1} \,.$$

Hence there are constants $0 < c_5 < c_6 < \infty$ such that

$$c_5 \leq \|u_n\|_X \leq c_6 \tag{4.15}$$

holds for all $n \in \mathbb{N}$. By (4.15) and due to the uniform convexity of X we can find a subsequence of $\{u_n\}_{n=1}^\infty$, still denoted by $\{u_n\}_{n=1}^\infty$, and an element $u \in X$, $u \geq 0$ in $\mathbb{R}^N$, such that

$$u_n \rightharpoonup u \quad \text{in} \quad X\,, \qquad \|u\|_X \leq c_6\,. \tag{4.16}$$

Due to Lemma 4.1 and (4.15), for each $\varepsilon > 0$ there exists a constant $K_\varepsilon > 0$ such that

$$\int_{|x|\geq K_\varepsilon} F(x, u_n(x))\,dx \leq \varepsilon\,, \quad \int_{|x|\geq K_\varepsilon} F(x, u(x))\,dx \leq \varepsilon \tag{4.17}$$

holds for each $n \in \mathbb{N}$. On the other hand, it follows from (4.9) and (4.11) that the Nemytskij operator

$$u \mapsto F(x, u(x))$$

is continuous from $L^{(\gamma+1)(\gamma_1+\delta)'}(B_{K_\varepsilon}(0))$ into $L^1(B_{K_\varepsilon}(0))$. This together with Theorem 1.2 (ii) and (4.16) yields

$$\int_{|x|<K_\varepsilon} F(x, u_n(x))\,dx \to \int_{|x|<K_\varepsilon} F(x, u(x))\,dx \tag{4.18}$$

as $n \to \infty$. However, (4.17) and (4.18) imply

$$\lim_{n\to\infty} \int_{\mathbb{R}^N} F(x, u_n(x))\,dx = \int_{\mathbb{R}^N} F(x, u(x))\,dx\,. \tag{4.19}$$

From (4.14), (4.15) and (4.19) it follows that

$$\int_{\mathbb{R}^N} F(x, u_n(x))\,dx \geq c_3(c_5^p + c_5^{p^*}) > 0$$

and therefore $u \not\equiv 0$. From the uniform boundedness principle for weakly convergent sequences in a uniformly convex Banach space (see e.g. E. Hewitt, K. Stromberg [33]), we obtain

$$\|u\|_X^p + \|u\|_X^{p^*} \leq \liminf_{n\to\infty} (\|u_n\|_X^p + \|u_n\|_X^{p^*})\,. \tag{4.20}$$

Now (4.19) and (4.20) imply that

$$\Phi(u) \geq \limsup_{n\to\infty} \Phi(u_n) = S\,,$$

and consequently

$$\Phi(u) = S\,.$$

We shall prove that there is $\lambda > 0$ such that λ and u solve (4.1) in the weak sense. Indeed, due to the fact that $u \not\equiv 0$, for any $\varphi \in X$ we can find $\varepsilon_\varphi > 0$ such that $\|u + \varepsilon\varphi\|_X > 0$ holds for all $\varepsilon \in (-\varepsilon_\varphi, \varepsilon_\varphi)$.

Define the function

$$\eta(\varepsilon) = \Phi(u + \varepsilon\varphi), \quad \varepsilon \in (-\varepsilon_\varphi, \varepsilon_\varphi).$$

Then $\eta'(0) = 0$ due to the fact that supremum of Φ is achieved for u, which implies that

$$\int_{\mathbb{R}^N} a(x)|\nabla u|^{p-2}\nabla u \nabla\varphi \, dx = \lambda \int_{\mathbb{R}^N} f(x, u(x))\varphi(x)\, dx,$$

where

$$\lambda = \frac{\|u\|_X^p + \|u\|_X^{p^*}}{(p + p^*\|u\|_X^{p^*-p}) \int_{\mathbb{R}^N} F(x, u(x))\, dx} > 0.$$

Step 2 We proceed in a similar way as in the proof of Lemma 3.2. For $M > 0$ define

$$v_M(x) = \min\{u(x), M\}$$

and choose $\varphi = v_M^{\kappa p+1}$ ($\kappa \geq 0$) as a test function in (4.7). Then obviously $\varphi \in X \cap L^\infty(\mathbb{R}^N)$ and it follows from (4.7) that

$$(\kappa p + 1) \int_{\mathbb{R}^N} a(x) v_M^{\kappa p} |\nabla v_M|^p \, dx = \lambda \int_{\mathbb{R}^N} f(x, u) v_M^{\kappa p+1}\, dx. \tag{4.21}$$

Due to (4.2) and (4.6) the left-hand side of (4.21) can be estimated from below as follows:

$$\begin{aligned}(\kappa p + 1) \int_{\mathbb{R}^N} a(x) v_M^{\kappa p} |\nabla v_M|^p \, dx &= \frac{\kappa p + 1}{(\kappa + 1)^p} \int_{\mathbb{R}^N} a(x) |\nabla (v_M)^{\kappa+1}|^p \, dx \\ &\geq \frac{1}{c_1^p} \frac{\kappa p + 1}{(\kappa + 1)^p} \left(\int_{\mathbb{R}^N} v_M^{(\kappa+1)p^*} \, dx \right)^{\frac{p}{p^*}}.\end{aligned} \tag{4.22}$$

By using the assumption (f3) and the Hölder inequality, the right-hand side of (4.21) can be formally estimated from above and we obtain

$$\begin{aligned}\lambda \int_{\mathbb{R}^N} f(x, u) v_M^{\kappa p+1}\, dx &\leq \lambda \left(\int_{\mathbb{R}^N} \rho(x)^{\gamma_1+\delta}\, dx \right)^{\frac{1}{\gamma_1+\delta}} \\ &\quad \cdot \left(\int_{\mathbb{R}^N} u^{(\kappa+1)p(\gamma_1+\delta)'} u^{(\gamma+1-p)(\gamma_1+\delta)'}\, dx \right)^{\frac{1}{(\gamma_1+\delta)'}} \\ &\leq \lambda \left(\int_{\mathbb{R}^N} \rho(x)^{\gamma_1+\delta}\, dx \right)^{\frac{1}{\gamma_1+\delta}} \left(\int_{\mathbb{R}^N} u^{(\kappa+1)q}\, dx \right)^{\frac{p}{q}} \\ &\quad \cdot \left(\int_{\mathbb{R}^N} u^{(\gamma+1-p)(\gamma_1+\delta)' \frac{q}{q-p(\gamma_1+\delta)'}}\, dx \right)^{\frac{q-p(\gamma_1+\delta)'}{q(\gamma_1+\delta)'}},\end{aligned} \tag{4.23}$$

where $q = \dfrac{pp^*(\gamma_1+\delta)'}{p^* - (\gamma+1-p)(\gamma_1+\delta)'}$. Obviously

$$q < p^*, \quad 1 < \frac{q}{p(\gamma_1+\delta)'} \quad \text{and} \quad (\gamma+1-p)(\gamma_1+\delta)' \frac{q}{q - p(\gamma_1+\delta)'} = p^*,$$

and hence (4.23) yields

$$\lambda \int_{\mathbb{R}^N} f(x,u) v_M^{\kappa p+1}\, dx \le \lambda \Big(\int_{\mathbb{R}^N} \rho(x)^{\gamma_1+\delta}\, dx \Big)^{\frac{1}{\gamma_1+\delta}} \cdot \Big(\int_{\mathbb{R}^N} u^{p^*}\, dx \Big)^{\frac{q-p(\gamma_1+\delta)'}{q(\gamma_1+\delta)'}} \Big(\int_{\mathbb{R}^N} u^{(\kappa+1)q}\, dx \Big)^{\frac{p}{q}}. \tag{4.24}$$

Now it follows from (4.6), (4.16), (4.21), (4.22) and (4.24) that there exists a constant $c_7 > 0$ (independent of $M > 0$ and $\kappa > 0$) such that

$$\Big(\int_{\mathbb{R}^N} v_M^{(\kappa+1)p^*}\, dx \Big)^{\frac{p}{p^*}} \le c_7 \frac{(\kappa+1)^p}{\kappa p+1} \Big(\int_{\mathbb{R}^N} u^{(\kappa+1)q}\, dx \Big)^{\frac{p}{q}},$$

i.e.

$$\|v_M\|_{(\kappa+1)p^*} \le c_8^{\frac{1}{\kappa+1}} \left(\frac{\kappa+1}{(\kappa p+1)^{\frac{1}{p}}} \right)^{\frac{1}{\kappa+1}} \|u\|_{(\kappa+1)q} \tag{4.25}$$

with $c_8 = c_7^{\frac{1}{p}}$. Since $u \in X$ and hence $u \in L^{p^*}(\mathbb{R}^N)$ we can choose $\kappa := \kappa_1$ in (4.25) such that $(\kappa_1+1)q = p^*$, i.e. $\kappa_1 = \dfrac{p^*}{q} - 1$. Then we have

$$\|v_M\|_{(\kappa_1+1)p^*} \le c_8^{\frac{1}{\kappa_1+1}} \left(\frac{\kappa_1+1}{(\kappa_1 p+1)^{\frac{1}{p}}} \right)^{\frac{1}{\kappa_1+1}} \|u\|_{p^*}$$

for any $M > 0$. Repeating the same iteration process as in the proof of Lemma 3.2 we arrive at

$$\|u\|_{(\kappa_n+1)p^*} \le c_9 \|u\|_{p^*}$$

with $\kappa_n \to \infty$, and taking into account (4.6), (4.16) we have

$$\|u\|_{r_n} \le c_{10} \tag{4.26}$$

for $r_n = (\kappa_n+1)p^* \to \infty$ with some $c_{10} > 0$. Repeating again the same argument as in the proof of Lemma 3.2 we derive from (4.26) that

$$\|u\|_\infty \le c_{10}. \tag{4.27}$$

Now it follows from (4.26) and (4.27) that $u \in L^r(\mathbb{R}^N)$ for all $p^* \le r \le \infty$. Moreover, we have a uniform bound for the corresponding norms independently of r.

Step 3 The assumptions (a), (f1), (f2) and f(3) allow to apply a result of J. Serrin [60] in order to prove the decay of the function u. Let $x \in \mathbb{R}^N$ be an arbitrary point and let $B_r(x)$ be the ball centered at x with radius $r > 0$. Then by Theorem 1.10 there exists a constant $c_{11} = c_{11}(N, p, \gamma_1) > 0$ such that

$$\|u\|_{L^\infty(B_1(x))} \leq c_{11}\left(\|u\|_{L^{p^*}(B_2(x))} + \|f(\cdot, u)\|_{L^{\gamma_1}(B_2(x))}\right) \tag{4.28}$$

holds independently of $x \in \mathbb{R}^N$. Note that $\gamma_1 > \frac{N}{p}$, which is important for the application of Serrin's result. Letting $|x| \to \infty$ the uniform decay of $u = u(x)$ follows directly from (4.28) (cf. Remark 1.4). □

If the coefficient $a = a(x)$ is smooth then we can obtain more than the nonnegativity and L^∞-boundedness of the weak solution: We get a regularity result and the strict positivity of the solution.

Theorem 4.2 *Suppose that the assumptions of Theorem* 4.1 *are satisfied and, moreover, let $\rho \in L^\infty(\mathbb{R}^N)$ and $a \in C^1(\mathbb{R}^N)$. Then the assertion of Theorem* 4.1 *holds with u positive in $\mathbb{R}^N$, $u \in C^{1,\alpha}(B_R(0))$ for any $R > 0$ with some $\alpha = \alpha(R) \in (0, 1)$.*

The idea of the proof is the following. We can apply the Harnack-type inequality (see Theorem 1.9) to prove that $u > 0$ in $\mathbb{R}^N$. Indeed, assume that this is not true. Since u is nonnegative and nontrivial there must be a point $x_0 \in \mathbb{R}^N$ such that $u(x_0) = 0$ and $u(x) > 0$ for x arbitrarily close to x_0. But this contradicts Theorem 1.9. Regularity of the solution u follows directly from Theorem 1.11.

Remark 4.2 Let us consider the problem

$$\begin{cases} Au = \lambda f(x, u) \quad \text{in} \quad \mathbb{R}^N, \\ u > 0 \quad \text{in} \quad \mathbb{R}^N, \qquad \lim_{|x| \to \infty} u(x) = 0, \end{cases} \tag{4.29}$$

where

$$(Au)(x) = -\sum_{i,j=1}^{N} \frac{\partial}{\partial x_j}\left(a_{ij}(x)\left|\frac{\partial u}{\partial x_i}\right|^{p-2} \frac{\partial u}{\partial x_i}\right).$$

Assume that $a_{ij} \in L^\infty(\mathbb{R}^N)$ (or $a_{ij} \in C^1(\mathbb{R}^N) \cap L^\infty(\mathbb{R}^N)$) and that for some $a_0 > 0$,

$$\sum_{i,j=1}^{N} a_{ij}(x)|\xi_i|^{p-2}\xi_i\xi_j \geq a_0|\xi|^p$$

holds for all $\xi \in \mathbb{R}^N$ and a.e. $x \in \mathbb{R}^N$ while the function f is supposed to satisfy (f1), (f2) and

(f3)′ $$0 \le f(x,s) \le \sum_{i=1}^{I} f_i(x) s^{\gamma_i},$$

where f_i and γ_i fulfil the same conditions as ρ and γ, respectively, in (f3) with some $0 < \delta_i < \infty$, $i = 1, \dots, I$.

Then obvious modifications of the proof of Theorem 4.1 yield the same assertion concerning the problem (4.29). Also the regularity result as in Theorem 4.2 applies for (4.29). Thus we have a direct generalization of earlier results of E. S. Noussair, C. A. Svanson [54] and W. Rother [58] concerning the Emden–Fowler equation.

Remark 4.3 The same approach as explained in this section applies to the nonlinear problem on the *exterior* domain $\Omega \subset \mathbb{R}^N$ with $C^{1,\alpha}$ boundary, $0 \le \alpha \le 1$:

$$\begin{cases} -\operatorname{div}(a(x)|\nabla u|^{p-2}\nabla u) = \lambda f(x,u) \quad \text{in} \quad \Omega, \\ u > 0 \quad \text{in} \quad \Omega, \quad u|_{\partial\Omega} = 0, \quad \lim_{|x|\to\infty} u(x) = 0. \end{cases}$$

We thus obtain a result parallel to Lao Sen Yu [43] but under rather more general assumptions on f.

Remark 4.4 Note that related problems were studied also in Li Gongbao, Yuo Shusen [45] and Li Gongbao [44] where a different approach was used.

4.2 Bifurcation problem for the p-Laplacian in $\mathbb{R}^N$

The results in this and the forthcoming section follow the lines of P. Drábek, Y. X. Huang [18, 19] and are related to those in Section 3.6, where analogous problems were studied on the bounded domain. In this section we will be concerned with the existence of a positive weak solution of the nonlinear problem

$$-\operatorname{div}(a(x)|\nabla u|^{p-2}\nabla u) = \lambda g(x)|u|^{p-2}u + f(\lambda, x, u) \quad \text{in} \quad \mathbb{R}^N, \tag{4.30}$$

where a is as in Section 4.1. We assume that $\lambda \in \mathbb{R}$ is a bifurcation parameter and consider first the case $1 < p < N$. Then we have $p' \in (1,\infty)$, $p^* \in (p,\infty)$ for $p' = \dfrac{p}{p-1}$ and $p^* = \dfrac{Np}{N-p}$, respectively. We will write the function $g = g(x)$ in the form

$$g(x) = g^+(x) - g^-(x). \tag{4.31}$$

Concerning functions a, g and f which appear in (4.30) we assume (a) (see p. 160) and the following:

(g1) $$g^+ \in L^\infty(\mathbb{R}^N) \cap L^{\frac{N}{p}}(\mathbb{R}^N),\ g^+ \not\equiv 0,$$

(g2) $g^- \in L^\infty(\mathbb{R}^N)$.

Recall that

$$\omega(x) = \frac{1}{(1+|x|)^p}, \quad x \in \mathbb{R}^N$$

is the weight function which appears in the Hardy inequality

$$\int_{\mathbb{R}^N} \frac{|u|^p}{(1+|x|)^p}\,dx \le \left(\frac{p}{N-p}\right)^p \int_{\mathbb{R}^N} |\nabla u|^p\,dx \tag{4.32}$$

(see Example 1.7, inequality (1.45)). Let

$$w(x) = \max\{g^-(x), \omega(x)\} > 0, \quad x \in \mathbb{R}^N.$$

Then we assume that f satisfies

($\hat{\text{f1}}$) $f \in \mathrm{CAR}$ (cf. Section 3.6);

($\hat{\text{f2}}$) $|f(\lambda, x, s)| \le c(\lambda)(\sigma(x) + \rho(x)|s|^\gamma)$

for a.e. $x \in \mathbb{R}^N$ and all $s, \lambda \in \mathbb{R}$, where $c(\lambda)$ is nonnegative and bounded on bounded subsets of $\mathbb{R}$, $p < \gamma + 1 < p^*$,

$$0 \le \rho \in L^{\gamma_1}(\mathbb{R}^N)$$

with

$$\gamma_1 = \frac{p^*}{p^* - (\gamma+1)}, \quad 0 \le \sigma \in L^{\frac{N}{p}}(\mathbb{R}^N, w^{\frac{N}{p}}),$$

and either

(i) $\sigma \in L^{(p^*)'}(\mathbb{R}^N)$, $(p^*)' = \dfrac{Np}{Np-(N-p)}$,

or

(ii) $\sigma \in L^{p'}(\mathbb{R}^N, w^{-\frac{1}{p-1}})$;

($\hat{\text{f3}}$) let $\lambda \in \Lambda \subset \mathbb{R}$, where Λ is a bounded interval, then for any $\varepsilon > 0$ there exists $\delta > 0$ such that

$$f(\lambda, x, s) \le \varepsilon w(x)|s|^{p-1}$$

holds for every $x \in \mathbb{R}^N$ and $w(x)|s|^{p-1} < \delta$.

In this section we shall work in the uniformly convex Banach space X which is defined as the completion of $C_0^\infty(\mathbb{R}^N)$ with respect to the norm

$$\|u\|_X = \left(\int_{\mathbb{R}^N} a(x)|\nabla u|^p\,dx + \int_{\mathbb{R}^N} w(x)|u|^p\,dx\right)^{\frac{1}{p}}.$$

As usual, we denote by X^* the dual space and by $\langle\cdot,\cdot\rangle$ the pairing between X^* and X.

We define operators $J, G, F_\lambda: X \to X^*$ as follows: for $u, \varphi \in X$,

$$\langle J(u), \varphi\rangle = \int_{\mathbb{R}^N} a(x)|\nabla u|^{p-2}\nabla u \nabla\varphi\, dx\,,$$

$$\langle G(u), \varphi\rangle = \int_{\mathbb{R}^N} g(x)|u|^{p-2}u\varphi\, dx\,,$$

$$\langle F_\lambda(u), \varphi\rangle = \int_{\mathbb{R}^N} f(\lambda, x, u)\varphi\, dx\,.$$

Sometimes we split G to $G = G^+ - G^-$, where according to (4.31) we have

$$\langle G^\pm(u), \varphi\rangle = \int_{\mathbb{R}^N} g^\pm(x)|u|^{p-2}u\varphi\, dx\,.$$

In order to define a proper functional setting for (4.30) we need some properties of J, G and F_λ.

Lemma 4.3 *The operators J, G, F_λ: $X \to X^*$ are well defined, J and G are $(p-1)$-homogeneous, and F_λ satisfies*

$$\lim_{\|u\|_X \to 0} \frac{\|F_\lambda(u)\|_{X^*}}{\|u\|_X^{p-1}} = 0 \tag{4.33}$$

uniformly for λ in a bounded subset of $\mathbb{R}$.

Proof. Homogenity of J and G is obvious. Observe that by the Hölder inequality

$$\begin{aligned}|\langle J(u), \varphi\rangle| &= \left|\int_{\mathbb{R}^N} a^{\frac{1}{p'}}(x)|\nabla u|^{p-2}\nabla u\, a^{\frac{1}{p}}(x)\nabla\varphi\, dx\right| \\ &\le \left(\int_{\mathbb{R}^N} a(x)|\nabla u|^p\, dx\right)^{\frac{1}{p'}} \left(\int_{\mathbb{R}^N} a(x)|\nabla\varphi|^p\, dx\right)^{\frac{1}{p}},\end{aligned}$$

so that J is well defined. Using the Hölder inequality again, we have

$$|\langle G(u), \varphi\rangle| \le \left(\int_{\mathbb{R}^N} |g||u|^p\, dx\right)^{\frac{1}{p'}} \left(\int_{\mathbb{R}^N} |g||\varphi|^p\, dx\right)^{\frac{1}{p}}.$$

Note that for any $u \in X$, $\int_{\mathbb{R}^N} g^-(x)|u|^p\, dx < \infty$ and by the Hölder inequality, (g1) and the Sobolev imbedding theorem, we have

$$\int_{\mathbb{R}^N} g^+(x)|u|^p\, dx \le \left(\int_{\mathbb{R}^N} g^+(x)^{\frac{N}{p}}\, dx\right)^{\frac{p}{N}} \left(\int_{\mathbb{R}^N} |u|^{p^*}\, dx\right)^{\frac{p}{p^*}} < \infty\,.$$

Hence G is well defined. For F_λ, we have from (f2)

$$|\langle F_\lambda(u), \varphi\rangle| \le c(\lambda)\left(\int_{\mathbb{R}^N} \sigma(x)|\varphi|\, dx + \int_{\mathbb{R}^N} \rho(x)|u|^\gamma|\varphi|\, dx\right).$$

Now, by $(\hat{f2})$, either

$$\int_{\mathbb{R}^N} \sigma(x)|\varphi|\,dx \le \left(\int_{\mathbb{R}^N} \sigma^{(p^*)'}\,dx\right)^{\frac{1}{(p^*)'}} \left(\int_{\mathbb{R}^N} |\varphi|^{p^*}\,dx\right)^{\frac{1}{p^*}},$$

or

$$\int_{\mathbb{R}^N} \sigma(x)|\varphi|\,dx \le \left(\int_{\mathbb{R}^N} w^{\frac{1}{1-p}}\sigma^{p'}\,dx\right)^{\frac{1}{p'}} \left(\int_{\mathbb{R}^N} w|\varphi|^p\,dx\right)^{\frac{1}{p}},$$

and

$$\begin{aligned}\int_{\mathbb{R}^N} \rho(x)|u|^\gamma|\varphi|\,dx &\le \left(\int_{\mathbb{R}^N} |u|^{p^*}\,dx\right)^{\frac{\gamma}{p^*}} \left(\int_{\mathbb{R}^N} \rho^{\frac{p^*}{p^*-\gamma}}|\varphi|^{\frac{p^*}{p^*-\gamma}}\,dx\right)^{\frac{p^*-\gamma}{p^*}} \\ &\le \left(\int_{\mathbb{R}^N} |u|^{p^*}\,dx\right)^{\frac{\gamma}{p^*}} \left(\int_{\mathbb{R}^N} |\varphi|^{p^*}dx\right)^{\frac{1}{p^*}} \left(\int_{\mathbb{R}^N} \rho^{\gamma_1}\,dx\right)^{\frac{1}{\gamma_1}} \\ &\le c_1\|u\|_X^\gamma\|\varphi\|_X,\end{aligned} \tag{4.34}$$

thus F_λ is well defined.

Let us now prove (4.33). By definition,

$$\begin{aligned}\lim_{\|u\|_X\to 0} \frac{\|F_\lambda(u)\|_{X^*}}{\|u\|_X^{p-1}} &= \lim_{\|u\|_X\to 0} \sup_{\|\varphi\|_X\le 1} \frac{1}{\|u\|_X^{p-1}}|\langle F_\lambda(u),\varphi\rangle| \\ &= \lim_{\|u\|_X\to 0} \sup_{\|\varphi\|_X\le 1} \frac{1}{\|u\|_X^{p-1}}\left|\int_{\mathbb{R}^N} f(\lambda,x,u)\varphi\,dx\right| \\ &\le \lim_{\|u\|_X\to 0} \sup_{\|\varphi\|_X\le 1} \int_{\mathbb{R}^N} \frac{|f(\lambda,x,u)|}{w|u|^{p-1}}|\tilde{u}|^{p-1}|\varphi|w\,dx,\end{aligned} \tag{4.35}$$

where $\tilde{u} = \dfrac{u}{\|u\|_X}$. We now estimate the last integral in (4.35). We first define, for $\delta > 0$, the set

$$\Omega_\delta(u) = \{x \in \mathbb{R}^N;\ w(x)|u(x)|^{p-1} \ge \delta\}$$

(for w see the beginning of Section 4.2). We claim that meas $\Omega_\delta(u) \to 0$ as $\|u\|_X \to 0$. Assume, on the contrary, meas $\Omega_\delta(u) \ge c_2 > 0$ while $\|u\|_X \to 0$. Let

$$\Omega_K = \Omega_\delta(u) \cap B_K(0) \quad \text{for} \quad K > 0.$$

Then meas $\Omega_K \ge \frac{1}{2}c_2$ provided $K > 0$ is large enough. Hence we have

$$\begin{aligned}0 < \delta \operatorname{meas}\Omega_K &\le \int_{\Omega_K} w(x)|u(x)|^{p-1}\,dx \\ &\le \left(\int_{\Omega_K} w(x)\,dx\right)^{\frac{1}{p}} \left(\int_{\Omega_K} w(x)|u(x)|^p\,dx\right)^{\frac{1}{p'}} \\ &\le \left(\int_{\Omega_K} w(x)\,dx\right)^{\frac{1}{p}} \|u\|_X^{p-1} \le c_3(\operatorname{meas}\Omega_K)^{\frac{1}{p}}\|u\|_X^{p-1}.\end{aligned}$$

That is,

$$0 < \delta \left(\frac{c_2}{2}\right)^{\frac{1}{p'}} < \delta (\operatorname{meas} \Omega_K)^{\frac{1}{p'}} \le c_3 \|u\|_X^{p-1},$$

a contradiction.

Due to ($\hat{f}3$), for any given $\varepsilon > 0$ there exists $\delta > 0$ such that

$$\frac{|f(\lambda, x, u)|}{w(x)|u|^{p-1}} \le \varepsilon$$

uniformly for $w(x)|u|^{p-1} < \delta$. We split the integral in (4.35) into integrals on $\mathbb{R}^N \setminus \Omega_\delta(u)$ and $\Omega_\delta(u)$. For the first integral we have

$$\int_{\mathbb{R}^N \setminus \Omega_\delta(u)} \frac{|f(\lambda, x, u)|}{w(x)|u|^{p-1}} |\tilde{u}|^{p-1} |\varphi| w \, dx \le \varepsilon \int_{\mathbb{R}^N \setminus \Omega_\delta(u)} |\tilde{u}|^{p-1} |\varphi| w \, dx \le c_4 \varepsilon ,$$

where we have used the inequality

$$\int_{\mathbb{R}^N} |\tilde{u}|^{p-1} |\varphi| w \, dx \le \left(\int_{\mathbb{R}^N} w |\tilde{u}|^p \, dx\right)^{\frac{1}{p'}} \left(\int_{\mathbb{R}^N} w |\varphi|^p \, dx\right)^{\frac{1}{p}} .$$

By ($\hat{f2}$), we have for the second integral

$$\begin{aligned} \int_{\Omega_\delta(u)} \frac{|f(\lambda, x, u)|}{w(x)|u|^{p-1}} |\tilde{u}|^{p-1} |\varphi| w \, dx \le \int_{\Omega_\delta(u)} & \frac{c(\lambda)\sigma(x)}{w(x)|u|^{p-1}} |\tilde{u}|^{p-1} |\varphi| w \, dx \\ & + \frac{c(\lambda)}{\|u\|_X^{p-1}} \int_{\Omega_\delta(u)} \rho(x) |u|^\gamma |\varphi| \, dx . \end{aligned}$$

We denote the last expression by $c(\lambda)(I_1 + I_2)$. Observe that

$$\begin{aligned} I_1 &\le \frac{1}{\delta} \int_{\Omega_\delta(u)} \sigma(x) w(x) |\tilde{u}|^{p-1} |\varphi| \, dx \\ &\le \frac{1}{\delta} \left(\int_{\Omega_\delta(u)} |\tilde{u}|^{p^*} dx\right)^{\frac{p-1}{p^*}} \left(\int_{\Omega_\delta(u)} (\sigma(x) w(x) |\varphi|)^{\frac{p^*}{p^*-(p-1)}} dx\right)^{\frac{p^*-(p-1)}{p^*}} \\ &\le \frac{1}{\delta} \left(\int_{\Omega_\delta(u)} |\tilde{u}|^{p^*} dx\right)^{\frac{p-1}{p^*}} \left(\int_{\Omega_\delta(u)} |\varphi|^{p^*} dx\right)^{\frac{1}{p^*}} \left(\int_{\Omega_\delta(u)} (\sigma(x) w(x))^{\frac{p^*}{p^*-p}} dx\right)^{\frac{p^*-p}{p^*}} \\ &\le c_5 \|\tilde{u}\|_X^{p-1} \|\varphi\|_X \left(\int_{\Omega_\delta(u)} (\sigma(x) w(x))^{\frac{N}{p}} dx\right)^{\frac{P}{N}} \to 0 \end{aligned}$$

since $\operatorname{meas} \Omega_\delta(u) \to 0$ and $\sigma \in L^{\frac{N}{p}}(\mathbb{R}^N, w^{\frac{N}{p}})$. The estimate (4.2) and $\|\varphi\|_X \le 1$ implies

$$I_2 \le c_1 \|u\|_X^{\gamma - p + 1} \|\varphi\|_X \to 0 .$$

We thus conclude that (4.33) holds. □

Lemma 4.4 *The operators J and G^- are continuous. The operators G^+ and F_λ are compact.*

Proof. Continuity of J follows from the continuity of the Nemytskij operator from $L^p(\mathbb{R}^N)$ into $L^{p'}(\mathbb{R}^N)$, continuity of G^- follows from the continuity of the Nemytskij operator from $L^p(\mathbb{R}^N, w)$ into $L^{p'}(\mathbb{R}^N, w^{1-p'})$ (see Section 1.3).

Let us prove the compactness of G^+. We first claim that, for any $\varepsilon > 0$ and $u \in X$, there exists $K > 0$ such that

$$\sup_{\|\varphi\|_X \le 1} \int_{|x|>K} g^+(x)|u|^{p-1}|\varphi|\,dx \le \varepsilon \|u\|_X^{p-1}.$$

Indeed, using twice the Hölder inequality, we have

$$\begin{aligned}
&\sup_{\|\varphi\|_X \le 1} \int_{|x|>K} g^+(x)|u|^{p-1}|\varphi|\,dx \\
&\le \sup_{\|\varphi\|_X \le 1} \left(\int_{|x|>K} g^+|u|^p\,dx\right)^{\frac{1}{p'}} \left(\int_{|x|>K} g^+|\varphi|^p\,dx\right)^{\frac{1}{p}} \\
&\le \sup_{\|\varphi\|_X \le 1} \left(\int_{|x|>K} (g^+)^{\frac{N}{p}}\,dx\right)^{\frac{p-1}{N}} \left(\int_{|x|>K} |u|^{p^*}\,dx\right)^{\frac{p-1}{p^*}} \\
&\quad \cdot \left(\int_{|x|>K} (g^+)^{\frac{N}{p}}\,dx\right)^{\frac{1}{N}} \left(\int_{|x|>K} |\varphi|^{p^*}\,dx\right)^{\frac{1}{p^*}} \\
&\le c_6 \sup_{\|\varphi\|_X \le 1} \left(\int_{|x|>K} (g^+)^{\frac{N}{p}}\,dx\right)^{\frac{p}{N}} \left(\int_{|x|>K} |u|^{p^*}\,dx\right)^{\frac{p-1}{p^*}} \|\varphi\|_X \le \varepsilon \|u\|_X^{p-1}
\end{aligned}$$

since $g^+ \in L^{\frac{N}{p}}(\mathbb{R}^N)$.

Now, suppose $u_n \rightharpoonup u_0$ in X. We estimate

$$\begin{aligned}
\|G^+(u_n) - G^+(u_0)\|_X &= \sup_{\|\varphi\|_X \le 1} |\langle G^+(u_n) - G^+(u_0), \varphi\rangle| \\
&= \sup_{\|\varphi\|_X \le 1} \left| \int_{\mathbb{R}^N} g^+(x)(|u_n|^{p-2}u_n - |u_0|^{p-2}u_0)\varphi\,dx \right| \\
&\le \sup_{\|\varphi\|_X \le 1} \left| \int_{|x|\le K} g^+(x)(|u_n|^{p-2}u_n - |u_0|^{p-2}u_0)\varphi\,dx \right| \\
&\quad + \sup_{\|\varphi\|_X \le 1} \left| \int_{|x|>K} g^+(x)(|u_n|^{p-2}u_n - |u_0|^{p-2}u_0)\varphi\,dx \right|.
\end{aligned}$$

Observe that for any $\varepsilon > 0$, we can choose $K > 0$ such that the second term is smaller than $\frac{\varepsilon}{2}$ for all $n \in \mathbb{N}$, while for fixed K, by strong convergence $u_n \to u_0$ in L^p on any bounded region, the first term is smaller than $\frac{\varepsilon}{2}$ for n large enough. We thus have proved that $G^+(u_n) \to G^+(u_0)$ in X^*, i.e. G^+ is compact.

To prove compactness of F_λ, let us assume again $u_n \rightharpoonup u_0$ in X. We have

$$\begin{aligned} \|F_\lambda(u_n) - F_\lambda(u_0)\|_X &= \sup_{\|\varphi\|_X \le 1} |\langle F_\lambda(u_n) - F_\lambda(u_0), \varphi\rangle| \\ &= \sup_{\|\varphi\|_X \le 1} \left| \int_{\mathbb{R}^N} (f(\lambda, x, u_n) - f(\lambda, x, u_0))\varphi\, dx \right| \\ &\le \sup_{\|\varphi\|_X \le 1} \left| \int_{|x| \le K} (f(\lambda, x, u_n) - f(\lambda, x, u_0))\varphi\, dx \right| \\ &\quad + \sup_{\|\varphi\|_X \le 1} \left| \int_{|x| > K} (f(\lambda, x, u_n) - f(\lambda, x, u_0))\varphi\, dx \right| . \end{aligned} \tag{4.36}$$

It is easy to see that the first term tends to zero as $n \to \infty$ by the continuity of the Nemytskij operator associated with f and acting from $L^p(B_K(0))$ into $L^{p'}(B_K(0))$ and by $W^{1,p}(B_K(0)) \hookrightarrow\hookrightarrow L^p(B_K(0))$. We estimate the integral over $\mathbb{R}^N \setminus B_K(0)$ similarly as for G^+:

$$\begin{aligned} &\sup_{\|\varphi\|_X \le 1} \left| \int_{|x|>K} f(\lambda, x, u)\varphi\, dx \right| \\ &\quad \le \sup_{\|\varphi\|_X \le 1} \int_{|x|>K} c(\lambda)\sigma|\varphi|\, dx + \sup_{\|\varphi\|_X \le 1} \int_{|x|>K} c(\lambda)\rho|u|^\gamma|\varphi|\, dx . \end{aligned}$$

The first term is estimated by

$$c_7\left(\int_{|x|>K} \sigma^{(p^*)'}\, dx\right)^{\frac{1}{(p^*)'}} \quad \text{or by} \quad c_7\left(\int_{|x|>K} w^{\frac{1}{1-p}}\sigma^{p'}\, dx\right)^{\frac{1}{p'}},$$

while the second term is estimated by

$$c_8\left(\int_{\mathbb{R}^N} |u|^{p^*}\, dx\right)^{\frac{\gamma}{p^*}} \left(\int_{\mathbb{R}^N} |\varphi|^{p^*}\, dx\right)^{\frac{1}{p^*}} \left(\int_{|x|>K} \rho^{\gamma_1}\, dx\right)^{\frac{1}{\gamma_1}}$$

as in the proof of Lemma 4.3. Hence for any $\varepsilon > 0$ and $u \in X$, there exists $K > 0$ such that

$$\sup_{\|\varphi\|_X \le 1} \left| \int_{|x|>K} f(\lambda, x, u)\varphi\, dx \right| \le \varepsilon \left(\|u\|_X^\gamma + 1\right) .$$

Thus $F_\lambda(u_n) \to F_\lambda(u_0)$ in X^*, i.e. F is compact. □

Definition 4.2 We say that $\lambda \in \mathbb{R}$ and $u \in X$ *solve* the problem (4.30) *weakly* if

$$J(u) - \lambda G(u) - F_\lambda(u) = 0 \quad \text{in} \quad X^* . \tag{4.37}$$

In the forthcoming two lemmas we prove some useful properties of the eigenvalue problem

$$-\operatorname{div}(a(x)|\nabla u|^{p-2}\nabla u) = \lambda g(x)|u|^{p-2}u \quad \text{in } \mathbb{R}^N, \quad \int_{\mathbb{R}^N} g|u|^p\,dx > 0. \tag{4.38}$$

Note that g changes sign in general and so the last inequality is essential in order to define the eigenvalue problem correctly.

Lemma 4.5 *The eigenvalue problem* (4.38) *has a pair* (λ_1, u_1) *of a principal eigenvalue* λ_1 *and an eigenfunction* u_1 *with* $\lambda_1 > 0$ *and* $0 < u_1 \in X \cap L^\infty(\mathbb{R}^N)$. *Moreover,* λ_1 *is simple and* $u_1(x)$ *decays uniformly as* $|x| \to \infty$.

Proof. Let us start with two observations. First, realize that

$$\int_{\mathbb{R}^N} g^+(x)|v_n|^p\,dx \to \int_{\mathbb{R}^N} g^+(x)|u_1|^p\,dx \tag{4.39}$$

whenever $v_n \rightharpoonup u_1$ in X. Indeed, for any $K > 0$, the Hölder inequality implies

$$\begin{aligned}&\int_{|x|\ge K} g^+(x)|v_n|^p\,dx \\ &\quad\le \left(\int_{|x|\ge K} (g^+(x))^{\frac{N}{p}}\,dx\right)^{\frac{p}{N}} \left(\int_{\mathbb{R}^N} |v_n|^{p^*}\,dx\right)^{\frac{p}{p^*}} \le \varepsilon(K)\end{aligned} \tag{4.40}$$

and $\varepsilon(K) \to 0$ for $K \to \infty$ due to the boundedness of $\|v_n\|_X$. Simultaneously, we have

$$\int_{|x|<K} g^+(x)|v_n|^p\,dx \to \int_{|x|<K} g^+(x)|u_1|^p\,dx \tag{4.41}$$

for any $K > 0$ due to Theorem 1.2 (ii). The convergence (4.39) now follows from (4.40) and (4.41). The second observation concerns the fact that the mappings

$$u \mapsto \int_{\mathbb{R}^N} a(x)|\nabla u|^p\,dx, \quad u \mapsto \int_{\mathbb{R}^N} g^-(x)|u|^p\,dx \tag{4.42}$$

are weakly lower semicontinuous functionals on X. This is a consequence of continuity and convexity (see e.g. S. Fučík, A. Kufner [27]).

Set

$$\lambda_1 = \inf\left\{\int_{\mathbb{R}^N} a(x)|\nabla v|^p\,dx\right\} \tag{4.43}$$

where the infimum is taken over all $v \in X$ satisfying $\int_{\mathbb{R}^N} g^+(x)|v|^p\,dx = 1$. We shall prove (similarly as in the proof of Lemma 3.1) that λ_1 is the principal eigenvalue of (4.38). The expression (4.43) will be refered to as its *variational characterization* (cf. Section 3 for the case of bounded domain). Clearly we have $\lambda \geq 0$. Let $\{v_n\}_{n=1}^{\infty} \subset X$ be the minimizing sequence for λ_1, i.e.

$$\int_{\mathbb{R}^N} g(x)|v_n|^p\,dx = 1 \quad \text{and} \quad \int_{\mathbb{R}^N} a(x)|\nabla v_n|^p\,dx = \lambda_1 + \delta_n \tag{4.44}$$

with $\delta_n \to 0+$ for $n \to 0$. It follows from (4.44) and from the Hardy inequality (4.32) that $\|v_n\|_X$ is bounded. Hence we can assume that $v_n \rightharpoonup u_1$ in X for some $u_1 \in X$. Now, it follows from our observations at the beginning of the proof and from (4.44) that

$$\int_{\mathbb{R}^N} a(x)|\nabla u_1|^p\,dx = \lambda_1 \tag{4.45}$$

and

$$\int_{\mathbb{R}^N} g(x)|u_1|^p\,dx \geq 1\,. \tag{4.46}$$

But the strict inequality in (4.46) is excluded due to the variational characterization of λ_1. In particular, we can assume that $u_1 \geq 0$ and it follows from (4.43), (4.45), (4.46) that $u_1 \not\equiv 0$, $\lambda_1 > 0$ and u_1 is the eigenfunction of (4.38) corresponding to the principal eigenvalue λ_1, i.e.

$$\int_{\mathbb{R}^N} a(x)|\nabla u_1|^{p-2}\nabla u_1 \nabla\varphi\,dx = \lambda_1 \int_{\mathbb{R}^N} g(x)|u_1|^{p-2}u_1\varphi\,dx \tag{4.47}$$

for any $\varphi \in X$. Let us choose $\varphi = v_M^{\kappa p+1}$ $(\kappa \geq 0)$ as a test function in (4.47), where v_M was defined in Step 2 of the proof of Theorem 4.1. Proceeding in a similar way as there we arrive at the inequality

$$\left(\int_{\mathbb{R}^N} v_M^{(\kappa+1)p^*}\,dx\right)^{\frac{1}{(\kappa+1)p^*}} \leq c_9^{\frac{1}{\kappa+1}} \left(\frac{\kappa+1}{(\kappa p+1)^{\frac{1}{p}}}\right)^{\frac{1}{\kappa+1}} \left(\int_{\mathbb{R}^N} u_1^{(\kappa+1)p}\,dx\right)^{\frac{1}{(\kappa+1)p}},$$

where $c_9 > 0$ depends only on λ_1 and $\|g^+\|_\infty$. Repeating the iteration procedure from the proof of Lemma 3.2 we get

$$\|u_1\|_{r_n} \leq c_{10}\|u_1\|_{p^*} \tag{4.48}$$

for $r_n \to \infty$ and hence

$$\|u_1\|_\infty \leq c_{10}\|u_1\|_{p^*}\,.$$

Hence we can apply the Harnack-type inequality (see Theorem 1.9) to get $u_1 > 0$ in $\mathbb{R}^N$. In particular, it follows that every eigenfunction corresponding to λ_1 is strictly of the same sign in $\mathbb{R}^N$. Indeed, if v is an eigenfunction corresponding to λ_1 then by its variational characterization also $|v|$ is an eigenfunction corresponding to λ_1. But then $|v| > 0$ in $\mathbb{R}^N$ by the above considerations.

Repeating the same arguments as in the proofs of Proposition 3.1 and Theorem 3.1 we can prove the simplicity of λ_1 (realize that the proof can be adopted considering $\mathbb{R}^N$ instead of Ω).

Finally, we apply Theorem 1.10 as in Step 3 of the proof of Theorem 4.1 in order to prove the decay of $u_1(x)$ as $|x| \to \infty$. Note that $\gamma_1 > \dfrac{N}{p}$ can be chosen arbitrary in the estimate

$$\|u_1\|_{L^\infty(B_1(x))} \le c_{11} \left(\|u_1\|_{L^{p^*}(B_2(x))} + \|g|u_1|^{p-2}u_1\|_{L^{\gamma_1}(B_2(x))} \right)$$

due to our assumptions on g and the uniform bound (4.48). □

Remark 4.5 Let us assume that $g^- \not\equiv 0$ and $g^\pm \in L^\infty(\mathbb{R}^N) \cap L^{\frac{N}{p}}(\mathbb{R}^N)$. Then by symmetry there is also a principal eigenpair (λ_1^*, u_1^*) of

$$-\operatorname{div}(a(x)|\nabla u|^{p-2}\nabla u) = \lambda g(x)|u|^{p-2}u \quad \text{in } \mathbb{R}^N, \quad \int_{\mathbb{R}^N} g|u|^p\, dx < 0 \qquad (4.38')$$

with $\lambda_1^* < 0$ and $0 < u_1^* \in X$ which has the same properties as stated in Lemma 4.5. This is the reason why we use the expression "the principal eigenvalue" instead of "the least eigenvalue" in this chapter. Further we will concentrate on properties of $\lambda_1 > 0$ while the analogous properties of $\lambda_1^* < 0$ are left to the reader.

Remark 4.6 Essentially the same eigenvalue problem as (4.38) was studied in W. Allegretto, Y. X. Huang [3], where also the existence of "higher" eigenvalues of (4.38) was proved using the Ljusternik–Schnirelmann variational approach.

The following assertion is an analogue of Lemma 3.9.

Lemma 4.6 *The principal eigenvalue* $\lambda_1 > 0$ *of* (4.38) *is isolated and every eigenfunction corresponding to the eigenvalue* $\lambda_0 > 0$, $\lambda_0 \ne \lambda_1$ *changes sign in* $\mathbb{R}^N$.

Proof. Similarly as in the proof of Lemma 3.9 we get $\lambda_0 > \lambda_1$ (by the variational characterization of λ_1) and the second part of the assertion of this lemma.

Let us suppose that (λ_0, u_0) is an eigenpair of (4.38) with $\lambda_0 > \lambda_1$, λ_0 belonging to a certain bounded neighbourhood of λ_1. Denote

$$\Omega_0^- = \{x \in \mathbb{R}^N;\ u_0(x) < 0\}.$$

Then

$$\int_{\mathbb{R}^N} a(x)|\nabla u_0|^{p-2}\nabla u_0 \nabla\varphi\, dx = \lambda_0 \int_{\mathbb{R}^N} g(x)|u_0|^{p-2}u_0\varphi\, dx$$

with $\varphi = u_0^-$ implies that

$$\int_{\Omega_0^-} a(x)|\nabla u_0^-|^p\, dx + \lambda_0 \int_{\Omega_0^-} g^-(x)|u_0^-|^p\, dx = \lambda_0 \int_{\Omega_0^-} g^+(x)|u_0^-|^p\, dx.$$

By the Hardy and Hölder inequalities, the definition of the norm in X, and the assumption on g^+, we derive from this that

$$c_{12}\|u_0^-\|_X^p \le \left(\int_{\Omega_0^-} (g^+(x))^{\frac{N}{p}}\,dx\right)^{\frac{p}{N}} \|u_0^-\|_{p^*}^p .$$

Consequently, we obtain, by the Sobolev imbedding theorem,

$$\left(\int_{\Omega_0^-} (g^+(x))^{\frac{N}{p}}\,dx\right)^{\frac{p}{N}} \ge c_{13} > 0 \tag{4.49}$$

(where c_{13} is independent of λ_0 and u_0). In particular, it follows from (4.49) that taking $K_0 > 0$ large enough we have

$$\operatorname{meas}(\Omega_0^- \cap B_K(0)) \ge c_{14} \tag{4.50}$$

for any $K \ge K_0$, where $c_{14} > 0$ depends neither on λ_0 nor on u_0.

Suppose now that there exists a sequence of eigenpairs (λ_n, u_n) of (4.38) with $\lambda_n \to \lambda_1$, $\lambda_n \ne \lambda_1$. Then $\lambda_n > \lambda_1$ and without loss of generality we may assume that $\|u_n\|_X = 1$, $u_n \rightharpoonup \tilde{u}$ in X for some $\tilde{u} \in X$. By this weak convergence and the simplicity of λ_1 we get from (4.38) that either $\tilde{u} = u_1$ or $\tilde{u} = -u_1$. Assume further that $u_n \rightharpoonup u_1 > 0$ in $\mathbb{R}^N$. It follows from (4.38) that

$$\begin{aligned}\int_{\mathbb{R}^N} & a(x)(|\nabla u_n|^{p-2}\nabla u_n - |\nabla u_1|^{p-2}\nabla u_1)\nabla(u_n - u_1)\,dx \\ &= \int_{\mathbb{R}^N} \lambda_n g(x)(|u_n|^{p-2}u_n - |u_1|^{p-2}u_1)(u_n - u_1)\,dx \\ &\quad + (\lambda_n - \lambda_1)\int_{\mathbb{R}^N} g(x)|u_1|^{p-2}u_1(u_n - u_1)\,dx .\end{aligned} \tag{4.51}$$

We then deduce

$$\begin{aligned}\int_{\mathbb{R}^N} & a(x)(|\nabla u_n|^{p-2}\nabla u_n - |\nabla u_1|^{p-2}\nabla u_1)(\nabla u_n - \nabla u_1)\,dx \\ &\ge \left[\left(\int_{\mathbb{R}^N} a(x)|\nabla u_n|^p\,dx\right)^{\frac{p-1}{p}} - \left(\int_{\mathbb{R}^N} a(x)|\nabla u_1|^p\,dx\right)^{\frac{p-1}{p}}\right] \\ &\quad \cdot \left[\left(\int_{\mathbb{R}^N} a(x)|\nabla u_n|^p\,dx\right)^{\frac{1}{p}} - \left(\int_{\mathbb{R}^N} a(x)|\nabla u_1|^p\,dx\right)^{\frac{1}{p}}\right] \ge 0\end{aligned} \tag{4.52}$$

and

$$
\begin{aligned}
&\int_{\mathbb{R}^N} \lambda_n g(x)(|u_n|^{p-2}u_n - |u_1|^{p-2}u_1)(u_n - u_1)\,dx \\
&\qquad + (\lambda_n - \lambda_1)\int_{\mathbb{R}^N} g(x)|u_1|^{p-2}u_1(u_n - u_1)\,dx \\
&\quad \le \int_{|x|\le K} \lambda_n g^+(x)(|u_n|^{p-2}u_n - |u_1|^{p-2}u_1)(u_n - u_1)\,dx \\
&\qquad + \int_{|x|>K} \lambda_n g^+(x)(|u_n|^{p-2}u_n - |u_1|^{p-2}u_1)(u_n - u_1)\,dx \\
&\qquad + c_{15}|\lambda_n - \lambda_1|(\|u_n\|_X^p + \|u_1\|_X^p)\,.
\end{aligned} \tag{4.53}
$$

The right-hand side of (4.53) can be made arbitrarily small as $n \to \infty$. Hence it follows from (4.51)–(4.53) that

$$\int_{\mathbb{R}^N} a(x)|\nabla u_n|^p\,dx \to \int_{\mathbb{R}^N} a(x)|\nabla u_1|^p\,dx\,.$$

By the Hardy inequality and (4.38), (4.39) we have

$$\int_{\mathbb{R}^N} w(x)|u_n|^p\,dx \to \int_{\mathbb{R}^N} w(x)|u_1|^p\,dx\,,$$

and therefore $u_n \to u_1$ in $L^p(\mathbb{R}^N, w)$. Fix some $K \ge K_0$. Then $w(x) \ge w_0 > 0$ and hence $u_n \to u_1$ in $L^p(B_K(0))$. By the Egorov theorem, u_n converge uniformly to u_1 on $B_K(0)$ with the exception of a set of arbitrarily small measure. But this contradicts (4.50), where we consider $\Omega_n^- = \{x \in \mathbb{R}^N;\ u_n(x) < 0\}$ instead of Ω_0^-. Thus we have proved that λ_1 is isolated. □

Remark 4.7 The eigenvalue problem (4.38) is investigated in W. Allegretto, Y.X. Huang [3] also for $p \ge N$. By their result, in order for $\lambda_1 > 0$ to exist, we have to assume $g(x) = g_1(x) - g_2(x)$, $g_i(x) \ge 0$, $i = 1, 2$,

$$g_1(x) \not\equiv 0\,, \quad g_1 \in L^\infty(\mathbb{R}^N) \cap L^{\frac{N_0}{p}}(\mathbb{R}^N) \quad \text{for some} \quad N_0 > p\,, \tag{4.54}$$

and moreover,

$$g_2 \in L^\infty(\mathbb{R}^N) \quad \text{and} \quad g_2(x) \ge \varepsilon > 0 \quad \text{in} \quad \mathbb{R}^N\,. \tag{4.55}$$

One can prove (see W. Allegretto, Y. X. Huang [3]) that the assumption (4.55) implies, in particular, that $\lambda_1^* < 0$ (cf. Remark 4.5) does not exist.

Let us assume (4.54), (4.55). Then we can assume, without loss of generality, that $g_2(x) \ge \omega(x)$ and so the norm in X is

$$\|u\|_X = \left(\int_{\mathbb{R}^N} a(x)|\nabla u|^p\,dx + \int_{\mathbb{R}^N} g_2(x)|u|^p\,dx\right)^{\frac{1}{p}}.$$

In the assumptions we replace p^* by $p_0 = \dfrac{N_0 p}{N_0 - p}$, and $(p^*)'$ by p_0'; concerning σ we assume $\sigma \in L^{\frac{N_0}{p}}(\mathbb{R}^N)$. By the Sobolev imbedding theorem we have

$$\left(\int_{\mathbb{R}^N} |u|^{p_0}\,dx\right)^{\frac{1}{p_0}} \le c_{16}\|u\|_X\,.$$

Then the proofs in this section can be carried over and the conclusions remain valid also for the case $p \ge N$.

The following definition is an analogue to Definition 3.6.

Definition 4.3 Define the space $E := \mathbb{R} \times X$ with the norm

$$\|(\lambda, u)\|_E = (|\lambda|^2 + \|u\|_X^2)^{\frac{1}{2}}\,, \qquad (\lambda, u) \in E$$

as in Section 3.6. Let

$$C \subset \{(\lambda, u) \in E\,;\ (\lambda, u) \text{ solves (4.30) weakly,} \quad u \neq 0\}\,,$$

C be connected with respect to the topology induced by the norm on E. Then it is called a *continuum of nontrivial solutions* of (4.30). The point $\lambda_0 \in \mathbb{R}$ is said to be a *global bifurcation point* of (4.30) (in the sense of Rabinowitz) if there is a continuum of nontrivial solutions C of (4.30) such that $(\lambda_0, 0) \in \overline{C}$ and C is either unbounded in E or there is an eigenvalue $\hat{\lambda}$ of (4.38) such that $\hat{\lambda} \neq \lambda_0$ and $(\hat{\lambda}, 0) \in \overline{C}$.

Let us consider the operators $M_\lambda, \tilde{M}_\lambda \colon X \to X^*$ defined by

$$M_\lambda(u) := J(u) - \lambda G(u) - F_\lambda(u)\,, \quad \tilde{M}_\lambda(u) := J(u) - \lambda G(u)\,.$$

Since G^+ and F_λ are compact operators (see Lemma 4.4), M_λ and $\tilde{M}_\lambda$ will satisfy condition $\alpha(X)$ for $\lambda > 0$ if the following assertion holds.

Lemma 4.7 *The operator $J + \lambda G^-$ satisfies condition $\alpha(X)$ for $\lambda > 0$.*

Proof. Assume $u_n \rightharpoonup u_0$ in X and

$$\limsup_{n\to\infty} \langle J(u_n) + \lambda G^-(u_n), u_n - u_0\rangle \le 0\,.$$

Then we have

$$\begin{aligned}
0 &\ge \limsup_{n\to\infty} \big(\langle J(u_n) - J(u_0), u_n - u_0\rangle + \lambda\langle G^-(u_n) - G^-(u_0), u_n - u_0\rangle\big)\\
&= \limsup_{n\to\infty}\bigg(\int_{\mathbb{R}^N} a(x)(|\nabla u_n|^{p-2}\nabla u_n - |\nabla u_0|^{p-2}\nabla u_0)(\nabla u_n - \nabla u_0)\,dx\\
&\qquad + \lambda\int_{\mathbb{R}^N} g^-(x)(|u_n|^{p-2}u_n - |u_0|^{p-2}u_0)(u_n - u_0)\,dx\bigg).
\end{aligned}$$

Similarly as in the proof of Lemmas 3.12 and 4.6 we derive from here that

$$\int_{\mathbb{R}^n} a(x)|\nabla u_n|^p\,dx \to \int_{\mathbb{R}^n} a(x)|\nabla u_0|^p\,dx\,,$$
$$\int_{\mathbb{R}^n} g^-(x)|u_n|^p\,dx \to \int_{\mathbb{R}^n} g^-(x)|u_0|^p\,dx\,.$$

This together with the weak convergence $u_n \rightharpoonup u_0$ in X implies that $\nabla u_n \to \nabla u_0$ in $L^p(\mathbb{R}^N, a)$. Moreover, this also means that $u_n \to u_0$ in X if $p \geq N$. For the case $1 < p < N$, it follows from the Hardy inequality that

$$\int_{\mathbb{R}^N} \frac{|u_n - u_0|^p}{(1+|x|)^p}\,dx \leq \left(\frac{p}{N-p}\right)^p \int_{\mathbb{R}^N} |\nabla(u_n - u_0)|^p\,dx \to 0\,.$$

Due to the definition of the norm $\|\cdot\|_X$ we conclude again $u_n \to u_0$ in X. □

Lemmas 4.3, 4.4 and 4.7 imply that the degrees

$$\operatorname{Deg}[M_\lambda; D, 0] \quad \text{and} \quad \operatorname{Deg}[\tilde{M}_\lambda; D, 0]$$

are well defined for any $\lambda > 0$ and any bounded nonempty open set $D \subset X$ such that $M_\lambda(u) \neq 0$ and $\tilde{M}_\lambda(u) \neq 0$ for any $u \in \partial D$, respectively. If $g \in L^\infty(\mathbb{R}^N) \cap L^{\frac{N}{p}}(\mathbb{R}^N)$ $(1 < p < N)$ then G is compact (cf. Remark 4.5) and the degree can be defined also for $\lambda \leq 0$.

We can now prove the main result of this section.

Theorem 4.3 *Let* $1 < p < N$ *and suppose that* a, g *and* f *satisfy hypotheses* (a), (g1), (g2), $(\hat{f}1)$–$(\hat{f}3)$. *Then the principal eigenvalue* λ_1 *of* (4.38) *is a global bifurcation point of* (4.30).

Proof. The idea of the proof is the same as that of the proof of Theorem 3.7. That is why we only sketch it and point out the differences. Recall that the Steps 1 and 3 are standard, Step 2 is the main and new part of the proof.

Step 1 Consider the operators M_λ and $\tilde{M}_\lambda$ defined above. It follows from the variational characterization of λ_1 that for $\lambda \in (0, \lambda_1)$ and any $u \in X$ with $\|u\|_X \neq 0$, we have

$$\langle \tilde{M}_\lambda(u), u\rangle > 0$$

and henceforth due to Theorem 1.6 we get

$$\operatorname{Deg}[\tilde{M}_\lambda; B_r(0), 0] = 1\,, \quad \lambda \in (0, \lambda_1) \tag{4.56}$$

for any ball $B_r(0) \subset X$.

Step 2 According to Lemma 4.6 there exists $\delta > 0$ such that the interval $(\lambda_1, \lambda_1 + \delta)$ contains no eigenvalue of the problem (4.38) and we can evaluate $\mathrm{Ind}(\tilde{M}_\lambda, 0)$ for $\lambda \in (\lambda_1, \lambda_1 + \delta)$. Set

$$\hat{\Psi}_\lambda(u) = \frac{1}{p}\langle J(u), u\rangle - \frac{\lambda}{p}\langle G(u), u\rangle + \psi\left(\frac{1}{p}\langle J(u), u\rangle\right)$$

with the function $\psi: \mathbb{R} \to \mathbb{R}$ defined in the proof of Theorem 3.7. Then clearly the same argument applies showing that the derivative $\hat{\Psi}'_\lambda$ has precisely three isolated critical points:

$$-u_1, 0, u_1 .$$

To show that $\hat{\Psi}_\lambda$ is weakly lower semicontinuous we argue as follows. Let us assume $u_n \rightharpoonup u_0$ in X. Then

$$\langle G^+(u_n), u_n\rangle \to \langle G^+(u_0), u_0\rangle \tag{4.57}$$

due to the compactness of G^+ (see Lemma 4.4), and

$$\begin{aligned} \liminf_{n\to\infty} &\left[\frac{1}{p}\langle J(u_n), u_n\rangle + \frac{\lambda}{p}\langle G^-(u_n), u_n\rangle + \psi\left(\frac{1}{p}\langle J(u_n), u_n\rangle\right)\right] \\ &\geq \frac{1}{p}\langle J(u_0), u_0\rangle + \frac{\lambda}{p}\langle G^-(u_0), u_0\rangle + \psi\left(\frac{1}{p}\langle J(u_0), u_0\rangle\right) \end{aligned} \tag{4.58}$$

by the fact that $\liminf_{n\to\infty} \|u_n\|_X \geq \|u_0\|_X$, $\liminf_{n\to\infty} \|\nabla u_n\|_p \geq \|\nabla u_0\|_p$ and that ψ is nondecreasing. The relations (4.57) and (4.58) then imply

$$\liminf_{n\to\infty} \hat{\Psi}_\lambda(u_n) \geq \hat{\Psi}_\lambda(u_0) .$$

The same argument as that used to prove (3.121) applies also here and so we have that $\hat{\Psi}_\lambda$ is coercive, i.e.

$$\lim_{\|u\|_X\to\infty} \hat{\Psi}_\lambda(u) = \infty .$$

Thus it follows that there are precisely two points at which the minimum of $\hat{\Psi}_\lambda$ is achieved: $-u_1, u_1$, and that the isolated critical point 0 of $\hat{\Psi}_\lambda$ is of "saddle type". In virtue of Theorem 1.8 we have

$$\mathrm{Ind}(\hat{\Psi}'_\lambda, -u_1) = \mathrm{Ind}(\hat{\Psi}'_\lambda, u_1) = 1 . \tag{4.59}$$

The same argument as in the proof of Theorem 3.7 proves that

$$\langle \hat{\Psi}'_\lambda(u), u\rangle > 0$$

for any $u \in X$, $\|u\|_X = \kappa$, with κ large enough and so

$$\mathrm{Deg}[\hat{\Psi}'_\lambda; B_\kappa(0), 0] = 1 \tag{4.60}$$

by Theorem 1.6. The additivity property of the degree (see Theorem 1.7) and (4.59), (4.60) yield

$$\operatorname{Ind}(\hat{\Psi}'_\lambda, 0) = -1 \quad \text{for } \lambda \in (\lambda_1, \lambda_1 + \delta). \tag{4.61}$$

So, as in the proof of Theorem 3.7, we have the following conclusion from (4.56), (4.61):

$$\begin{aligned} \operatorname{Ind}(\tilde{M}_\lambda, 0) &= 1 \quad \text{for} \quad \lambda \in (0, \lambda_1), \\ \operatorname{Ind}(\tilde{M}_\lambda, 0) &= -1 \quad \text{for} \quad \lambda \in (\lambda_1 \lambda_1 + \delta). \end{aligned} \tag{4.62}$$

If follows now from (4.33), (4.62) and the homotopy invariance property of the degree (see Theorem 1.5 (iii)) that

$$\begin{aligned} \operatorname{Ind}(M_\lambda, 0) &= 1 \quad \text{for} \quad \lambda \in (0, \lambda_1), \\ \operatorname{Ind}(M_\lambda, 0) &= -1 \quad \text{for} \quad \lambda \in (\lambda_1, \lambda_1 + \delta). \end{aligned}$$

Step 3 Following the proof of Theorem 1.3 in P. H. Rabinowitz [57] we get the conclusion. □

As a consequence of our considerations in Remark 4.5 we have the following theorem.

Theorem 4.4 *Let* $1 < p < N$ *and assume* (a), ($\hat{\text{f1}}$)–($\hat{\text{f3}}$) *and* $g^\pm \in L^\infty(\mathbb{R}^N) \cap L^{\frac{N}{p}}(\mathbb{R}^N)$, $g^\pm \not\equiv 0$. *Then the conclusion of Theorem* 4.3 *remains valid.*

Moreover, the principal eigenvalue $\lambda_1^* < 0$ *of the eigenvalue problem*

$$-\operatorname{div}(a(x)|\nabla u|^{p-2}\nabla u) = \lambda g(x)|u|^{p-2}u \quad in \quad \mathbb{R}^N, \qquad \int_{\mathbb{R}^N} g|u|^p\,dx < 0,$$

is a global bifurcation point of (4.30).

Modifying the assumptions in the spirit of Remark 4.7 if $p \geq N$, we have the following analogue of Theorem 4.3.

Theorem 4.5 *Let* $p \geq N$. *Assume* (a), ($\hat{\text{f1}}$)–($\hat{\text{f3}}$) *(with* p^* *replaced by* p_0*) and* $g_1 \in L^\infty(\mathbb{R}^N) \cap L^{\frac{N_0}{p}}(\mathbb{R}^N)$, $g_1 \not\equiv 0$, $N_0 > p$, $g_2 \in L^\infty(\mathbb{R}^N)$, $g_2(x) \geq \varepsilon > 0$ *in* $\mathbb{R}^N$. *Then the principal eigenvalue* $\lambda_1 > 0$ *of the eigenvalue problem* (4.38) *is a global bifurcation point of* (4.30).

Note that by the reasons mentioned in Remark 4.7 we have no analogue of Theorem 4.4 if $p \geq N$.

We can provide more precise information about the structure and qualitative behaviour of the continuum of nontrivial solutions if we consider the nonlinearity f of more special form. Instead of ($\hat{\text{f2}}$) we will further assume

$(\hat{\text{f}}2)'$ there is a nonnegative $\tilde{\rho} \in L^\infty(\mathbb{R}^N) \cap L^{\gamma_1}(\mathbb{R}^N)$ such that

$$|f(\lambda, x, s)| \leq c(\lambda)\tilde{\rho}(x)|s|^\gamma$$

for all $\lambda, s \in \mathbb{R}$ and a.e. $x \in \mathbb{R}^N$, where $c(\lambda)$, γ_1 and γ are as in $(\hat{\text{f}}2)$.

Then we have the following result.

Proposition 4.1 *Let* $1 < p < N$, *and assume that* (a), (g1), (g2), $(\hat{\text{f}}1)$, $(\hat{\text{f}}2)'$, $(\hat{\text{f}}3)$ *hold. Then for any weak solution* u *of* (4.30) *with* $\lambda \geq 0$, *we have* $u \in L^r(\mathbb{R}^N)$, $p^* \leq r \leq \infty$ *and* $u(x)$ *decays uniformly as* $|x| \to \infty$. *Moreover,* $u \in C^{1,\alpha}(B_K(0))$ *for any* $K > 0$ *with some* $\alpha = \alpha(K) \in (0, 1)$ *if* $a \in C^1(\mathbb{R}^N)$.

Proof. Let u be a weak solution of (4.30). For $u^+(x) = \max\{u(x), 0\}$ and $M > 0$ define

$$v_M(x) = \min\{u^+(x), M\}.$$

Choose $\varphi = v_M^{\kappa p+1}$ $(\kappa \geq 0)$ as a test function in

$$\int_{\mathbb{R}^N} a(x)|\nabla u|^{p-2}\nabla u \nabla\varphi\, dx = \lambda \int_{\mathbb{R}^N} g(x)|u|^{p-2}u\varphi\, dx + \int_{\mathbb{R}^N} f(\lambda, x, u)\varphi\, dx.$$

We then get

$$\begin{aligned}(\kappa p + 1)\int_{\mathbb{R}^N} a(x)v_M^{\kappa p}|\nabla v_M|^p\, dx &+ \lambda \int_{\mathbb{R}^N} g^-(x)|u^+|^{p-2}u^+v_M^{\kappa p+1}\, dx \\ &= \lambda \int_{\mathbb{R}^N} g^+(x)|u^+|^{p-2}u^+v_M^{\kappa p+1}\, dx + \int_{\mathbb{R}^N} f(\lambda, x, u^+)v_M^{\kappa p+1}\, dx.\end{aligned} \tag{4.63}$$

The left-hand side of (4.63) is estimated from below similarly as in (4.22) by the expression

$$\frac{1}{c_1^p}\frac{\kappa p + 1}{(\kappa + 1)^p}\left(\int_{\mathbb{R}^N} v_M^{(\kappa+1)p^*}\, dx\right)^{\frac{p}{p^*}}. \tag{4.64}$$

For the estimate of the right-hand side of (4.63) we use the definition of $\delta > 0$ and $q \in (p, p^*)$ from the proof of Theorem 4.1. The first integral on the right-hand side in (4.63) is estimated from above by the expression

$$\lambda\left(\int_{\mathbb{R}^N} g^+(x)^{\frac{q}{q-p}}\, dx\right)^{\frac{q-p}{q}}\left(\int_{\mathbb{R}^N} u^+(x)^{(\kappa+1)q}\, dx\right)^{\frac{p}{q}}, \tag{4.65}$$

where the first integral is finite due to (g1) and $\dfrac{N}{p} < \dfrac{q}{q-p}$. The second integral on the right-hand side of (4.63) is estimated by

$$c(\lambda)\left(\int_{\mathbb{R}^N} \tilde{\rho}(x)^{\gamma_1+\delta}\, dx\right)^{\frac{1}{\gamma_1+\delta}}\left(\int_{\mathbb{R}^N} u^+(x)^{p^*}\, dx\right)^{\frac{q-p(\gamma_1+\delta)'}{q(\gamma_1+\delta)'}}\left(\int_{\mathbb{R}^N} u^+(x)^{(\kappa+1)q}\, dx\right)^{\frac{p}{q}}. \tag{4.66}$$

The first two integrals are finite due to $(\hat{f2})'$ and the Sobolev inequality (1.12). Hence it follows from (4.63)–(4.66) that there exists a constant $c_{17} = c_{17}(\lambda) > 0$ such that

$$\|v_M\|_{(\kappa+1)p^*} \le c_{17}^{\frac{1}{\kappa+1}} \left(\frac{\kappa+1}{(\kappa p+1)^{\frac{1}{p}}} \right)^{\frac{1}{\kappa+1}} \|u^+\|_{(\kappa+1)q} \,.$$

Then similarly as in the proof of Theorem 4.1 we get that $u^+ \in L^r(\mathbb{R}^N)$ for any $p^* \le r \le \infty$. The same approach yields $u^- \in L^r(\mathbb{R}^N)$.

The decay of $u(x)$ as $|x| \to \infty$ follows from the estimate in Theorem 1.10 and the regularity of u from Theorem 1.11. The proof of Proposition 4.1 is complete. □

Let $u^* \in X^*$ be fixed and satisfy

$$\langle u^*, u_1 \rangle = 1 \,,$$

where $u_1 > 0$ is the eigenfunction corresponding to the principal eigenvalue $\lambda_1 > 0$ of (4.38). For $\tau \in (0, 1)$ define the sets

$$K_\tau^+ = \left\{ (\lambda, u) \in E\,;\ \langle u^*, u \rangle > \tau \|u\|_X \right\} ,$$
$$K_\tau^- = \left\{ (\lambda, u) \in E\,;\ \langle u^*, u \rangle < -\tau \|u\|_X \right\} .$$

Note that the sets $K_\tau^\pm$ are cones with the vertex at the origin containing (λ, u_1), $(\lambda, -u_1)$, respectively, and the number τ is a measure of "the openness" of these cones. Furthermore, let $B_\eta(\lambda_1, 0)$ denote the ball in E centered at the point $(\lambda_1, 0)$ with radius $\eta > 0$.

Theorem 4.6 *Let the assumptions of Proposition* 4.1 *be fulfilled with* $a \in C^1(\mathbb{R}^N)$. *Moreover, suppose that the equation*

$$J(u) = \lambda_1 G(u) + F_{\lambda_1}(u) \tag{4.67}$$

has no nonzero solution in X. *Then there are maximal (in the sense of inclusion) connected sets* C^+, C^- *of* $\{(\lambda, u) \in E;\ (\lambda, u)$ *solves* (4.30) *weakly,* $u \ne 0\}$ *containing* $(\lambda_1, 0)$ *in their closures,*

$$C^\pm \cap B_\eta(\lambda_1, 0) \subset K_\tau^\pm \quad (\eta = \eta(\tau) \to 0 \quad as \quad \tau \to 1)\,, \tag{4.68}$$

and such that both $C^\pm$ *are unbounded in* E. *Moreover,* $\lambda > \lambda_1$ *for any* $(\lambda, u) \in C^\pm$ *and we have* $u > 0$ *in* $\mathbb{R}^N$ *if* $(\lambda, u) \in C^+$ *and* $u < 0$ *in* $\mathbb{R}^N$ *if* $(\lambda, u) \in C^-$.

Proof. We can apply the same method as that used in the proof of Theorem 2 in E. N. Dancer [13]. We get the existence of maximal connected sets $C^\pm \subset \{(\lambda, u) \in E;$ (λ, u) solves (4.30) weakly, $u \ne 0\}$ satisfying (4.68) and either being unbounded in E or containing in their closures a common point of E different from $(\lambda_1, 0) \in E$. We will show now that the latter case cannot occur.

Note that $\lambda > \lambda_1$ for any $(\lambda, u) \in C^{\pm}$ due to the fact that (4.67) has no nonzero solution for $\lambda = \lambda_1$. Let $(\lambda_n, u_n) \in C^+$ be a sequence such that $\lambda_n \to \lambda_1$ and $\|u_n\|_X \to 0$, $\|u_n\|_X \neq 0$. Denoting $\tilde{u}_n = \dfrac{u_n}{\|u_n\|_X}$ we can assume due to the simplicity of λ_1, $(\hat{f2})'$ and $(\hat{f3})$ that $\tilde{u}_n \rightharpoonup u_1$ in X. Due to (4.33) we derive that $\tilde{u}_n \to u_1$ using the same argument as in the proof of Lemma 4.6. We prove now that $u_n > 0$ for n large enough. Denote $\Omega_n^- = \{x \in \mathbb{R}^N ;\ u_n(x) < 0\}$. Then we have

$$\int_{\mathbb{R}^N} a(x)|\nabla u_n^-|^p\,dx + \lambda_n \int_{\mathbb{R}^N} g^-(x)|u_n^-|^p\,dx$$
$$= \lambda_n \int_{\mathbb{R}^N} g^+(x)|u_n^-|^p\,dx - \int_{\mathbb{R}^N} f(\lambda_n, x, u_n)u_n^-\,dx\,.$$

By the Hardy, Hölder, and Sobolev inequalities, and the assumptions on g and $(\hat{f2})'$, we get

$$c_{18}\|u_n^-\|_X^p \le c_{19}\left(\int_{\Omega_n^-}(g^+(x))^{\frac{N}{p}}\,dx\right)^{\frac{p}{N}}\|u_n^-\|_X^p + c_{20}\left(\int_{\Omega_n^-}(\tilde{\rho}(x))^{\gamma_1}\,dx\right)^{\frac{1}{\gamma_1}}\|u_n^-\|_X^{\gamma+1},$$

i.e.

$$c_{18} \le c_{19}\left(\int_{\Omega_n^-}(g^+(x))^{\frac{N}{p}}\,dx\right)^{\frac{p}{N}} + c_{20}\left(\int_{\Omega_n^-}(\tilde{\rho}(x))^{\gamma_1}\,dx\right)^{\frac{1}{\gamma_1}}\|u_n^-\|_X^{\gamma+1-p}\,.$$

Since $\|u_n\|_X \to 0$, $g^+ \in L^{\frac{N}{p}}(\mathbb{R}^N)$, $\tilde{\rho} \in L^{\gamma_1}(\mathbb{R}^N)$ and c_{18}, c_{19}, c_{20} do not depend on u_n, we derive from here that for $K_0 > 0$ large enough,

$$\operatorname{meas}(\Omega_n^- \cap B_K(0)) \ge c_{21}$$

for any $K > K_0$, where $c_{21} > 0$ depends on neither λ_n nor u_n. Now, using the same argument as in the proof of Lemma 4.6 based on the Egorov theorem we deduce that $\tilde{u}_n$ (and hence u_n) is nonnegative in $\mathbb{R}^N$ for n large enough. It then follows that $u \ge 0$ for any $(\lambda, u) \in C^+ \cap B_\eta(\lambda_1, 0)$ with $\eta > 0$ small enough.

Assume now that there exists $(\lambda^*, v^*) \in C^+$ such that $v^* \not\equiv 0$, we have $v^* \ge 0$, and $v^*(x_0) = 0$ for some $x_0 \in \mathbb{R}^N$. This violates the Harnack inequality (see Theorem 1.9). Hence $u > 0$ for any $(\lambda, u) \in C^+$ due to the fact that C^+ is connected set in E and similarly $u < 0$, $\lambda > \lambda_1$ for all $(\lambda, u) \in C^-$. This excludes the second possibility of Theorem 2 in E. N. Dancer [13] and thus $C^{\pm}$ must be unbounded in E. □

Remark 4.8 For the case $p > N$, any solution u of BVP (4.30) is in $L^\infty(\mathbb{R}^N)$, while for $p = N$, a solution of (4.30) belongs to $L^r(\mathbb{R}^N)$ with any $r \ge N$ by the Sobolev imbedding theorem. Observe that in either case changing the assumptions in the spirit of Remark 4.7, the arguments in the proofs of Proposition 4.1 and Theorem 4.6 are still valid, so conclusions remain true also for $p \ge N$.

Remark 4.9 For $1 < p < N$, if $(\hat{f1})$, $(\hat{f2})'$, $(\hat{f3})$ hold, $g \in L^\infty(\mathbb{R}^N) \cap L^{\frac{N}{p}}(\mathbb{R}^N)$ and $g^\pm \not\equiv 0$, the same conclusions as in Proposition 4.1 and Theorem 4.6 also hold for $\lambda_1^* < 0$.

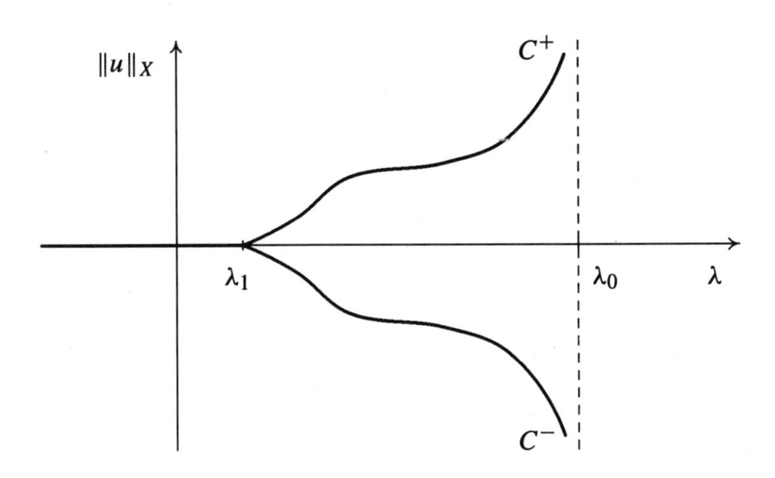

Figure 4.1: Unbounded branches C^+, C^-.

Example 4.1 Let us consider the equation

$$-\operatorname{div}(a(x)|\nabla u|^{p-2}\nabla u) = \lambda g(x)|u|^{p-2}u - c(\lambda)\tilde{\rho}(x)|u|^{\gamma-1}u \quad \text{in} \quad \mathbb{R}^N ,$$

where $a(x)$ satisfies (a), $a \in C^1(\mathbb{R}^N)$, $g(x)$ satisfies (g1), (g2),γ, $\tilde{\rho}(x) > 0$ are as in $(\hat{f2})$' and $c(\lambda) > 0$ is a continuous function defined on $\mathbb{R}$. Then the assumptions of Theorem 4.6 are satisfied. Indeed, due to the variational characterization of λ_1, we have

$$\int_{\mathbb{R}^N} a(x)|\nabla u|^p \, dx - \lambda \int_{\mathbb{R}^N} g(x)|u|^p \, dx > 0$$

and

$$\int_{\mathbb{R}^N} f(\lambda, x, u)u \, dx = -c(\lambda) \int_{\mathbb{R}^N} \tilde{\rho}(x)|u|^{\gamma+1} \, dx < 0$$

for any $\lambda \in [\lambda_1 - \delta, \lambda_1)$ and any $u \in X$ with $u \not\equiv 0$ and some $\delta > 0$. Note that in this case we have $(\lambda, u) \in C^+$ if and only if $(\lambda, -u) \in C^-$.

Let us suppose that $c(\lambda_0) = 0$ at some point $\lambda_0 > \lambda_1$. Then due to Theorem 4.6 the set C^+ (and similarly C^-) "blows up" in $\|u\|_X$. In fact, since

$$-\operatorname{div}(a(x)|\nabla u|^{p-2}\nabla u) = \lambda_0 g(x)|u|^{p-2}u$$

cannot have a positive solution in $\mathbb{R}^N$ (the eigenfunction corresponding to any eigenvalue $\lambda_0 > \lambda_1$ changes sign in $\mathbb{R}^N$), the parameter λ cannot cross the value λ_0 and hence $\|u\|_X$ is unbounded along C^+ (or C^-). ◇

4.3 Bifurcation problem for the perturbed p-Laplacian in $\mathbb{R}^N$

In this section we are concerned with the existence of positive solutions of the nonlinear problem

$$-\operatorname{div}(a(x,u)|\nabla u|^{p-2}\nabla u) = \lambda g(x)|u|^{p-2}u + f(\lambda, x, u) \quad \text{in} \quad \mathbb{R}^N, \tag{4.69}$$

with $1 < p < N$, $x \in \mathbb{R}^N$, $g(x)$, $f(\lambda, x, u)$ satisfying similar conditions as in the preceding section but with a function $a(x,u)$ which (similarly as in Chapter 3) has no growth restrictions with respect to u and can possess degeneracies or singularities in x. The main result of this section is again the bifurcation theorem and even if the method is a direct extension of that used in Section 4.2, the method itself cannot be applied directly to (4.69). Note that nontrivial difficulties arise due to the nonlinear dependence of $a(x,u)$ on u and due to the above mentioned singularity or degeneracy of $a(x,u)$ in x described by a nontrivial weight function (see below). A very important role in dealing with these difficulties is played by the generalized weighted Sobolev inequality (see Example 1.8, inequality (1.46)).

Let us introduce two weight functions $\nu(x)$ and $\omega(x)$ as follows:

$$\nu(x) = \frac{1}{(1+|x|)^{\beta}}, \quad \omega(x) = \frac{1}{(1+|x|)^{\alpha}}, \quad x \in \mathbb{R}^N$$

with $\alpha = p + \beta$, $\alpha < N$. Recall that the following generalized Hardy inequality holds (see Example 1.7):

$$\int_{\mathbb{R}^N} \omega(x)|u|^p \, dx \le \left(\frac{p}{N-\alpha}\right)^p \int_{\mathbb{R}^N} \nu(x)|\nabla u|^p \, dx. \tag{4.70}$$

Let us assume that the function $g(x)$ satisfies

($\tilde{g1}$) $\quad g^+\nu^{-1} \in L^\infty(\mathbb{R}^N) \cap L^{\frac{N}{p}}(\mathbb{R}^N)$, $g^+ \not\equiv 0$,

($\tilde{g2}$) $\quad g^- \in L^\infty(\mathbb{R}^N)$.

Denote $w(x) = \max\{\omega(x),\ g^-(x)\}$. Then we define the norm

$$\|u\|_X = \left(\int_{\mathbb{R}^N} \nu(x)|\nabla u|^p \, dx + \int_{\mathbb{R}^N} w(x)|u|^p \, dx\right)^{\frac{1}{p}},$$

and the basic space X we will work in is defined as the completion of $C_0^\infty(\mathbb{R}^N)$ with respect to the norm $\|\cdot\|_X$.

We assume that the function $a(x,s)$ satisfies

(a1) $a(x,s)$ is defined on $\mathbb{R}^N \times \mathbb{R}$, $a \in \mathrm{CAR}$ and, for all $s \in \mathbb{R}$ and for a.e. $x \in \mathbb{R}^N$,

$$a(x,s) > c_1 \nu(x);$$

(a2) for any $s_0 \in \mathbb{R}$,

$$\lim_{s \to s_0} \frac{a(x,s) - a(x,s_0)}{\nu(x)} = 0 \tag{4.71}$$

uniformly for a.e. $x \in \mathbb{R}^N$. Moreover, there exists a nondecreasing function $\tilde{g}: [0,\infty) \to [1,\infty)$ such that

$$a(x,s) \le \tilde{g}(|s|)\nu(x) \tag{4.72}$$

for all $s \in \mathbb{R}$ and for a.e. $x \in \mathbb{R}^N$.

We assume that $f(\lambda, x, s)$ satisfies

($\tilde{\text{f1}}$) $f(\lambda, x, s) \in$ CAR (cf. Section 4.2);

($\tilde{\text{f2}}$) there is a nonnegative function $c(\lambda)$ defined on $\mathbb{R}$, bounded on bounded subsets of $\mathbb{R}$, such that

$$|f(\lambda,x,s)| \le c(\lambda)(\sigma(x) + \rho(x)|s|^{\gamma})$$

holds for a.e. $x \in \mathbb{R}^N$ and for all $s \in \mathbb{R}$, where $p - 1 < \gamma < p^* - 1, 0 \le \rho(x)$, $\rho\nu^{\frac{-(\gamma+1)}{p}} \in L^{\gamma_1}(\mathbb{R}^N)$ with $\gamma_1 = \dfrac{p^*}{p^* - (\gamma+1)}$, $0 \le \sigma \in L^{\frac{N}{p}}(\mathbb{R}^N, (\frac{\nu}{w})^{\frac{-N}{p}})$ and $\sigma\nu^{\frac{-1}{p}} \in L^{(p^*)'}(\mathbb{R}^N)$ with $(p^*)' = \dfrac{Np}{Np - (N-p)}$;

($\tilde{\text{f3}}$) let $\lambda \in \Lambda \subset \mathbb{R}$, where Λ is a bounded interval, then for any $\varepsilon > 0$ there exists $\delta > 0$ such that

$$|f(\lambda,x,s)| \le \varepsilon w(x)|s|^{p-1}$$

holds for every $x \in \mathbb{R}^N$ and $w(x)|s|^{p-1} < \delta$.

Define operators $J, G, G^{\pm}, F_\lambda: X \to X^*$ by the same formulas as at the beginning of Section 4.2. Moreover, define, for any real number $d > 0$,

$$\tilde{a}(x,s) = \begin{cases} a(x,-d) & \text{for } x \in \mathbb{R}^N, s < -d, \\ a(x,s) & \text{for } x \in \mathbb{R}^N, |s| < d, \\ a(x,d) & \text{for } x \in \mathbb{R}^N, s > d, \end{cases}$$

and an operator $A: X \to X^*$ by

$$\langle A(u), \varphi \rangle = \int_{\mathbb{R}^N} \tilde{a}(x,u)|\nabla u|^{p-2}\nabla u \nabla\varphi \, dx .$$

Now, we have the following Lemmas 4.8 and 4.9 which summarize some basic properties of operators defined above.

Lemma 4.8 *The operators* $A, J, G, F_\lambda : X \to X^*$ *are well defined,* G *and* J *are* $(p-1)$*-homogeneous,* A *is continuous, and* F_λ *satisfies*

$$\lim_{\|u\|_X \to 0} \frac{\|F_\lambda(u)\|_{X^*}}{\|u\|_X^{p-1}} = 0 \tag{4.73}$$

uniformly for λ *in a bounded subset of* $\mathbb{R}$.

Proof. The proof follows the lines of that of Lemma 4.3. That is why we only sketch it. Homogenity of J and G is obvious. By the Hölder inequality and (4.72) we have

$$\begin{aligned}
|\langle A(u), \varphi\rangle| &= \left| \int_{\mathbb{R}^N} \tilde{a}(x,u)|\nabla u|^{p-2}\nabla u \nabla\varphi \, dx \right| \\
&\le \int_{\mathbb{R}^N} \tilde{g}(d)\nu(x)|\nabla u|^{p-2}|\nabla u \nabla \varphi| \, dx \\
&\le \tilde{g}(d) \left(\int_{\mathbb{R}^N} \nu(x)|\nabla u|^p \, dx \right)^{\frac{1}{p'}} \left(\int_{\mathbb{R}^N} \nu(x)|\nabla\varphi|^p \, dx \right)^{\frac{1}{p}}.
\end{aligned}$$

So A is well defined. A similar procedure implies that also J is well defined.

Using the Hölder inequality again, we have

$$\begin{aligned}
|\langle G(u), \varphi\rangle| &\le \int_{\mathbb{R}^N} \left| g^{\frac{1}{p'}}(x)|u|^{p-2}u g^{\frac{1}{p}}(x)\varphi \right| dx \\
&\le \left(\int_{\mathbb{R}^N} |g||u|^p \, dx \right)^{\frac{1}{p'}} \left(\int_{\mathbb{R}^N} |g||\varphi|^p \, dx \right)^{\frac{1}{p}}.
\end{aligned}$$

Note that for $u \in X$ we have $\int_{\mathbb{R}^N} g^-(x)|u|^p \, dx < \infty$ by the definition of the norm in X. By the Hölder inequality, assumptions $(\tilde{g}1)$ and the generalized Sobolev inequality, we have

$$\int_{\mathbb{R}^N} g^+(x)|u|^p \, dx \le \left(\int_{\mathbb{R}^N} \nu^{-\frac{N}{p}} (g^+)^{\frac{N}{p}} \, dx \right)^{\frac{p}{N}} \left(\int_{\mathbb{R}^N} \nu^{N_1}|u|^{p^*} \, dx \right)^{\frac{p}{p^*}} < \infty .$$

Recall that $N_1 = \dfrac{p^*}{p} = \dfrac{N}{N-p}$. A similar argument applies for $\varphi \in X$. Hence G is well defined.

For F_λ, we have

$$\begin{aligned}
|\langle F_\lambda(u), \varphi\rangle| &= \left| \int_{\mathbb{R}^N} f(\lambda, x, u)\varphi \, dx \right| \\
&\le c(\lambda) \left(\int_{\mathbb{R}^N} \sigma(x)|\varphi| \, dx + \int_{\mathbb{R}^N} \rho(x)|u|^\gamma |\varphi| \, dx \right).
\end{aligned}$$

Similarly as in the proof of Lemma 4.3, by ($\tilde{f}2$) we obtain

$$\int_{\mathbb{R}^N} \sigma(x)|\varphi|\,dx \le \left(\int_{\mathbb{R}^N} (\nu^{\frac{-1}{p}}\sigma)^{(p^*)'}\,dx\right)^{\frac{1}{(p^*)'}} \left(\int_{\mathbb{R}^N} \nu^{N_1}|\varphi|^{p^*}\,dx\right)^{\frac{1}{p^*}},$$

$$\begin{aligned}\int_{\mathbb{R}^N} \rho(x)|u|^{\gamma}|\varphi|\,dx &\le \left(\int_{\mathbb{R}^N} \left(\nu^{\frac{-N_1\gamma}{p^*}}\rho|\varphi|\right)^{\frac{p^*}{p^*-\gamma}} dx\right)^{\frac{p^*-\gamma}{p^*}} \left(\int_{\mathbb{R}^N} \nu^{N_1}|u|^{p^*}\,dx\right)^{\frac{\gamma}{p^*}} \\ &\le \left(\int_{\mathbb{R}^N} \left(\nu^{\frac{-(\gamma+1)}{p}}\rho\right)^{\gamma_1} dx\right)^{\frac{1}{\gamma_1}} \left(\int_{\mathbb{R}^N} \nu^{N_1}|\varphi|^{p^*}\,dx\right)^{\frac{1}{p^*}} \qquad (4.74) \\ &\quad\cdot \left(\int_{\mathbb{R}^N} \nu^{N_1}|u|^{p^*}\,dx\right)^{\frac{\gamma}{p^*}} \le c_2\|\varphi\|_X\|u\|_X^{\gamma}.\end{aligned}$$

Thus F_λ is well defined.

Let us prove continuity of A. Assume $u \to v$ in X. Then

$$\begin{aligned}\|A(u)-A(v)\|_{X^*} &= \sup_{\|\varphi\|_X\le 1} |\langle A(u)-A(v),\varphi\rangle| \\ &= \sup_{\|\varphi\|_X\le 1}\left|\int_{\mathbb{R}^N} (\tilde{a}(x,u)|\nabla u|^{p-2}\nabla u - \tilde{a}(x,v)|\nabla v|^{p-2}\nabla v)\nabla\varphi\,dx\right| \\ &\le \sup_{\|\varphi\|_X\le 1}\left|\int_{\mathbb{R}^N} \tilde{a}(x,v)(|\nabla u|^{p-2}\nabla u - |\nabla v|^{p-2}\nabla v)\nabla\varphi\,dx\right| \\ &\quad + \sup_{\|\varphi\|_X\le 1}\left|\int_{\mathbb{R}^N} (\tilde{a}(x,u)-\tilde{a}(x,v))|\nabla u|^{p-2}\nabla u\nabla\varphi\,dx\right|. \end{aligned} \qquad (4.75)$$

The convergence $u \to v$ in X obviously implies that

$$(\tilde{a}(x,v))^{\frac{1}{p'}}|\nabla u|^{p-2}\nabla u \to (\tilde{a}(x,v))^{\frac{1}{p'}}|\nabla v|^{p-2}\nabla v$$

in $L^{p'}(\mathbb{R}^N)$ and hence

$$\begin{aligned}&\sup_{\|\varphi\|_X\le 1}\left|\int_{\mathbb{R}^N} \tilde{a}(x,v)(|\nabla u|^{p-2}\nabla u - |\nabla v|^{p-2}\nabla v)\nabla\varphi\,dx\right| \\ &\quad\le \sup_{\|\varphi\|_X\le 1} \tilde{g}(d)\left(\int_{\mathbb{R}^N} \tilde{a}(x,v)\left|(|\nabla u|^{p-2}\nabla u - |\nabla v|^{p-2}\nabla v)\right|^{p'} dx\right)^{\frac{1}{p'}} \\ &\quad\cdot\left(\int_{\mathbb{R}^N} \nu(x)|\nabla\varphi|^p\,dx\right)^{\frac{1}{p}} \to 0.\end{aligned}$$

Let us consider the last integral in (4.75),

$$I = \int_{\mathbb{R}^N} (\tilde{a}(x,u)-\tilde{a}(x,v))|\nabla u|^{p-2}\nabla u\nabla\varphi\,dx.$$

It follows from (4.71) that for any $\varepsilon > 0$ there exists a $\delta > 0$ such that

$$|\tilde{a}(x,u) - \tilde{a}(x,v)| \le \varepsilon \nu(x)$$

provided $|u(x) - v(x)| < \delta$. Define

$$\Omega_\delta^*(u) = \{x \in \mathbb{R}^N\,;\ |u(x) - v(x)| \ge \delta\}\,.$$

Let $K > 0$ and let $B_K(0) \subset \mathbb{R}^N$ be the ball centered at the origin with radius K. Then $u \to v$ in X implies $u \to v$ in $L^p(B_K(0))$ and hence we get

$$\operatorname{meas}(\Omega_\delta^*(u) \cap B_K(0)) \to 0\,.$$

Now, the integral I on $\mathbb{R}^N \setminus \Omega_\delta^*(u)$ can be estimated by

$$\varepsilon \int_{\mathbb{R}^N \setminus \Omega_\delta^*(u)} \nu(x)|\nabla u|^{p-1}|\nabla\varphi|\,dx \le \varepsilon \|u\|_X^{p-1}\|\varphi\|_X\,.$$

On the other hand, the integral I on $\Omega_\delta^*(u)$ is estimated by

$$c_3\left[\left(\int_{\Omega_\delta^*(u)\cap B_K(0)} \nu(x)|\nabla u|^p\,dx\right)^{\frac{p-1}{p}} + \left(\int_{|x|>K} \nu(x)|\nabla u|^p\,dx\right)^{\frac{p-1}{p}}\right]\|\varphi\|_X\,.$$

Observe that the latter integral can be made arbitrarily small for $K > 0$ sufficiently large. Due to $u \to v$ in X, the former integral approaches zero for any fixed $K > 0$. This implies that $I \to 0$ as $u \to v$ in X. Continuity of A then follows from the estimate (4.75).

To prove (4.73), we proceed in a similar way as in the proof of Lemma 4.3. We again have (4.35); let us estimate the integral on the right-hand side of (4.35). For $\delta > 0$ define the set

$$\Omega_\delta(u) = \{x \in \mathbb{R}^N\,;\ w(x)|u(x)|^{p-1} \ge \delta\}\,.$$

Then we have

$$\lim_{\|u\|_X \to 0} \operatorname{meas}\Omega_\delta(u) = 0\,. \tag{4.76}$$

Now for any given $\varepsilon > 0$, by ($\tilde{\mathrm{f}}3$), there exists $\delta > 0$ such that

$$\frac{|f(\lambda,x,u)|}{w(x)|u|^{p-1}} \le \varepsilon$$

uniformly for $w(x)|u|^{p-1} < \delta$. Splitting again the integral in (4.35) into integrals on $\mathbb{R}^N \setminus \Omega_\delta(u)$ and $\Omega_\delta(u)$, we have the same estimate for the first integral:

$$\int_{\mathbb{R}^N\setminus\Omega_\delta(u)} \frac{|f(\lambda,x,u)|}{w(x)|u|^{p-1}}|\tilde{u}|^{p-1}|\varphi|w\,dx \le \varepsilon\int_{\mathbb{R}^N\setminus\Omega_\delta(u)} |\tilde{u}|^{p-1}|\varphi|w\,dx \le c_4\varepsilon.$$

For the estimate of the second integral we use ($\tilde{f}2$):

$$\begin{aligned}
&\int_{\Omega_\delta(u)} \frac{|f(\lambda,x,u)|}{w(x)|u|^{p-1}}|\tilde{u}|^{p-1}|\varphi|w\,dx \\
&\quad\le c(\lambda)\int_{\Omega_\delta(u)} \frac{\sigma(x)}{w(x)|u|^{p-1}}|\tilde{u}|^{p-1}|\varphi|w\,dx + \frac{c(\lambda)}{\|u\|_X^{p-1}}\int_{\Omega_\delta(u)} \rho(x)|u|^{\gamma}|\varphi|\,dx\,.
\end{aligned}$$

We denote the last expression by $c(\lambda)(I_1+I_2)$ and in this case we have

$$\begin{aligned}
I_1 &\le \frac{1}{\delta}\int_{\Omega_\delta(u)} \sigma(x)w(x)|\tilde{u}|^{p-1}|\varphi|\,dx \\
&\le \frac{1}{\delta}\left(\int_{\Omega_\delta(u)} \nu^{N_1}|\tilde{u}|^{p^*}\,dx\right)^{\frac{p-1}{p^*}} \\
&\quad\cdot\left(\int_{\Omega_\delta(u)} \left(\nu^{-\frac{N_1(p-1)}{p^*}}w\sigma|\varphi|\right)^{\frac{p^*}{p^*-(p-1)}}dx\right)^{\frac{p^*-(p-1)}{p^*}} \\
&\le \frac{1}{\delta}\left(\int_{\Omega_\delta(u)} \nu^{\frac{-N_1(N-p)}{p}}(w\sigma)^{\frac{N}{p}}\,dx\right)^{\frac{p}{N}}\left(\int_{\Omega_\delta(u)} \nu^{N_1}|\tilde{u}|^{p^*}\,dx\right)^{\frac{p-1}{p^*}} \qquad (4.77)\\
&\quad\cdot\left(\int_{\Omega_\delta(u)} \nu^{N_1}|\varphi|^{p^*}\,dx\right)^{\frac{1}{p^*}} \\
&\le c_5\left(\int_{\Omega_\delta(u)} \nu^{\frac{-N}{p}}(w\sigma)^{\frac{N}{p}}\,dx\right)^{\frac{p}{N}}\|\tilde{u}\|_X^{p-1}\|\varphi\|_X \\
&\le c_5\left[\left(\int_{\Omega_\delta(u)\cap B_K(0)} (\nu^{-1}w\sigma)^{\frac{N}{p}}\,dx\right)^{\frac{p}{N}} + \left(\int_{|x|>K} (\nu^{-1}w\sigma)^{\frac{N}{p}}\,dx\right)^{\frac{p}{N}}\right] \\
&\quad\cdot\|\tilde{u}\|_X^{p-1}\|\varphi\|_X\,.
\end{aligned}$$

We have that

$$\int_{|x|>K} (\nu^{-1}w\sigma)^{\frac{N}{p}}\,dx$$

can be arbitrarily small if $K>0$ is large enough due to the assumption on σ, and for any fixed $K>0$ we have

$$\int_{\Omega_\delta(u)\cap B_K(0)} (\nu^{-1}w\sigma)^{\frac{N}{p}}\,dx \to 0$$

due to (4.76). Thus it follows from (4.77) that $I_1 \to 0$ as $\|u\|_X \to 0$. The estimate (4.74) implies

$$I_2 \le c_2\|u\|_X^{\gamma-p+1}\|\varphi\|_X \to 0$$

as $\|u\|_X \to 0$. Hence (4.73) holds. □

Remark 4.10 Note that if $\sigma(x) \equiv 0$ then (4.73) follows directly from (4.74).

Lemma 4.9 *The operators J and G^- are continuous and the operators G^+, F_λ are compact.*

The proof of this lemma proceeds in the same way as that of Lemma 4.4.

The following assertion is an analogue of Lemma 3.13 and will be used to establish the admissible homotopy in the proof of the main result of this section.

Lemma 4.10 *Let $d > 0$ be given, $\varphi \in X$, $t_n \in [0, 1]$, $\|u_n\|_X \to 0$, $\|u_n\|_X \neq 0$. Then*

$$\lim_{n\to\infty} \frac{1}{\|u_n\|_X^{p-1}} \int_{\mathbb{R}^N} (\tilde{a}(x, t_n u_n) - a(x, 0))|\nabla u_n|^{p-2} \nabla u_n \nabla \varphi dx = 0 . \tag{4.78}$$

Proof. We again estimate the term

$$\begin{aligned} &\frac{1}{\|u_n\|_X^{p-1}} \left| \int_{\mathbb{R}^N} (\tilde{a}(x, t_n u_n) - a(x, 0))|\nabla u_n|^{p-2} \nabla u_n \nabla \varphi \, dx \right| \\ &\qquad = \left| \int_{\mathbb{R}^N} (\tilde{a}(x, t_n u_n) - a(x, 0))|\nabla \tilde{u}_n|^{p-2} \nabla \tilde{u}_n \nabla \varphi \, dx \right| , \end{aligned} \tag{4.79}$$

where $\tilde{u}_n = \dfrac{u_n}{\|u_n\|_X}$. Let us define sets

$$\tilde{\Omega}_\delta(u_n) = \{x \in \mathbb{R}^N ;\ t_n |u_n(x)| \geq \delta\} .$$

By (4.71), for any $\varepsilon > 0$ there exists $\delta > 0$ such that

$$|\tilde{a}(x, t_n u_n) - a(x, 0)| \leq \varepsilon \nu(x)$$

provided $t_n|u_n(x)| < \delta$. We now split the integral into integrals on $\mathbb{R}^N \setminus \tilde{\Omega}_\delta(u_n)$ and $\tilde{\Omega}_\delta(u_n)$. On $\mathbb{R}^N \setminus \tilde{\Omega}_\delta(u_n)$ we have

$$\begin{aligned} &\int_{\mathbb{R}^N \setminus \tilde{\Omega}_\delta(u_n)} |\tilde{a}(x, t_n u_n) - a(x, 0)||\nabla \tilde{u}_n|^{p-1} |\nabla \varphi| \, dx \\ &\qquad \leq \varepsilon \int_{\mathbb{R}^N \setminus \tilde{\Omega}_\delta(u_n)} \nu(x) |\nabla \tilde{u}_n|^{p-1} |\nabla \varphi| \, dx \leq \varepsilon \|\tilde{u}_n\|_X^{p-1} \|\varphi\|_X = \varepsilon \|\varphi\|_X . \end{aligned}$$

On $\tilde{\Omega}_\delta(u_n)$, due to (4.72), the definition of $\tilde{a}(x, s)$ and the generalized Sobolev inequality (1.46), the integral is estimated by

$$\begin{aligned} &\int_{\tilde{\Omega}_\delta(u_n)} (|\tilde{a}(x, t_n u_n)| + |a(x, 0)|)|\nabla \tilde{u}_n|^{p-1} |\nabla \varphi| \, dx \\ &\qquad \leq 2\tilde{g}(d) \int_{\tilde{\Omega}_\delta(u_n)} \nu(x) |\nabla \tilde{u}_n|^{p-1} |\nabla \varphi| \, dx \\ &\qquad \leq 2\tilde{g}(d) \|\tilde{u}_n\|_X^{p-1} \left(\int_{\tilde{\Omega}_\delta(u_n)} \nu(x) |\nabla \varphi|^p \, dx \right)^{\frac{1}{p}} . \end{aligned}$$

Now, repeating the same argument as in the proof of Lemma 4.8, we can estimate the last integral separately on $\tilde{\Omega}_\delta(u_n) \cap B_K(0)$ and on $\mathbb{R}^N \setminus B_K(0)$ with $K > 0$ large enough to prove that it is arbitrarily small. Thus (4.79) can be made arbitrarily small and so (4.78) is established. □

Let us define the weak solution of (4.69).

Definition 4.4 We say that $\lambda \in \mathbb{R}$ and $u \in X$ *solve* the problem (4.69) *weakly* if

$$\int_{\mathbb{R}^N} a(x,u)|\nabla u|^{p-2}\nabla u \nabla\varphi\, dx = \lambda \int_{\mathbb{R}^N} g(x)|u|^{p-2}u\varphi\, dx + \int_{\mathbb{R}^N} f(\lambda, x, u)\varphi\, dx \tag{4.80}$$

holds for any $\varphi \in X$.

Remark 4.11 Observe that if $u \in X$ satisfies $|u(x)| < d$ for a.e. $x \in \Omega$ then the solvability of the operator equation

$$A(u) - \lambda G(u) - F_\lambda(u) = 0 \tag{4.81}$$

is equivalent to the (weak) solvability of (4.69) in the sense of Definition 4.4.

The following two assertions are estimates of the weak solution of (4.69) in $L^\infty(\mathbb{R}^N)$. Due to these results we can deal with (4.81) instead of (4.80) in the spirit of Remark 4.11.

Proposition 4.2 *Let us assume* (a1), (a2), ($\tilde{\text{g}}$1), ($\tilde{\text{g}}$2), ($\tilde{\text{f}}$1), ($\tilde{\text{f}}$2), ($\tilde{\text{f}}$3) *and recall that* $N_1 = \frac{p^*}{p}$. *Assume, in addition,*

(i) *for some* $\delta > 0$ *we have* $\rho \in L^{\gamma_1+\delta}(\mathbb{R}^N, \nu^{-N_1(\gamma_1+\delta-1)})$;

(ii) *for* $q = \dfrac{pp^*(\gamma_1+\delta)'}{p^* - (\gamma+1-p)(\gamma_1+\delta)'}$ *we have* $g^+\nu^{-\frac{p^*}{q}} \in L^{\frac{q}{q-p}}(\mathbb{R}^N)$;

(iii) *for all* μ *with* $\frac{q}{p} < \mu < q$ *we have* $\sigma\nu^{-\frac{N_1}{\mu}} \in L^{\mu'}(\mathbb{R}^N)$.

Then, for $\lambda \in \Lambda$, *where* Λ *is a bounded set, any weak solution of* (4.69) *satisfies*

$$\|u\|_\infty \le c(\|u\|_X), \tag{4.82}$$

where $c(s)$ *is a positive function bounded on bounded sets.*

Proof. We proceed in a similar way as in the proof of apriori estimates in previous sections. Let λ and u solve weakly (4.69) and let λ belong to a bounded set. Without loss of generality we assume $u \ge 0$ (otherwise we can deal with u^+ and u^- separately — cf. the proof of Proposition 4.1). For $M > 0$ define $v_M(x) = \min\{u(x), M\}$

and for $\kappa > 0$ take $\varphi = v_M^{\kappa p+1}$ as a test function in (4.80). Then $\varphi \in L^\infty(\mathbb{R}^N) \cap X$ and we have

$$\begin{aligned}(\kappa p+1)&\int_{\mathbb{R}^N} a(x,u)v_M^{\kappa p}|\nabla v_M|^p\,dx\\ &=\lambda\int_{\mathbb{R}^N} g(x)u^{p-1}v_M^{\kappa p+1}\,dx+\int_{\mathbb{R}^N} f(\lambda,x,u)v_M^{\kappa p+1}\,dx\,.\end{aligned}\tag{4.83}$$

Using (a1) and the generalized Sobolev inequality (1.46), the left-hand side of (4.83) can be estimated from below as

$$\begin{aligned}(\kappa p+1)&\int_{\mathbb{R}^N} a(x,u)v_M^{\kappa p}|\nabla v_M|^p\,dx\\ &=\frac{(\kappa p+1)}{(\kappa+1)^p}\int_{\mathbb{R}^N} a(x,u)|\nabla(v_M^{\kappa+1})|^p\,dx\\ &\geq\frac{(\kappa p+1)}{(\kappa+1)^p}\int_{\mathbb{R}^N} c_1\nu(x)|\nabla(v_M^{\kappa+1})|^p\,dx\\ &\geq c_9\frac{(\kappa p+1)}{(\kappa+1)^p}\left(\int_{\mathbb{R}^N}\nu^{N_1}v_M^{(\kappa+1)p^*}\,dx\right)^{\frac{p}{p^*}}.\end{aligned}\tag{4.84}$$

The right-hand side of (4.83) can be estimated from above by

$$\lambda\int_{\mathbb{R}^N} g(x)u^{(\kappa+1)p}\,dx+c(\lambda)\int_{\mathbb{R}^N}\sigma(x)u^{\kappa p+1}\,dx+c(\lambda)\int_{\mathbb{R}^N}\rho(x)u^{\kappa p+1+\gamma}\,dx\,.$$

Let us estimate these integrals one by one. We have

$$\begin{aligned}\int_{\mathbb{R}^N} g(x)u^{(\kappa+1)p}\,dx&\leq\int_{\mathbb{R}^N} g^+(x)u^{(\kappa+1)p}\,dx\\ &\leq\left(\int_{\mathbb{R}^N}(g^+\nu^{-\varepsilon})^{\frac{q}{q-p}}\,dx\right)^{\frac{q-p}{q}}\left(\int_{\mathbb{R}^N}u^{(\kappa+1)q}\nu^{\varepsilon\frac{q}{p}}\,dx\right)^{\frac{p}{q}},\end{aligned}$$

where $q=\dfrac{pp^*(\gamma_1+\delta)'}{p^*-(\gamma+1-p)(\gamma_1+\delta)'}$, $\varepsilon=\dfrac{p^*}{q}$. We note here that

$$p^*>q>p(\gamma_1+\delta)',\quad p^*=(\gamma+1-p)(\gamma_1+\delta)'\frac{q}{q-p(\gamma_1+\delta)'}\,.\tag{4.85}$$

Thus by condition (ii) we have

$$\int_{\mathbb{R}^N} g(x)u^{(\kappa+1)p}\,dx\leq c_{10}\left(\int_{\mathbb{R}^N}u^{(\kappa+1)q}\nu^{N_1}\,dx\right)^{\frac{p}{q}}.\tag{4.86}$$

The second integral can be estimated (using condition (iii)) as

$$\int_{\mathbb{R}^N} \sigma(x) u^{\kappa p+1}\,dx \le \left(\int_{\mathbb{R}^N} (\nu^{-\frac{N_1}{\mu}}\sigma)^{\mu'}\,dx\right)^{\frac{1}{\mu'}} \left(\int_{\mathbb{R}^N} \nu^{N_1} u^{(\kappa+1)q}\,dx\right)^{\frac{1}{\mu}}, \tag{4.87}$$

where $\mu = \dfrac{(\kappa+1)q}{\kappa p+1}$ and $\dfrac{q}{p} < \mu < q$. For the last integral we have

$$\begin{aligned}\int_{\mathbb{R}^N} \rho(x) u^{\kappa p+1+\gamma}\,dx &\le \left(\int_{\mathbb{R}^N} \nu^{-\varepsilon_1(\gamma_1+\delta)} \rho^{\gamma_1+\delta}\,dx\right)^{\frac{1}{\gamma_1+\delta}} \\ &\quad \cdot \left(\int_{\mathbb{R}^N} u^{(\kappa+1)p(\gamma_1+\delta)'} u^{(\gamma+1-p)(\gamma_1+\delta)'} \nu^{\varepsilon_1(\gamma_1+\delta)'}\,dx\right)^{\frac{1}{(\gamma_1+\delta)'}} \\ &\le \left(\int_{\mathbb{R}^N} \nu^{-\varepsilon_1(\gamma_1+\delta)} \rho^{\gamma_1+\delta}\,dx\right)^{\frac{1}{\gamma_1+\delta}} \left(\int_{\mathbb{R}^N} \nu^{N_1} u^{(\kappa+1)q}\,dx\right)^{\frac{p}{q}} \\ &\quad \cdot \left(\int_{\mathbb{R}^N} \nu^{N_1} [u^{(\gamma+1-p)(\gamma_1+\delta)'}]^{\frac{q}{q-p(\gamma_1+\delta)'}}\,dx\right)^{\frac{q-p(\gamma_1+\delta)'}{q(\gamma_1+\delta)'}},\end{aligned}$$

where ε_1 is chosen in such a way that $\varepsilon_1 = \dfrac{N}{(\gamma_1+\delta)'}$. Thus we have by (4.85) and condition (i),

$$\begin{aligned}\int_{\mathbb{R}^N} \rho(x) u^{\kappa p+1+\gamma}\,dx &\le \left(\int_{\mathbb{R}^N} \nu^{-N_1(\gamma_1+\delta-1)} \rho^{\gamma_1+\delta}\,dx\right)^{\frac{1}{\gamma_1+\delta}} \\ &\quad \cdot \left(\int_{\mathbb{R}^N} \nu^{N_1} u^{(\kappa+1)q}\,dx\right)^{\frac{p}{q}} \left(\int_{\mathbb{R}^N} \nu^{N_1} u^{p^*}\,dx\right)^{\frac{q-p(\gamma_1+\delta)'}{q(\gamma_1+\delta)'}}.\end{aligned} \tag{4.88}$$

We then derive from (4.83), (4.84), (4.86)–(4.88) that

$$\begin{aligned}\left(\int_{\mathbb{R}^N} \nu^{N_1} v_M^{(\kappa+1)p^*}\,dx\right)^{\frac{p}{p^*}} &\le c_{11} \frac{(\kappa+1)^p}{\kappa p+1} \left[\left(\int_{\mathbb{R}^N} \nu^{N_1} u^{(\kappa+1)q}\,dx\right)^{\frac{p}{q}}\right. \\ &\quad + \left(\int_{\mathbb{R}^N} (\sigma\nu^{-\frac{N_1}{\mu}})^{\mu'}\,dx\right)^{\frac{1}{\mu'}} \left(\int_{\mathbb{R}^N} \nu^{N_1} u^{(\kappa+1)q}\,dx\right)^{\frac{1}{\mu}} \\ &\quad + \left(\int_{\mathbb{R}^N} \nu^{-N_1(\gamma_1+\delta-1)} \rho^{\gamma_1+\delta}\,dx\right)^{\frac{1}{\gamma_1+\delta}} \\ &\quad \left.\cdot \left(\int_{\mathbb{R}^N} \nu^{N_1} u^{p^*}\,dx\right)^{\frac{q-p(\gamma_1+\delta)'}{q(\gamma_1+\delta)'}} \left(\int_{\mathbb{R}^N} \nu^{N_1} u^{(\kappa+1)q}\,dx\right)^{\frac{p}{q}}\right].\end{aligned}$$

This together with our assumptions (i)–(iii) and the generalized Sobolev inequality (1.46) implies that

$$\left(\int_{\mathbb{R}^N} \nu^{N_1} v_M^{(\kappa+1)p^*}\,dx\right)^{\frac{p}{p^*}} \le c_{11}\frac{(\kappa+1)^p}{\kappa p+1}\left[c_{12}\left(\int_{\mathbb{R}^N} \nu^{N_1} u^{(\kappa+1)q}\,dx\right)^{\frac{1}{\mu}} + \bar{c}(\|u\|_X)\left(\int_{\mathbb{R}^N} \nu^{N_1} u^{(\kappa+1)q}\,dx\right)^{\frac{p}{q}}\right].$$

Observing that $\dfrac{1}{\mu} < \dfrac{p}{q}$, we further obtain that

$$\left(\int_{\mathbb{R}^N} \nu^{N_1} v_M^{(\kappa+1)p^*}\,dx\right)^{\frac{p}{p^*}} \le c_{11}\frac{(\kappa+1)^p}{\kappa p+1}\left[c_{13}+\hat{c}(\|u\|_X)\left(\int_{\mathbb{R}^N} \nu^{N_1} u^{(\kappa+1)q}\,dx\right)^{\frac{p}{q}}\right].$$

Thus we conclude

$$\begin{aligned}\left(\int_{\mathbb{R}^N} \nu^{N_1} v_M^{(\kappa+1)p^*}\,dx\right)^{\frac{1}{(\kappa+1)p^*}} & \\ \le c_{11}^{\frac{1}{p(\kappa+1)}}\left(\frac{\kappa+1}{(\kappa p+1)^{\frac{1}{p}}}\right)^{\frac{1}{\kappa+1}}&\left[c_{14}+\tilde{c}(\|u\|_X)\left(\int_{\mathbb{R}^N} \nu^{N_1} u^{(\kappa+1)q}\,dx\right)^{\frac{1}{(\kappa+1)q}}\right].\end{aligned} \tag{4.89}$$

Note that $\bar{c}(\|u\|_X)$, $\hat{c}(\|u\|_X)$ and $\tilde{c}(\|u\|_X)$ depend in general on $\|u\|_X$.

Now we can use a similar argument as in the proof of Theorem 3.5, Step 1. Actually, we derive that for $\kappa \to \infty$ the following inequality holds:

$$\|u\|_{L^{(\kappa_n+1)p^*}(\mathbb{R}^N,\nu^{N_1})} \le c(\|u\|_X), \tag{4.90}$$

where the function $c = c(s)$ is bounded on bounded sets.

This implies (4.82). Indeed, assume

$$\|u\|_\infty > c(\|u\|_X). \tag{4.91}$$

Then due to (4.91) on a bounded set $\mathcal{A} \subset \mathbb{R}^N$ of positive measure we have

$$\begin{aligned}\left(\int_{\mathbb{R}^N} (\nu^{\frac{N_1}{r}} u)^r\,dx\right)^{\frac{1}{r}} &\ge \left(\int_{\mathcal{A}} (\nu^{\frac{N_1}{r}} u)^r\,dx\right)^{\frac{1}{r}} \\ &> c(\|u\|_X)\left(\int_{\mathcal{A}} \nu^{N_1}\,dx\right)^{\frac{1}{r}}.\end{aligned} \tag{4.92}$$

For $r \to \infty$, the last expression tends to $c(\|u\|_X)$, which contradicts (4.90). □

Since the conditions in Proposition 4.2 are in general rather complicated to check, we will present one modification of the assumptions on $g(x)$ and $f(\lambda, x, s)$ when these conditions are satisfied readily. Recall that $\nu(x) = \dfrac{1}{(1+|x|)^\beta}$.

Let us introduce

$(\tilde{f2})^-$ $\quad \sigma\nu^{-\frac{1}{p}} \in L^\infty(\mathbb{R}^N) \cap L^{(p^*)'}(\mathbb{R}^N)$, $\rho\nu^{-\frac{\gamma+1}{p}} \in L^\infty(\mathbb{R}^N) \cap L^{\gamma_1}(\mathbb{R}^N)$;

$(\tilde{f2})^+$ $\quad \sigma\nu^{-N_1} \in L^\infty(\mathbb{R}^N) \cap L^{(p^*)'}(\mathbb{R}^N)$, $\rho\nu^{-N_1} \in L^\infty(\mathbb{R}^N) \cap L^{\gamma_1}(\mathbb{R}^N)$;

$(\tilde{g1,2})^+$ $\quad g^+\nu^{-N_1} \in L^\infty(\mathbb{R}^N) \cap L^{\frac{N}{p}}(\mathbb{R}^N)$, $g^- \in L^\infty(\mathbb{R}^N)$.

Proposition 4.3 *Assume, for $\beta \le 0$, that* $(\tilde{g1})$, $(\tilde{g2})$ *and* $(\tilde{f2})^-$ *hold with nonnegative σ and ρ, and for $\beta > 0$, that* $(\tilde{g1}, 2)^+$ *and* $(\tilde{f2})^+$ *hold. Then conditions* (i)–(iii) *in Proposition* 4.2 *are satisfied and so its conclusion remains valid.*

Proof. In the proof we use essentially the pointwise estimates of the weight function $\nu(x)$ and its powers.

Condition (i). Observe that $\gamma_1 = \left(\dfrac{p^*}{\gamma+1}\right)'$. Then we have

$$1 < (\gamma_1 + \delta)' < \frac{p^*}{\gamma+1}. \tag{4.93}$$

For $\beta \le 0$ it follows from (4.93) that

$$\nu(x)^{-\frac{N_1}{(\gamma_1+\delta)'}} \le \nu(x)^{-\frac{N_1(\gamma+1)}{p^*}} \quad \text{in} \quad \mathbb{R}^N$$

and hence

$$\begin{aligned}\int_{\mathbb{R}^N} \nu^{-N_1(\gamma_1+\delta-1)}\rho^{\gamma_1+\delta}\,dx &= \int_{\mathbb{R}^N} (\nu^{-\frac{N_1}{(\gamma_1+\delta)'}}\rho)^{\gamma_1+\delta}\,dx \\ &\le \int_{\mathbb{R}^N} (\nu^{-\frac{N_1(\gamma+1)}{p^*}}\rho)^{\gamma_1+\delta}\,dx = \int_{\mathbb{R}^N} (\nu^{-\frac{\gamma+1}{p}}\rho)^{\gamma_1+\delta}\,dx \\ &\le \|\nu^{-\frac{\gamma+1}{p}}\rho\|_\infty^\delta \int_{\mathbb{R}^N} (\nu^{-\frac{\gamma+1}{p}}\rho)^{\gamma_1}\,dx < \infty,\end{aligned}$$

due to $(\tilde{f2})^-$. For $\beta > 0$, we have (using (4.93) again)

$$\nu(x)^{-\frac{N_1}{(\gamma_1+\delta)'}} \le \nu(x)^{-N_1} \quad \text{in} \quad \mathbb{R}^N$$

and so

$$\begin{aligned}\int_{\mathbb{R}^N} (\nu^{-\frac{N_1}{(\gamma_1+\delta)'}}\rho)^{\gamma_1+\delta}\,dx &\le \int_{\mathbb{R}^N} (\nu^{-N_1}\rho)^{\gamma_1+\delta}\,dx \\ &\le \|\nu^{-N_1}\rho\|_\infty^\delta \int_{\mathbb{R}^N} (\nu^{-N_1}\rho)^{\gamma_1}\,dx < \infty,\end{aligned}$$

due to $(\tilde{f2})^+$. Thus condition (i) is satisfied.

Condition (ii). Observe that, since $q < p^*$, we have

$$\frac{q}{q-p} = \left(\frac{q}{p}\right)' > \left(\frac{p^*}{p}\right)' = \frac{N}{p}, \tag{4.94}$$

$$-N_1 = -\frac{p^*}{p} < -\frac{p^*}{q} < -1. \tag{4.95}$$

Hence for $\beta \leq 0$ it follows from (4.94), (4.95) that

$$\begin{aligned}\int_{\mathbb{R}^N} \left(g^+\nu^{-\frac{p^*}{q}}\right)^{\frac{q}{q-p}} dx &\leq \int_{\mathbb{R}^N} (g^+\nu^{-1})^{\frac{q}{q-p}} dx \\ &\leq \|g^+\nu^{-1}\|_\infty^{\frac{q}{q-p}-\frac{N}{p}} \int_{\mathbb{R}^N} (g^+\nu^{-1})^{\frac{N}{p}} dx < \infty,\end{aligned}$$

since $g^+\nu^{-1} \in L^\infty(\mathbb{R}^N) \cap L^{\frac{N}{p}}(\mathbb{R}^N)$ due to ($\tilde{g}1$) and

$$\nu(x)^{-\frac{p^*}{q}} \leq \nu(x)^{-1} \quad \text{in} \quad \mathbb{R}^N.$$

For $\beta > 0$,

$$\begin{aligned}\int_{\mathbb{R}^N} (g^+\nu^{-\frac{p^*}{q}})^{\frac{q}{q-p}} dx &\leq \int_{\mathbb{R}^N} (g^+\nu^{-N_1})^{\frac{q}{q-p}} dx \\ &\leq \|g^+\nu^{-N_1}\|_\infty^{\frac{q}{q-p}-\frac{N}{p}} \int_{\mathbb{R}^N} (g^+\nu^{-N_1})^{\frac{N}{p}} dx < \infty\end{aligned}$$

since, by $(\tilde{g}1, 2)^+$, $g^+\nu^{-N_1} \in L^\infty(\mathbb{R}^N) \cap L^{\frac{N}{p}}(\mathbb{R}^N)$, and by (4.95),

$$\nu(x)^{-\frac{p^*}{q}} \leq \nu^{-N_1}(x) \quad \text{in} \quad \mathbb{R}^N.$$

Hence condition (ii) is fulfilled.

Condition (iii). Since $q > \mu > \frac{q}{p}$, we have

$$-\frac{1}{p} = -\frac{N_1}{p^*} > -\frac{N_1}{q} > -\frac{N_1}{\mu} > -\frac{pN_1}{q} > -N_1. \tag{4.96}$$

For $\beta \leq 0$ we have, due to (4.96),

$$\nu(x)^{-\frac{N_1}{\mu}} \leq \nu(x)^{-\frac{1}{p}} \quad \text{in} \quad \mathbb{R}^N$$

and so

$$\begin{aligned}\int_{\mathbb{R}^N} (\nu^{-\frac{N_1}{\mu}}\sigma)^{\mu'} dx &\leq \int_{\mathbb{R}^N} (\nu^{-\frac{1}{p}}\sigma)^{\mu'} dx \\ &\leq \|\nu^{-\frac{1}{p}}\sigma\|_\infty^{\mu'-(p^*)'} \int_{\mathbb{R}^N} (\nu^{-\frac{1}{p}}\sigma)^{(p^*)'} dx < \infty,\end{aligned}$$

since $\mu' > (p^*)'$, and by $(\tilde{f}2)^-$ we obtain $\nu^{-\frac{1}{p}}\sigma \in L^\infty(\mathbb{R}^N) \cap L^{(p^*)'}(\mathbb{R}^N)$. For $\beta > 0$, (4.96) implies

$$\nu(x)^{-\frac{N_1}{\mu}} \le \nu(x)^{-N_1} \quad \text{in} \quad \mathbb{R}^N ,$$

and so

$$\begin{aligned} \int_{\mathbb{R}^N} (\nu^{-\frac{N_1}{\mu}}\sigma)^{\mu'}\, dx &\le \int_{\mathbb{R}^N} (\nu^{-N_1}\sigma)^{\mu'}\, dx \\ &\le \|\nu^{-N_1}\sigma\|_\infty^{\mu'-(p^*)'} \int_{\mathbb{R}^N} (\nu^{-N_1}\sigma)^{(p^*)'}\, dx < \infty , \end{aligned}$$

since $\nu^{-N_1}\sigma \in L^\infty(\mathbb{R}^N) \cap L^{(p^*)'}(\mathbb{R}^N)$ by $(\tilde{f}2)^+$. Condition (iii) then follows. □

Let us denote $a(x) := a(x, 0)$, $x \in \mathbb{R}$. Concerning the homogeneous eigenvalue problem

$$-\operatorname{div}(a(x)|\nabla u|^{p-2}\nabla u) = \lambda g(x)|u|^{p-2}u \quad \text{in} \quad \mathbb{R}^N \tag{4.97}$$

we have the following assertion which is an analogue of Lemmas 4.5 and 4.6 from the previous section. Its proof can be carried out along the same lines as the proofs of Lemmas 4.5 and 4.6 with obvious modifications subject to the assumptions on $a(x)$, $g(x)$ and the definition of the norm in X. Note that in this section $a(x)$ and $g(x)$ may degenerate for $|x| \to \infty$.

Lemma 4.11 *Let us assume* (a1), (a2), $(\tilde{g}1)$, $(\tilde{g}2)$. *Then*

(i) *the problem* (4.97) *has a pair* (λ_1, u_1) *of a principal eigenvalue and an eigenfunction with* $\lambda_1 > 0$ *and* $0 < u_1 \in X \cap L^\infty(\mathbb{R}^N)$. *Moreover, such* λ_1 *is simple and unique.*

(ii) *If* $g^- \not\equiv 0$ *and* $g^-\nu^{-1} \in L^\infty(\mathbb{R}^N) \cap L^{\frac{N}{p}}(\mathbb{R}^N)$, *then there is also a principal eigenpair* (λ_1^*, u_1^*) *with* $\lambda_1^* < 0$ *and* $0 < u_1^* \in X$ *with analogous properties.*

(iii) *Every eigenfunction corresponding to an eigenvalue* λ_0 *with* $0 < \lambda_0 \ne \lambda_1$ *changes sign in* $\mathbb{R}^N$. *Moreover, every such eigenvalue* $\lambda_0 > 0$ *satisfies* $\lambda_0 > \lambda_1$. *(Similarly for* $0 > \lambda_0 \ne \lambda_1^*$.*)*

(iv) *The principal eigenvalue* $\lambda_1 > 0$ $(\lambda_1^* < 0)$ *is isolated.*

Let us consider the operator $T_\lambda : X \to X^*$,

$$T_\lambda(u) = A(u) - \lambda G(u) - F_\lambda(u) , \qquad u \in X .$$

Since G^+ and F_λ are compact operators (see Lemma 4.9), it follows from the next assertion and from Lemma 1.2 that T_λ satisfies condition $\alpha(X)$ for any $\lambda > 0$.

Lemma 4.12 *The operator* $A + \lambda G^- : X \to X^*$ *satisfies condition* $\alpha(X)$ *for* $\lambda > 0$.

Proof. Assume $u_n \rightharpoonup u_0$ in X and

$$\limsup_{n\to\infty} \langle A(u_n) + \lambda G^-(u_n), u_n - u_0\rangle \le 0 .$$

Then we have

$$\begin{aligned}
0 &\ge \limsup_{n\to\infty} (\langle A(u_n) - A(u_0), u_n - u_0\rangle + \lambda \langle G^-(u_n) - G^-(u_0), u_n - u_0\rangle) \\
&= \limsup_{n\to\infty} \bigg(\int_{\mathbb{R}^N} (\tilde{a}(x, u_n) - \tilde{a}(x, u_0)) |\nabla u_0|^{p-2} \nabla u_0 (\nabla u_n - \nabla u_0)\, dx \\
&\quad + \int_{\mathbb{R}^N} \tilde{a}(x, u_n)(|\nabla u_n|^{p-2}\nabla u_n - |\nabla u_0|^{p-2}\nabla u_0)(\nabla u_n - \nabla u_0)\, dx \qquad (4.98) \\
&\quad + \lambda \int_{\mathbb{R}^N} g^-(x)(|u_n|^{p-2}u_n - |u_0|^{p-2}u_0)(u_n - u_0)\, dx \bigg).
\end{aligned}$$

We will estimate the first integral similarly as the analogous expression in (4.75). It follows from (4.71) that for any $\varepsilon > 0$ there exists $\delta > 0$ such that

$$|\tilde{a}(x, u_n) - \tilde{a}(x, u_0)| \le \varepsilon \nu(x)$$

provided $|u_n(x) - u_0(x)| < \delta$. Define

$$\Omega_\delta(n) = \{x \in \mathbb{R}^N ;\ |u_n(x) - u_0(x)| \ge \delta\} .$$

Let $K > 0$. Then $u_n \to u_0$ in $L^p(B_K(0))$ due to $W^{1,p}(B_K(0)) \hookrightarrow\hookrightarrow L^p(B_K(0))$ and hence we get

$$\operatorname{meas}(\Omega_\delta(n) \cap B_K(0)) \to 0$$

(cf. the proof of Lemma 4.8). Now the integral on $\mathbb{R}^N \setminus \Omega_\delta(n)$ can be estimated by

$$\varepsilon \int_{\mathbb{R}^N \setminus \Omega_\delta(n)} \nu |\nabla u_0|^{p-1} |\nabla u_n - \nabla u_0|\, dx \le \varepsilon \|u_0\|_X^{p-1} (\|u_n\|_X + \|u_0\|_X) ,$$

while the corresponding integral on $\Omega_\delta(n)$ is estimated by

$$c_{17} \left[\left(\int_{\Omega_\delta(n) \cap B_K(0)} \nu |\nabla u_0|^p\, dx \right)^{\frac{1}{p'}} + \left(\int_{|x|>K} \nu |\nabla u_0|^p\, dx \right)^{\frac{1}{p'}} \right] (\|u_n\|_X + \|u_0\|_X) .$$

The first integral in the brackets obviously approaches zero as $n \to \infty$ for any fixed $K > 0$ while the other can be made arbitrarily small for $K > 0$ sufficiently large. (Note that $\|u_n\|_X$ is bounded due to the weak convergence of u_n in X.) This implies

$$\int_{\mathbb{R}^N} (\tilde{a}(x, u_n) - \tilde{a}(x, u_0)) |\nabla u_0|^{p-2} \nabla u_0 (\nabla u_n - \nabla u_0)\, dx \to 0$$

as $n \to \infty$. Then it follows from (4.98) and (a1) that

$$\int_{\mathbb{R}^N} \nu(x) |\nabla u_n|^p\, dx \to \int_{\mathbb{R}^N} \nu(x) |\nabla u_0|^p\, dx \qquad (4.99)$$

and

$$\int_{\mathbb{R}^N} g^-(x)|u_n|^p\,dx \to \int_{\mathbb{R}^N} g^-(x)|u_0|^p\,dx\,. \tag{4.100}$$

The convergence (4.99) together with $u_n \rightharpoonup u_0$ in X implies that $\nabla u_n \to \nabla u_0$ in $L^p(\mathbb{R}^N, \nu)$. It follows from the generalized Hardy inequality (4.70) that

$$\int_{\mathbb{R}^N} \omega(x)|u_n - u_0|^p\,dx \le \left(\frac{p}{N-\alpha}\right)^p \int_{\mathbb{R}^N} \nu(x)|\nabla(u_n - u_0)|^p\,dx \to 0\,. \tag{4.101}$$

So, using the fact that $u_n \rightharpoonup u_0$ in X, (4.100), (4.101) and the fact that $\nabla u_n \to \nabla u_0$ in $L^p(\mathbb{R}^N, \nu)$, we conclude that

$$u_n \to u_0 \quad \text{in} \quad X\,.$$

□

Remark 4.12 Straightforward modification of the proof of Lemma 4.7 implies that also $J + \lambda G^-$ (and hence $\tilde{M}_\lambda = J - \lambda G$) satisfies condition $\alpha(X)$ for $\lambda > 0$. In particular, it follows from here and from Lemma 4.12 that the degrees

$$\operatorname{Deg}[T_\lambda; D, 0] \quad \text{and} \quad \operatorname{Deg}[\tilde{M}_\lambda; D, 0]$$

are well defined for any $\lambda > 0$ and any bounded nonempty open set $D \subset X$ such that $T_\lambda(u) \neq 0$ and $\tilde{M}_\lambda(u) \neq 0$ for any $u \in \partial D$, respectively.

If $g\nu^{-1} \in L^\infty(\mathbb{R}^N) \cap L^{\frac{N}{p}}(\mathbb{R}^N)$ $(1 < p < N)$ then G is compact (cf. Lemma 4.9 and the proof of compactness of G^+) and the degree can be defined also for $\lambda \le 0$.

Definition 4.5 Let $C \subset E$ be a continuum of nontrivial solutions of (4.69) defined as in Definition 4.3. We say that $\lambda_0 \in \mathbb{R}$ is a *local bifurcation point* of (4.69) if there is a continuum of nontrivial solutions C of (4.69) such that $(\lambda_0, 0) \in \overline{C}$.

We have the following bifurcation result.

Theorem 4.7 *Let* $1 < p < N$ *and let the assumptions of Proposition* 4.2 *be satisfied. Then the principal eigenvalue* $\lambda_1 > 0$ *of the eigenvalue problem* (4.97) *is a local bifurcation point of* (4.69).

Proof. Proposition 4.2 implies that we can choose $d > 0$ large enough so that in a small neighbourhood B of $(\lambda_1, 0)$ in the space E, any weak solution (λ, u) of (4.69) satisfies $\|u\|_\infty < d$. (Note that λ_1 does not depend on d.) We will investigate now the topological degree of the mapping T_λ for λ "near" λ_1. Let us introduce the homotopy

$$\mathcal{H}(t, \lambda, u) = A(t, u) - \lambda G(u) - t F_\lambda(u)\,, \tag{4.102}$$

where $t \in [0, 1]$ and $A(t, u)$ is defined by

$$\langle A(t, u), \varphi \rangle = \int_{\mathbb{R}^N} \tilde{a}(x, tu)|\nabla u|^{p-2}\nabla u \nabla \varphi \, dx$$

for any $u, \varphi \in X$. Then $\mathcal{H}(1, \lambda, u) = T_\lambda(u)$ and $\mathcal{H}(0, \lambda, u) = \tilde{M}_\lambda(u)$. We claim that, for any $\varepsilon > 0$ small enough, there exists $\delta > 0$ such that for $\lambda = \lambda_1 \pm \varepsilon$, $\mathcal{H}(t, \lambda, u)$ is an admissible homotopy between T_λ and $\tilde{M}_\lambda$ in $B_\delta(0)$. Suppose this is not true. Then there exist $u_n \in X$ with $\|u_n\|_X \to 0$, $\|u_n\|_X \neq 0$, $t_n \in [0, 1]$, such that $\mathcal{H}(t_n, \lambda, u_n) = 0$ (for some $\lambda \neq \lambda_1$, λ in a "small" neighbourhood of λ_1). That is, for $\varphi \in X$ we have

$$\langle A(t_n, u_n), \varphi \rangle - \lambda \langle G(u_n), \varphi \rangle - t_n \langle F_\lambda(u_n), \varphi \rangle = 0 . \tag{4.103}$$

Dividing (4.103) by $\|u_n\|_X^{p-1}$ and denoting $\tilde{u}_n = \dfrac{u_n}{\|u_n\|_X}$ and

$$\langle \tilde{A}(t_n, \tilde{u}_n), \varphi \rangle = \int_{\mathbb{R}^N} \tilde{a}(x, t_n u_n)|\nabla \tilde{u}_n|^{p-2}\nabla \tilde{u}_n \nabla \varphi \, dx ,$$

we obtain for all $\varphi \in X$ and $n \in \mathbb{N}$,

$$\langle \tilde{A}(t_n, \tilde{u}_n), \varphi \rangle - \lambda \langle G(\tilde{u}_n), \varphi \rangle - t_n \frac{\langle F_\lambda(u_n), \varphi \rangle}{\|u_n\|_X^{p-1}} = 0 . \tag{4.104}$$

By Lemma 4.8 we have

$$\lim_{n \to \infty} \sup_{\|\varphi\|_X \le 1} \frac{|\langle F_\lambda(u_n), \varphi \rangle|}{\|u_n\|^{p-1}} = 0 ,$$

and we thus conclude that

$$\lim_{n \to \infty} \sup_{\|\varphi\|_X \le 1} \left| \langle \tilde{A}(t_n, \tilde{u}_n) - \lambda G(\tilde{u}_n), \varphi \rangle \right| = 0 ,$$

i.e.

$$\tilde{A}(t_n, \tilde{u}_n) - \lambda G(\tilde{u}_n) \to 0 \quad \text{in} \quad X^* .$$

Since $\tilde{u}_n$ is bounded in X, we can assume that $\tilde{u}_n \rightharpoonup u_0$ in X for some $u_0 \in X$. Lemma 4.9 then implies that $\lambda G^+(\tilde{u}_n) \to \lambda G^+(u_0)$ in X^*. We thus infer that

$$\tilde{A}(t_n, \tilde{u}_n) + \lambda G^-(\tilde{u}_n) \to \lambda G^+(u_0) \quad \text{in} \quad X^* . \tag{4.105}$$

It then follows from (4.105) and from $u_n \rightharpoonup u_0$ in X that

$$\lim_{n \to \infty} \langle \tilde{A}(t_n, \tilde{u}_n) + \lambda G^-(\tilde{u}_n), \tilde{u}_n - u_0 \rangle = 0 .$$

Obvious modification of the proof of Lemma 4.12 implies that $\tilde{u}_n \to u_0$ in X and hence $\|u_0\|_X = 1$. This strong convergence, (4.104), continuity of G (see Lemmma 4.9) and Lemma 4.10 imply that $u_0 \in X$ satisfies

$$J(u_0) - \lambda G(u_0) = 0 .$$

This contradicts the fact that λ_1 is an isolated eigenvalue and $\lambda \neq \lambda_1$. Hence we have the following property of the topological degree of T_λ : for any $\varepsilon > 0$ small enough, there exists $\delta > 0$ such that

$$\operatorname{Deg}[T_\lambda; B_\delta(0), 0] = \operatorname{Deg}[\tilde{M}_\lambda; B_\delta(0), 0] . \tag{4.106}$$

Now, we can apply the same technique as in the proof of Theorem 4.3 and show that

$$1 = \operatorname{Deg}[J - (\lambda_1 - \varepsilon)G; B_\delta(0), 0] \neq \operatorname{Deg}[J - (\lambda_1 + \varepsilon)G; B_\delta(0), 0] = -1 . \tag{4.107}$$

It follows from (4.106), (4.107) that we have a "jump" of $\operatorname{Deg}[T_\lambda; B_\delta(0), 0]$ as λ crosses λ_1. Hence $(\lambda_1, 0)$ is a bifurcation point of the equation

$$T_\lambda(u) = 0$$

(cf. the proof of Theorem 4.3). It follows from the considerations in Remark 4.11 that $(\lambda_1, 0)$ is also a bifurcation point of (4.69). □

Similarly, we have the following "dual" version of the previous theorem.

Theorem 4.8 *Let the assumptions of Theorem* 4.7 *be satisfied and assume, in addition, that* $g^- \not\equiv 0$, $g^- \nu^{-1} \in L^\infty(\mathbb{R}^N) \cap L^{\frac{N}{p}}(\mathbb{R}^N)$. *Then the conclusion of Theorem* 4.7 *remains valid and, moreover, the principal eigenvalue* $\lambda_1^* < 0$ *of the eigenvalue problem* (4.97) *is also a local bifurcation point of* (4.69).

Similarly as in Section 4.2 we can obtain more precise information about the bifurcating solutions of BVP (4.69) by strengthening the assumptions on $a(x, s)$ and $f(\lambda, x, s)$.

Theorem 4.9 *Let* $1 < p < N$, *and assume* $a \in C^1(\mathbb{R}^{N+1})$, $\sigma(x) \equiv 0$. *Assume, moreover,* (a1), (a2), $(\tilde{g}1)$, $(\tilde{g}2)$, $(\tilde{f}1)$, $(\tilde{f}2)$, $(\tilde{f}2)^-$, $(\tilde{f}3)$, *if* $\beta \le 0$ *and* (a1), (a2), $(\tilde{g}1)$, $(\tilde{g}2)$, $(g1,2)^+$, $(\tilde{f}2)$, $(\tilde{f}2)^+$, $(\tilde{f}3)$, *if* $\beta > 0$. *Then the bifurcating solutions of* (4.69) *belonging to sufficiently small neighbourhood* B *of* $(\lambda_1, 0)$ *in* E *are strictly of the same sign in* $\mathbb{R}^N$. *Moreover, every such solution satisfies* $u \in C^{1,\alpha}(B_K(0))$ *for any* $K > 0$ *with some* $\alpha = \alpha(K) \in (0, 1)$.

Proof. First, let us show that the bifurcating solutions guaranteed by Theorem 4.7 do not change sign in $\mathbb{R}^N$ if they belong to a sufficiently small neighbourhood B of $(\lambda_1, 0)$. Assume that (λ_n, u_n) solves (4.69) weakly and $\lambda_n \to \lambda_1$, $\|u_n\|_X \to 0$, $\|u_n\|_X \neq 0$.

Then due to Lemmas 4.8, 4.9 and 4.10 we can assume $\tilde{u}_n \rightharpoonup u_1$ in X for $\tilde{u}_n = \dfrac{u_n}{\|u_n\|_X}$, where u_1 is positive principal eigenfunction associated with λ_1. The same method used in the proof of Theorem 4.7 yields that $\tilde{u}_n \to u_1$ in X. Now, repeating the same argument as in the proof of Theorem 4.6 (but using the corresponding properties (a1), (a2) of $a(x, s)$ and ($\tilde{\text{g}}$1), ($\tilde{\text{g}}$2), or ($\tilde{\text{g}}1, 2)^+$ of $g(x)$) we can show that $u_n \geq 0$ in $\mathbb{R}^N$ for n large enough. Realize that our weight functions ν and ω are bounded from above and below by positive constants on any bounded ball in $\mathbb{R}^N$, and the bifurcating solutions are uniformly bounded in $L^\infty(\mathbb{R}^N)$ when they belong to a small neighbourhood B of $(\lambda_1, 0)$. Using these facts and the assumption $\sigma(x) \equiv 0$ we can deduce from the Harnack-type inequality (applied in any bounded ball in $\mathbb{R}^N$ — see Theorem 1.9), similarly as in the proof of Theorem 4.6, that $u_n > 0$ in $\mathbb{R}^N$.

The result of Theorem 1.11 applied in any bounded ball in $\mathbb{R}^N$ yields the local $C^{1,\alpha}$-regularity of the bifurcating solutions. □

Remark 4.13 The proof of Theorem 4.7 runs even more easily if T_λ and A are replaced by M_λ and J, respectively. Also the assertions of Theorems 4.8 and 4.9 remain true if we replace $a(x, s)$ by $a(x)$ and consider the bifurcation problem (4.30) instead of (4.69). We thus have a straighforward generalization of the results in Section 4.2 to the case of a degenerated (or singular) coefficient $a(x)$. Note that for the problem (4.30) we can formulate, in a more general setting of this section, also a bifurcation result in the spirit of Theorem 4.6.

Remark 4.14 Let us emphasize once again that the function $a(x, s)$ in the principal part of our perturbed problem (4.69) depends on x and s in a rather general way — no growth restrictions with respect to s are required and $a(x, s)$ may be degenerated (or singular) in x as $|x| \to \infty$. These facts justify the somewhat complicated assumptions which we have to pose on g and f and also the delicacy of our apriori estimate which we derived for the weak solutions of (4.69).

Example 4.2 Let us consider the equation with a blowing-up weight

$$\begin{aligned} &-\operatorname{div}((1 + |x|)^\tau b(u)|\nabla u|^{p-2}\nabla u) \\ &\quad = \lambda g(x)|u|^{p-2}u + c(\lambda)(\sigma(x) + \rho(x)|u|^{\gamma-1}u) \quad \text{in} \quad \mathbb{R}^N, \end{aligned} \tag{4.108}$$

where $\tau > 0$, $b(s) \geq \text{const} > 0$ and $c(\lambda)$ are continuous functions bounded on bounded intervals in $\mathbb{R}$. The weight functions take the forms

$$\nu(x) = (1 + |x|)^\tau \quad \text{and} \quad \omega(x) = (1 + |x|)^{\tau - p}.$$

Let us assume $g^+(x) \leq O(|x|^{k_0})$, $|x| \to \infty$, with $k_0 < \tau - p$, $g^-(x) \leq \text{const}$, g^+, g^- measurable,

$$\sigma(x) = O(|x|^{k_1}), \quad |x| \to \infty, \quad k_1 < \min\left\{0, \tau - p, \frac{\tau - p + N - Np}{p}\right\},$$

$$\rho(x) = O(|x|^{k_2}), \quad |x| \to \infty, \quad k_2 < \frac{(\gamma + 1)(\tau - p + N) - Np}{p}.$$

Then $\lambda_1 > 0$ is a local bifurcation point of (4.108) by Theorem 4.7, where λ_1 is the principal eigenvalue of the problem

$$-\operatorname{div}((1+|x|)^\tau b(0)|\nabla u|^{p-2}\nabla u) = \lambda g(x)|u|^{p-2}u \quad \text{in} \quad \mathbb{R}^N$$

with $\int_{\mathbb{R}^N} g|u|^p\,dx > 0$. We note that $b(s)$ is a rather general function. Two examples of such $b(s)$ are

$$b_1(s) = \sqrt{1+|s|^2}\,, \quad b_2(s) = e^{|s|^k}\,, \quad k \in \mathbb{R}\,.$$

◇

Example 4.3 Let us consider the equation with a decaying weight

$$\begin{aligned} &-\operatorname{div}\left(\frac{1}{(1+|x|)^\tau} b(u)|\nabla u|^{p-2}\nabla u\right) \\ &\quad = \lambda g(x)|u|^{p-2}u + c(\lambda)(\sigma(x) + \rho(x)|u|^{\gamma-1}u) \quad \text{in} \quad \mathbb{R}^N\,, \end{aligned} \tag{4.109}$$

where $\tau > 0$, $b(s)$ and $c(\lambda)$ are as in Example 4.2. In this case

$$\nu(x) = \frac{1}{(1+|x|)^\tau} \quad \text{and} \quad \omega(x) = \frac{1}{(1+|x|)^{p+\tau}}\,.$$

Assume $g^+(x) \le O(|x|^{k_3})$, $|x| \to \infty$, with $k_3 < -p - \tau N_1$, $g^-(x) \le \text{const}$, g^+, g^- measurable,

$$\sigma(x) = O(|x|^{k_4})\,, \quad |x| \to \infty\,, \quad k_4 < \min\left\{-p-\tau, -\frac{Np - N + p + \tau p^*}{p}\right\}$$

and

$$\rho(x) = O(|x|^{k_5})\,, \quad |x| \to \infty\,, \quad k_5 < -\frac{Np - (\gamma+1)(N-p) + \tau p^*}{p}\,.$$

Then $\hat{\lambda}_1 > 0$ is a local bifurcation point of (4.109) by Theorem 4.7, where $\hat{\lambda}_1$ is the principal eigenvalue of

$$-\operatorname{div}\left(\frac{b(0)}{(1+|x|)^\tau}|\nabla u|^{p-2}\nabla u\right) = \lambda g(x)|u|^{p-2}u \quad \text{in} \quad \mathbb{R}^N\,,$$

with $\int_{\mathbb{R}^N} g|u|^p\,dx > 0$.

◇

Example 4.4 Consider the equation

$$\begin{aligned}&-\operatorname{div}\Big((1+|x|)^{\tau}b(u)|\nabla u|^{p-2}\nabla u\Big)\\&\quad=\lambda g(x)|u|^{p-2}u+c(\lambda)(\sigma(x)+\rho(x)|u|^{\gamma-1}u)\quad\text{in}\quad\mathbb{R}^N,\end{aligned}\tag{4.110}$$

where $b(s)$ and $c(\lambda)$ are as above, $\tau > 0$ and $g(x) = \cos(|x|)g_0(x)$ with g_0 measurable and $g_0(x) > 0$, $x \in \mathbb{R}^N$. If $g_0(x) = O(|x|^{k_6})$, $|x| \to \infty$, where $k_6 < \min\{0, \tau - p\}$, and σ, ρ are as in Example 4.2, then (4.109) has two bifurcation points $\lambda^+ > 0$ and $\lambda^- < 0$, where $\lambda^{\pm}$ are the principal eigenvalues of

$$-\operatorname{div}((1+|x|)^{\tau}b(0)|\nabla u|^{p-2}\nabla u)=\lambda g(x)|u|^{p-2}u\quad\text{in}\quad\mathbb{R}^N,$$

with positive eigenfunctions $u^{\pm}$ in $\mathbb{R}^N$ satisfying $\pm\int_{\mathbb{R}^N} g(u^{\pm})^p\,dx > 0$.

Note that also σ and ρ can change signs, provided $|\sigma|$ and $|\rho|$ satisfy the above conditions. ◇

Bibliography

[1] R. A. Adams, *Sobolev Spaces,* Academic Press, Inc., New York 1975.

[2] S. Agmon, *Lectures on Elliptic Boundary Value Problems*, Van Nostrand Comp., New York 1965.

[3] W. Allegretto, Y. X. Huang, *Eigenvalues of the indefinite weight p-Laplacian in weighted* $\mathbb{R}^N$ *spaces*, Funkcial. Ekvac. 38 (1995), 233–242.

[4] H. Amann, *A note on degree theory for gradient mappings*, Proc. Amer. Math. Soc. 85 (1982), 591–595.

[5] A. Anane, *Simplicité et isolation de la premiére valeur propre du p-laplacien avec poids,* C. R. Acad. Sci. Paris Sér. I Math. 305 (1987), 725–728.

[6] J. Appell, P. P. Zabrejko, *Nonlinear Superposition Operators*, Cambridge University Press, Cambridge 1990.

[7] G. Barles, *Remarks on uniqueness results of the first eigenvalue of the p-Laplacian,* Ann. Fac. Sci. Toulouse Math. IX, no 1 (1988), 65–75.

[8] T. Bhattacharya, *Radial symmetry of the first eigenfunction for the p-Laplacian in the ball,* Proc. Amer. Math Soc. 104 (1988), 169–174.

[9] L. Boccardo, *Positive eigenfunctions for a class of quasi-linear operators,* Boll. Un. Mat. Ital. (5) 18–B (1981), 951–959.

[10] F. E. Browder, *Nonlinear elliptic boundary value problems and the generalized topological degree,* Bull. Amer. Math. Soc. 76 (1970), 999–1005.

[11] F. E. Browder, W. V. Petryshin, *Approximation methods and the generalized topological degree for nonlinear mappings in Banach spaces,* J. Funct. Anal. 3, (1969), 217–245.

[12] R. C. Brown, B. Opic, *Embeddings of weighted Sobolev spaces into spaces of continuous functions,* Proc. Roy. Soc. London Ser. A 439 (1992), 279–296.

[13] E. N. Dancer, *On the structure of the solutions of non-linear eigenvalue problems*, Indiana Univ. Math. J. 23 (11), (1974), 1069–1076.

[14] M. A. Del Pino, R. Manásevich, *Global bifurcation from the eigenvalues of the p-Laplacian*, J. Differential Equations 92 (1991), 226–251.

[15] P. Drábek, *Nonlinear eigenvalue problem for the p-Laplacian in* $\mathbb{R}^N$, Math. Nachr. 173 (1995), 131–139.

[16] P. Drábek, *On the bifurcation for a class of degenerate equations*, Ann. Mat. Pura Appl. 159 (1991), 1–16.

[17] P. Drábek, *Solvability and Bifurcations of Nonlinear Equations,* Pitman Res. Notes Math. Ser. 232, Longman Scientific & Technical, Harlow 1992.

[18] P. Drábek, Y. X. Huang, *Bifurcation problems for the p-Laplacian in* $\mathbb{R}^N$, Trans. Amer. Math. Soc. 349 (1997), 171–188.

[19] P. Drábek, Y. X. Huang, *Perturbed p-Laplacian in* $\mathbb{R}^N$*: Bifurcation from the principal eigenvalue*, J. Math. Anal. Appl. 204 (1996), 582–608.

[20] P. Drábek, M. Kučera, *Generalized eigenvalue and bifurcations of second order boundary value problems with jumping nonlinearities,* Bull. Austral. Math. Soc. 37 (1988), 179–187.

[21] P. Drábek, A. Kufner, F. Nicolosi, *On the solvability of degenerated quasilinear elliptic equations of higher order,* J. Differential Equations 109 (1994), 325–347.

[22] P. Drábek, F. Nicolosi, *Existence of bounded solutions for some degenerated quasilinear elliptic equations*, Ann. Mat. Pura Appl. (IV), CLXV (1993), 217–238.

[23] P. Drábek, F. Nicolosi, *Solubilité des problémes elliptiques dégénérés d' ordere supérieur a l' aide du théoreme de Leray–Lions,* C. R. Acad. Sci. Paris Sér. I Math. 315 (1992), 689–692.

[24] N. Dunford, J. T. Schwartz, *Linear Operators*, Interscience Publ., New York 1958.

[25] D. E. Edmunds, B. Opic, *Weighted Poincaré and Friedrichs inequalities,* J. London Math. Soc. (2) 47 (1993), 79–96.

[26] J. Fleckinger, J. Hernández, F. De Thélin, *Principe du maximum pour un systéme elliptique non linéaire,* C. R. Acad. Sci. Paris Sér. I Math. 314, (1992), 665–668.

[27] S. Fučík, A. Kufner, *Nonlinear Differential Equations,* Elsevier, Amsterdam–Oxford–New York 1980.

[28] J. García Azorero, I. Peral Alonso, *Existence and non-uniqueness for the p-Laplacian: Non-linear eigenvalues,* Comm. Partial Differential Equations 12 (1987), 1389–1430.

[29] V. P. Glushko, *Linear Degenerated Differential Equations* (Russian), Voronezh, 1972.

[30] F. Guglielmino, F. Nicolosi, *Sulle W-soluzioni dei problemi al contorno per operatori ellittici degeneri,* Ricerche Mat. 36 (1987), 59–72.

[31] F. Guglielmino, F. Nicolosi, *Teoremi di esistenza per i problemi al contorno relativi alle equazioni ellittiche quasilineari,* Ricerche Mat. 37 (1988), 157–176.

[32] B. Hanouzet, *Espaces de Sobolev avec poids. Application en problème de Dirichlet dans une demiespace*, Rend. Sem. Math. Univ. Padova 46 (1971), 227–272.

[33] E. Hewitt, K. Stromberg, *Real and Abstract Analysis,* Grad. Texts in Math. 25, Springer-Verlag, Berlin–Heidelberg 1975 (Third printing).

[34] I. A. Kiprijanov, *A certain class of singular elliptic operators I, II*, Differentsial'nye. Uravneniya 7 (1971), 2066–2077; Sibirsk. Mat. Zh. 14 (1973), 560–568 (Russian); Siberian Math. J. 14 (1973), (1974), 388–394.

[35] M. A. Krasnoselskij, *Positive Solutions of Operator Equations* (Russian), Moscow 1962. English translation: P. Noordhoff, Groningen.

[36] M. G. Krein, M. A. Rutman, *Linear operators leaving invariant a cone in a Banach space,* Amer. Math. Soc. Transl. Ser. 1 10 (1950), 199–325.

[37] A. Kufner, *Weighted Sobolev Spaces,* 2nd edition J. Wiley & Sons, Chichester–New York–Brisbane–Toronto–Singapore 1985.

[38] A. Kufner, O. John, S. Fučík, *Function Spaces,* Academia, Prague 1977.

[39] A. Kufner, S. Leonardi, *Solvability of degenerate elliptic boundary value problems: Another approach,* Math. Bohem. 119 (1994), 255–274.

[40] A. Kufner, B. Opic, *How to define reasonably weighted Sobolev space,* Comment. Math. Univ. Carolin. 25 (1984), 537–554.

[41] A. Kufner, B. Opic, *The Dirichlet problem and weighted spaces I, II,* Čas. Pěst. Mat. 108 (1983), 381–408; 111 (1986), 242–253.

[42] A. Kufner, A. M. Sändig, *Some Applications of Weighted Sobolev Spaces,* Teubner, Leipzig 1987.

[43] Lao Sen Yu, *Nonlinear p-Laplacian problems on unbounded domains*, Proc. Amer. Math. Soc. 115 (4) (1992), 1037–1045.

[44] Li Gongbao, *Some properties of weak solutions of nonlinear scalar field equations*, Ann. Acad. Sci. Fenn. Ser. A. I Math. 14 (1989), 27–36.

[45] Li Gongbao, You Shusen, *Eigenvalue problems for quasilinear elliptic equations on* $\mathbb{R}^N$, Comm. Partial Differential Equations 14 (1989), 1291–1314.

[46] P. Linqvist, *On the equation* $\operatorname{div}(|\nabla u|^{p-2}\nabla u) + \lambda |u|^{p-2}u = 0$, Proc. Amer. Math. Soc. 109 (1990), 157–164.

[47] J.-L. Lions, *Équations differentielles operationelles et problémes aux limites,* Springer-Verlag, Berlin–Göttingen–Heidelberg 1961.

[48] J.-L. Lions, *Quelques méthodes de résolution des problèmes aux limites non linéaires*, Dunod Gauthier–Villars, Paris 1969.

[49] J.-L. Lions, E. Magenes, *Problèmes aux limites non homogènes et applications*, Vol. 1, Dunod, Paris 1968.

[50] C. Miranda, *Istituzioni di analisi funzionale lineare*, Vol. I e II, Un. Mat. Ital., Gubbio, Bologna 1978/1979.

[51] M. K. V. Murthy, G. Stampacchina, *Boundary value problems for same degenerate elliptic operators*, Ann. Mat. Pura Appl. (4), 80 (1968), 1–122.

[52] J. Nečas, *Les méthodes directes en théorie des équations elliptiques*, Masson, Paris 1967.

[53] S. M. Nikolskij, *On a boundary value problem of the first kind with a strong degeneracy*, Dokl. Akad. Nauk SSSR 222 (1975), 281–283 (Russian); Soviet Math. Dokl. 16 (1975), 624–627.

[54] E. S. Noussair, C. A. Swanson, *An* $L^q(\mathbb{R}^N)$*-theory of subcritical semilinear elliptic problems*, J. Differential Equations 84 (1990), 52–61.

[55] B. Opic, A. Kufner, *Hardy-Type Inequalities,* Pitman Research Notes in Mathematics Series 219, Longman Scientific & Technical, Harlow 1990.

[56] M. Otani, T. Teshima, *The first eigenvalue of some quasilinear elliptic equations,* Proc. Japan Acad. Ser. A Math. Sci. 64 A (1988), 8–10.

[57] P. H. Rabinowitz, *Some global results for nonlinear eigenvalue problems,* J. Funct. Anal. 7 (1971), 487–513.

[58] W. Rother, *Generalized Emden–Fowler equations of subcritical growth*, J. Austral. Math. Soc. (Series A) 54 (1993), 254–262.

[59] A. Rumbos, A. Edelson, *Bifurcation properties of semilinear elliptic equations in* $\mathbb{R}^N$, Differential Integral Equations 7 (1994), 399–410.

[60] J. Serrin, *Local behavior of solutions of quasilinear equations*, Acta Math. 111 (1964), 247–302.

[61] I. V. Skrypnik, *Nonlinear Elliptic Boundary Value Problems*, Teubner, Leipzig 1986.

[62] P. Tolksdorf, *Regularity for a more general class of quasilinear elliptic equations,* J. Differential Equations 51 (1984), 126–150.

[63] N. S. Trudinger, *On Harnack type inequalities and their application to quasilinear elliptic equations,* Comm. Pure Appl. Math. 20 (1967), 721–747.

[64] M. M. Vainberg, *Variational Methods for the Study of Nonlinear Operators,* Holden-Day, Inc., San Francisco 1964.

Index